Nanotechnology in Environmental Remediation: Perspectives and Prospects

Edited by

Neha Agarwal
Department of Chemistry
Navyug Kanya Mahavidyalaya
University of Lucknow, Lucknow, India

Vijendra Singh Solanki
Department of Chemistry
IPS Academy, Indore, India

Neetu Singh
Department of Physics
Maharaja Bijli Pasi Government P.G. College
Ashiana, Lucknow, India

&

Maulin P. Shah
Environmental Microbiology Consultant
Ankleshwar, Gujrat, India

Nanotechnology in Environmental Remediation: Perspectives and Prospects

Editors: Neha Agarwal, Vijendra Singh Solanki, Neetu Singh and Maulin P. Shah

ISBN (Online): 978-981-5322-94-1

ISBN (Print): 978-981-5322-95-8

ISBN (Paperback): 978-981-5322-96-5

Published by Bentham Science Publishers Pte. Ltd. Singapore, in collaboration with Eureka Conferences, USA. All Rights Reserved.

First published in 2025.

need for a court order if at any point you breach any terms of this License Agreement. In no event will any delay or failure by Bentham Science Publishers in enforcing your compliance with this License Agreement constitute a waiver of any of its rights.
3. You acknowledge that you have read this License Agreement, and agree to be bound by its terms and conditions. To the extent that any other terms and conditions presented on any website of Bentham Science Publishers conflict with, or are inconsistent with, the terms and conditions set out in this License Agreement, you acknowledge that the terms and conditions set out in this License Agreement shall prevail.

Bentham Science Publishers Pte. Ltd.
No. 9 Raffles Place
Office No. 26-01
Singapore 048619
Singapore
Email: subscriptions@benthamscience.net

BENTHAM SCIENCE

CONTENTS

FOREWORD

In recent years, the rapid growth in the domain of nanotechnology has opened up new horizons to address some of the most serious environmental issues. The fusion of nanotechnology and nanotechnological methods carries enormous potential to enhance the cost-effectiveness, potency, efficiency, and sustainability of efforts of environmental protection and pollution control for environmental remediation. This book, "Nanotechnology in Environmental Remediation: Perspectives and Prospects," exhibits a comprehensive analysis of the applications, recent advancements, and prospects of nanotechnology in the field of environmental remediation and protection.

The book starts with an insightful introduction to the origins of several contaminants and pollutants across distinct environmental zones. This deep and basic knowledge contributes to developing a better understanding of the necessity and scope of nanotechnological advancements related to environmental protection. The following chapters scrutinize the synthesis, characterizations, and properties of different types of nanomaterials for the remediation of the environment and also highlight the uniqueness and potency of nanomaterials.

Nanomaterials in the wastewater treatment section of the book will surely create awareness amongst the readers about the advanced materials that effectively remove impurities from wastewater, thereby improving water quality and availability. The application of nanotechnology and nanomaterials for the remediation of soil and air has also been thoroughly discussed, presenting unique solutions to tackle pollution in the environment.

A commendable feature of this book is to focus on several types of specialized nanomaterials, such as carbon nanomaterials, nanocomposites, metal oxides, polymers, and materials for the degradation of toxic organic pollutants. Each chapter of this book offers deep insights into the advantages, mechanisms, and effectiveness of these materials, supported by recent advanced research studies.

Further chapters show the utilization of different kinds of nanomaterials in environmental detection and remediation along with their applications, such as nanobiosensors and photocatalytic properties. These aforementioned technologies contribute to the early detection of several types of environmental pollutants, which proactively helps in environmental protection strategies.

The book also draws attention to its broader themes like the social and environmental implications of different nanomaterials, green nanotechnology, sustainability, and the risk of nanotechnological applications. These analyses are very necessary to certify that the utilization and growth of nanotechnologies are ethical, safe, and sustainable for the environment.

Ultimately, the book offers a wide range of perspectives regarding the future directions, challenges, and applications of nanotechnology in the remediation of the environment. This commendable section targets a boarder range of readers like scientists, researchers, practitioners, and policymakers to carry on their research in this innovative and dynamic field.

I appreciate the editors and contributors for their meticulous and comprehensive approach to this multifaceted field. Their collective expertise and thorough analyses make this book an extremely useful resource for scientists, researchers, professionals, students, and technologists interested in the deep analysis of nanotechnology and environmental science.

As we go through the issues of environmental degradation and search for sustainable solutions, "Nanotechnology in Environmental Remediation: Perspectives and Prospects" proves itself a beacon of knowledge that directs us toward a safer, cleaner, and better planet.

Khac-Uan Do
School of Chemistry and Life Sciences
Hanoi University of Science and Technology
Vietnam

PREFACE

As an emerging field, nanotechnology plays a potential role in environmental remediation. Due to the continuous increase in the level of toxic pollutants, advanced eco-friendly detection and remediation technologies are required. The book focuses on nanotechnology based approaches that offer easier, faster and economical processes in environmental monitoring and remediation. The aim of the book "*Nanotechnology in Environmental Remediation: Perspectives and Prospects*" is to provide comprehensive knowledge related to the use and applications of nanotechnology/nanomaterials for the remediation of environmental contaminants, along with other applications.

The book covers several aspects of nanotechnology in the remediation of air, water, and soil, along with prominent biological applications, and presents recent advances in these fields. Furthermore, the book provides a deep insight into the role of nanotechnology in the decontamination of the environment: tools, methods, and approaches for detection and remediation. It has also addressed the social and economic aspects related to nanotechnology and the toxicological footprint of advanced functional nanomaterials. The safety and sustainability aspects of the use of nanomaterials and future directions in multifaceted aspects of using nanomaterials have also been discussed which will facilitate in formulation of strategies of environmental restoration.

In chapter one, the author has given an overview of environmental nanotechnology by exploring the techniques utilized in the biogenic synthesis of nanoparticles along with their characterization. He has also highlighted how environmental nanotechnology has profoundly influenced all facets of the field, spanning from identification to remediation.

In chapter two, the author has presented an insight into the potential role of nanomaterials in the identification and treatment of wastewater effluents. She has also taken into account the opportunities and risks, highlighting the imperative need for the responsible and cautious application of nanomaterials.

The Authors in chapter three have demonstrated a broad range of prospective nanotechnologies that have been tried to treat and enhance the organoleptic properties of drinking water and wastewater, supplying a safe and harmless liquid to society and the environment in a responsible manner.

The Authors in chapter four have examined the recent advancements achieved in the elimination of contaminants from contaminated water utilizing diverse forms of carbon NMs as adsorptive agents, including graphene, carbon nanotubes, activated carbon, and fullerenes.

The Authors in chapter five have outlined the latest developments in synthetic techniques for the production of copolymer nanocomposites and highlighted their potential uses in environmental remediation. They have also highlighted how there will be a significant increase in the potential applications of copolymer nanocomposites as innovative adsorbents for environmental remediation in the future.

In chapter six, the authors have given a comprehensive insight into biochar-based nanocomposites with precise preparation techniques and their efficacy in eliminating pollutants. According to them, embracing biochar-based nanocomposites represents a crucial step toward promoting cleaner and healthier ecosystems, contributing to a more sustainable future.

In chapter seven, the author has given a detailed account of the mechanism of environmental remediation by biofabricated nano-based adsorbents while also addressing the remediation of persistent organic pollutants. He has also highlighted the long-term development of environmentally benign biofabricated nanomaterial-based adsorbents, along with basic mechanisms as well as societal applications.

The Authors in chapter eight have shed light on nano-bioremediation and phytonanotechnology for the remediation of various categories of pollutants. They have also highlighted why these methods deliver great efficiency at a low cost when applied widely.

The Authors in chapter nine have explored the synthesis of bionanomaterials from various sources, their characterization, and diverse applications in the remediation of different environmental matrices such as water, air, and soil. Furthermore, they have also examined the challenges that need to be addressed and presented prospects for bio-nanomaterials in the ongoing battle against environmental pollution.

In chapter ten, the authors have discussed the principles of green nanotechnology, potential applications of green synthesised nanoparticles in the remediation of air, water, and soil, along with their superiority over other conventional treatment techniques. The authors have also highlighted the limitations and associated challenges so that, with continued research and development, green nanotechnology can ensure a brighter future for generations to come.

The Authors in chapter eleven have presented a comprehensive understanding of the photocatalytic activity and potential of NMs, paving the way for sustainable environmental remediation strategies. They have also discussed how the integration of nanomaterials in sustainable environmental management holds great promise for achieving cleaner air, water, and soil while minimizing the ecological footprint and safeguarding human health for future generations.

The Authors in chapter twelve have given a deep insight into the impact of nanocomposite TiO_2 photocatalyst in wastewater effluents. They have also tried to prove the idea of modulating the photocatalytic process and anticipated the potential for using this process to accomplish the utilization of wastewater effluent resources.

The Authors in chapter thirteen have given a comparative account of different types of nanomaterial based carbon-di-oxide sensors and their wide applications in various fields. Their discussion has highlighted the role of carbon dioxide nano-sensors in the agri-food sector, leading to more sustainable methods, less waste, and better use of available resources for environmental monitoring.

The Authors in chapter fourteen have specifically discussed the practical use of a range of nanomaterials for air pollution remediation applications. They have also discussed the pivotal role of nanomaterials as nano adsorbents, nanocatalysts, nanofilters, and nanosensors, showcasing the versatility and effectiveness of nanotechnological applications in this field.

The Authors in chapter fifteen have highlighted the importance of nanotechnology as an innovative technique for the remediation of degraded soil. They have emphasized that the promotion of efficient and sustainable use of nanomaterials can enhance the productivity and fertility of polluted soils to ensure a safe and healthy environment without degrading natural resources.

The Authors in chapter sixteen have highlighted the pivotal role of nanoparticles in the degradation of toxic organic materials by leveraging their unique properties, making nanomaterials a promising solution for addressing environmental pollution and promoting sustainable remediation practices.

Neha Agarwal
Department of Chemistry
Navyug Kanya Mahavidyalaya
University of Lucknow, Lucknow, India

Vijendra Singh Solanki
Department of Chemistry
IPS Academy, Indore, India

Neetu Singh
Department of Physics
Maharaja Bijli Pasi Government P.G. College
Ashiana, Lucknow, India

&

Maulin P. Shah
Environmental Microbiology Consultant
Ankleshwar, Gujrat, India

List of Contributors

Alisha	Department of Chemistry, National Council of Educational Research and Training (NCERT), New Delhi, India
Amlesh Yadav	Department of Botany, Govt. PG college, Hardoi, Uttar Pradesh, India
Amrit Mitra	Department of Chemistry, Government General Degree College, Singur, Hooghly, India
Anup K. Parmar	Department of Chemistry, C. J. Patel College, Tirora, Dist. Gondia, Maharashta, India
Anjali Mehta	Department of Chemistry, Banasthali Vidyapith, Rajasthan 304022, India
A. Ramesh Babu	Department of Chemistry, SVA Govt. Degree College, Srikalahasti A.P 517644, India
Ajay Kumar Tiwari	Department of Chemistry, School of Applied Sciences, Uttaranchal University, Dehradun, Uttarakhand 248007, India
A. K. Srivastava	Department of Applied Science, Institute of Engineering and Technology, Dr. Rammanohar Lohia Avadh University, Ayodhya 224000, Uttar Pradesh, India
Anjali Mehta	Department of Chemistry, Banasthali Vidyapith, Rajasthan 304022, India
Amrit Krishna Mitra	Department of Chemistry, Government General Degree College Singur, Hooghly, West Bengal, India
C. K. Kaithwas	Department of Applied Science, Institute of Engineering and Technology, Dr. Rammanohar Lohia Avadh University, Ayodhya 224000, Uttar Pradesh, India
Hemant Khambete	Faculty of Pharmacy, Medicaps University, Rau, Indore, India
Kamal Kant Sharma	Gurukula kangri (Deemed to be university), Haridwar, Uttarakhand, India
Karnica Srivastava	Department of Physics, Isabella Thoburn College, Lucknow 226007, Uttar Pradesh, India
Mishu Singh	Department of Chemistry, Pt. D. D. U. Govt. Girls P.G College, Lucknow, India
M.P. Laavanyaa shri	PG and Research Department of Chemistry, Bishop Heber College, Trichy - 17, Tamil Nadu, India
Mithlesh Kumar	Department of Botany, Rajkeeya Mahavidhyalaya Todarpur, Hardoi 241125, Uttar Pradesh, India
Nitisha Chakraborty	Department of Chemistry and Chemical Biology, Indian Institute of Technology (ISM), Dhanbad, India
Nilesh Gupta	Faculty of Pharmacy, Medicaps University, Rau, Indore, India
Neha Agarwal	Department of Chemistry, Navyug Kanya Mahavidyalaya, University of Lucknow, Lucknow, India
Priyanka Singh	Department of Botany, University of Lucknow, Lucknow, India
Rashmi R. Dubey	Department of Chemistry, Kamla Nehru Mahavidyalaya, Nagpur 440024, India
R. Margrate Thatcher	PG and Research Department of Chemistry, Bishop Heber College, Trichy - 17, Tamil Nadu, India

R. Sakthi Sri	PG and Research Department of Chemistry, Bishop Heber College, Trichy - 17, Tamil Nadu, India
Ratindra Gautam	Department of Applied Science, Institute of Engineering and Technology, Dr. Rammanohar Lohia Avadh University, Ayodhya 224000, Uttar Pradesh, India
Reenu Gill	Department of Botany, Rajkeeya Mahavidhyalaya Todarpur, Hardoi 241125, Uttar Pradesh, India
Sudhanshu Sharma	Department of Chemistry, Banasthali Vidyapith, Rajasthan 304022, India
Sudesh Kumar	Department of Chemistry, National Council of Educational Research and Training (NCERT), New Delhi 110016, India
Sourab Billore	Faculty of Pharmacy, Medicaps University, Rau, Indore, India
Sanjay Jain	Faculty of Pharmacy, Medicaps University, Rau, Indore, India
Sapna A. Kondalkar	Regional Ayurved Research Institute, Gwalior, Amkhoo, India
S. Ambika	PG and Research Department of Chemistry, Bishop Heber College, Trichy - 17, Tamil Nadu, India
Sankara Rao Miditana	Department of Chemistry, Government Degree College, Puttur, Tirupathi A.P 517583, India
Saivenkatesh Korlam	Department of Chemistry, SVA Govt. Degree College, Srikalahasti A.P 517644, India
Satheesh Ampolu	Department of Chemistry, Centurion University of Technology and Management, A.P, India
Sheerin Masroor	Department of Chemistry, A.N. College, Patliputra University, Patna 800013, Bihar, India
Shivani Chaudhary	Department of Physics, Deen Dayal Upadhyay Gorkhpur University, Gorakhpur, Uttar Pradesh, India
Tanisha Kathuria	Department of Chemistry, Banasthali Vidyapith, Rajasthan 304022, India
U. B. Singh	Department of Physics, Deen Dayal Upadhyay Gorkhpur University, Gorakhpur, Uttar Pradesh, India
Vijendra Singh Solanki	Department of Chemistry, SVA Govt. Degree College, Srikalahasti A.P 517644, India
W. B. Gurnule	Department of Chemistry, Kamla Nehru Mahavidyalaya, Nagpur 440024, India
Yashpal U. Rathod	Department of Chemistry, C. J. Patel College, Tirora, Dist. Gondia, Maharashta, India
Y. Manojkumar	PG and Research Department of Chemistry, Bishop Heber College, Trichy - 17, Tamil Nadu, India

Exploring Environmental Nanotechnology: Synthesis Techniques and Characterization Methods

Nitisha Chakraborty[1] and **Amrit Mitra**[2,*]

[1] *Department of Chemistry and Chemical Biology, Indian Institute of Technology (ISM), Dhanbad, Jharkhand, India*

[2] *Department of Chemistry, Government General Degree College, Singur, Hooghly, West Bengal, 712409, India*

Abstract: Environmental nanotechnology deals with environmental issues and plays a crucial role in contemporary science and engineering. These cutting-edge nanomaterials (NMs) are being used for a variety of purposes, with a primary focus on environmental preservation. Understanding matter has been made possible by nanoscience and nanotechnologies, which have significant effects on all industries and economies, including food and agriculture, energy production efficiency, the automobile industry, cosmetics, medicine and pharmaceuticals, computers, weapons, and household appliances. The environmental sector leverages nanotechnology to develop sensors for the detection, monitoring, and analysis of toxic contaminants, contributing to the protection of the environment. The field of nanotechnology is improving the detection of hazardous water-borne compounds and creating new avenues for water purification, desalination, and decontamination. For the purpose of detecting pesticides, NM-based unit-molecular and array types of biosensors are being developed. Environmental applications of nanotechnology include developing solutions to present environmental challenges as well as preventative measures for future problems caused by interactions between energy, materials, and the environment. These applications additionally seek to evaluate and alleviate any possible dangers associated with nanotechnology. Different physicochemical and biological techniques can be used to synthesize nanoparticles (NPs) for a variety of uses. The utilization of microorganisms in the biogenic synthesis of NPs offers several advantages over alternative methods and is increasingly gaining attention. This chapter provides an overview of environmental nanotechnology and explores the techniques utilized in the biogenic synthesis of NPs, along with their characterization.

Keywords: Characterization, Environment, Nanomaterials, Environmental nanotechnology, Nanoparticles, Nanosensors, Pollutants, Synthesis.

---
* **Corresponding author Amrit Mitra:** Department of Chemistry, Government General Degree College, Singur, Hooghly, West Bengal, 712409, India; E-mail: ambrosia12june@gmail.com

Neha Agarwal, Vijendra Singh Solanki, Neetu Singh & Maulin P. Shah (Eds.)

INTRODUCTION

The term 'nanos' signifies 'extremely small' and applies to both dwarfs and individuals of exceptionally short stature [1]. In the last decade, there has been an increasing utilization of the prefix 'nano' across various fields of expertise. Presently, it is a commonly employed term in a significant portion of contemporary research, leading to its more frequent appearance in scientific literature [1 - 5]. The fundamental characteristic distinguishing all NPs — with all materials possessing at least one dimension at the nanoscale— is their size, typically ranging from 1 to 100 nanometers (nm) [1]. Consequently, materials exhibiting at least one dimension within this range and possessing a spherical surface area per volume exceeding 60 m^2/cm^3 are classified as NMs. Nanoparticulate matter represents a distinct state of matter, which is different from all other states [1, 6 - 10].

Pinpointing the exact origins of human utilization of nanoscale entities is challenging. However, the use of NMs is not a recent phenomenon; humans have inadvertently employed them for various purposes over time [11]. As far back as approximately 4500 years ago, humans utilized asbestos nanofibers to reinforce ceramic mixes, albeit unknowingly [11, 12]. Ancient Egyptians, around 4000 years ago, incorporated PbS NPs into a traditional hair-dying recipe [11, 13, 14]. Another notable historical artefact is the Lycurgus Cup, crafted by the Romans in the fourth century AD. This cup exhibits a dichroic effect, appearing jade-like under direct light and translucent ruby under transmitted light, owing to the presence of Ag and Au NPs [11, 15]. In 1959, Richard Feynman, an American physicist and Nobel laureate, delivered what is considered the inaugural academic discourse on nanotechnology at the annual meeting of the American Chemical Society. Feynman's speech introduced the concept of nanotechnology [11, 16]. He emphasized that human's ability to manipulate matter is not constrained by natural laws but rather by a lack of tools and techniques. This notion laid the groundwork for modern nanotechnology, earning him the title of the father of this field [11, 17]. It is usually believed that Norio Taniguchi may have coined the term "nanotechnology" in 1974 [11, 16, 18]. Although nanotechnology was primarily a topic of discussion before the 1980s, its potential for future advancement was deeply embedded in the minds of scholars [11]. NPs have found extensive use in Ayurveda as well. Through scientific research, superhydrophobic surfaces have been developed for various applications, such as Lotusan self-cleaning paint (inspired by the lotus effect) and slippery liquid-infused porous surfaces (SLIPS) for refrigeration systems (inspired by nepenthes walls). Bhasma, a herbal-mineral metallic compound in Ayurveda, possesses nano dimensions typically ranging from 5 to 50 nm. Classical Indian alchemy, known as "Ayurveda Rasa Shastra," produces items utilized in the treatment of numerous chronic diseases [19].

Nanotechnology and nanodevices have found diverse applications across fields, such as cancer treatment, nanoscale electronics, hydrogen fuel cells, and nanographene batteries, owing to their remarkable versatility in altering physicochemical properties. By utilizing smaller-sized materials, nanotechnology enables precise adjustments at the molecular and substance levels, thereby enhancing material mechanical qualities or facilitating access to previously inaccessible regions of the body. NMs find commercial applications in various sectors, such as paints, cosmetics, electronics, environmental remediation, sensors, and energy storage devices. The untapped potential of these NMs poses a significant risk to society due to their wide array of applications, which could lead to unforeseen impacts on the environment and health. There is a concern that the waste generated by these innovative materials may surpass current waste management capabilities, rendering them inadequate for proper disposal [20].

Current environmental research and engineering have been significantly influenced by environmental nanotechnology, often referred to as E-nano. The production of NPs for nanotechnology holds the potential for generating substantially less waste, minimizing hazardous chemical synthesis, enhancing catalysis, and accelerating advancements in clean and sustainable technologies. Nanotechnology is revolutionizing the detection of harmful water-borne compounds and opening up new avenues for water purification, desalination, and decontamination. In the realm of pesticide detection, NM-based biosensors, both unit-molecular and array types, are under development [21]. Advanced treatment methods for pesticide-contaminated soil and water include innovative approaches such as ultrasound-promoted remediation and other sophisticated oxidation processes, alongside conventional methods like incineration, phytoremediation, and photochemical processes. Their hydrophobic nature, high reactivity, varied shapes and sizes, large surface area, biological interactions, durability, deformability, tendency to aggregate, and sensitivity to light are just a few of the distinctive characteristics of these NMs. These attributes significantly influence their interactions with other environmental contaminants [22 - 24]. As a result, they could potentially expedite and enhance the spread of these pollutants through air, soil, and water media. Recent studies indicate that NMs have the capacity to swiftly traverse aquifers and soils, potentially aiding in the rapid dissemination of hazardous substances over vast distances by acting as carriers for other pollutants.

This chapter aims to showcase a summary of environmental nanotechnology and examines the methods employed in the biogenic creation of NPs, as well as their analysis.

CLASSIFICATION OF NMS

Depending on their origin, size, structural arrangement, pore dimensions, and potential toxicity, NMs can be classified into five distinct classes. Fig. (**1**) provides a visual depiction outlining this classification of NMs into five groups [1]. These NMs possess unique attributes, such as a heightened surface area-t--volume ratio and diverse functionalities that markedly differ from those observed in bulk materials. These exceptional properties make NPs highly significant. Factors contributing to these distinctive NP features include electron confinement within the particles, modification of bandgap energies through size manipulation, and the substantial surface-to-volume ratio [22, 25]. Over the past decade, there has been significant interest in the development of innovative synthesis techniques and methodologies for various NMs, including metal NPs, carbon nanotubes (CNTs), graphene, quantum dots (QDs), and their composites, within the field of nanoscience and technology.

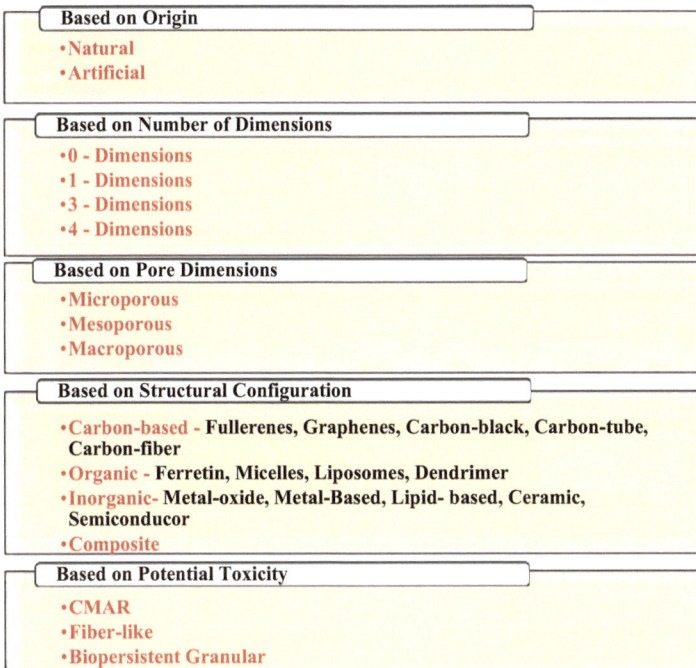

Based on Origin
- Natural
- Artificial

Based on Number of Dimensions
- 0 - Dimensions
- 1 - Dimensions
- 3 - Dimensions
- 4 - Dimensions

Based on Pore Dimensions
- Microporous
- Mesoporous
- Macroporous

Based on Structural Configuration
- Carbon-based - Fullerenes, Graphenes, Carbon-black, Carbon-tube, Carbon-fiber
- Organic - Ferretin, Micelles, Liposomes, Dendrimer
- Inorganic- Metal-oxide, Metal-Based, Lipid- based, Ceramic, Semiconducor
- Composite

Based on Potential Toxicity
- CMAR
- Fiber-like
- Biopersistent Granular

Fig. (1). General Classification of NMs [1].

PROPERTIES OF NMS

NMs can exhibit remarkable optical, magnetic, mechanical, electrical, and catalytic properties that differ significantly from their bulk counterparts. By precisely controlling parameters such as size, shape, synthesis conditions, and appropriate functionalization, it is possible to tailor NMs to achieve desired characteristics [1, 11]. This capability allows for fine-tuning of NM properties to meet specific requirements. Fig. (**2**) illustrates a broad spectrum of properties exhibited by NMs [19].

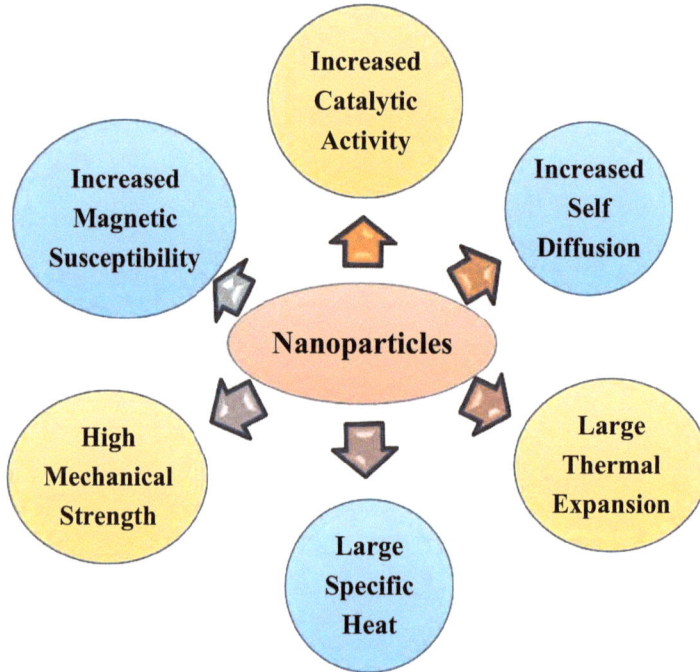

Fig. (2). NPs with novel properties [19].

Physical Properties

A bulk material's melting temperature is independent of its size; however, because NMs have unbounded surface atoms, their melting point lowers as particle size reduces [1, 26]. When a bulk substance is split into nanoscale components, its total volume stays the same, but its surface area grows, leading to an increase in the surface-to-volume ratio. The molecules or atoms on the surface are prone to aggregate and have a high surface energy [1, 27].

Magnetic Properties

The size of magnetic NPs can alter the magnetic behaviour of an element at the nanoscale. At critical grain sizes the size super-paramagnetic behaviour and coactivity is enhanced. At the nanoscale, nonmagnetic bulk materials can acquire magnetic properties [1]. For instance, while non-magnetic in bulk, gold and platinum are magnetic at the nanoscale [1, 28]. Magnetic NPs find application in biomedical fields, including magnetic fluid hyperthermia magnetic resonance imaging (MRI) and medication delivery [1, 29, 30].

Optical Properties

Localized surface plasmon resonance (LSPR) is an optical characteristic of NPs. Several investigations have demonstrated that the size of the NPs affects the line width. For instance, the emission light location shifts from the near-infrared (NIR) to the ultraviolet (UV) area when Au NP size decreases. Because NPs are so small, they have the potential to lose their LSPR and turn photoluminescent [1, 31]. By changing the nanoscale dimensions, visible light emission can be tuned due to quantum confinement in NMs. It has been found that the peak emission changes toward shorter wavelengths when the NMs' size decreases. Nanoscale color changes are possible for matter; for instance, gold nanospheres can change from yellow at 100 nm to red at 25 nm, greenish-yellow at 50 nm, and orange at 200 nm. Silver, on the other hand, can likewise change from orange at 200 nm to light blue at 90 nm and blue at 40 nm spherical thin film length [1, 32, 33].

Mechanical Properties

Elasticity, ductility, tensile strength, and flexibility are examples of mechanical qualities of materials that are crucial to their use. Impact on an NM's mechanical attributes, including its toughness, yield strength, elastic modulus, and hardness in comparison to bulk materials [1, 34]. Nanostructured materials get stronger and harder when grain size and grain boundary deformation decrease. All that is required for an improvement in mechanical strength is a decrease in the likelihood of flaws and an increase in imperfection. It increased the toughness and hardness of the alloy and the ceramic super plasticity [1, 11, 35].

Thermal and Electrical Properties

At the nanoscale level exceptional thermal and electrical conductivity can be seen according to the nature of the NM. For instance, Graphene derived from graphite is an important example that shows these properties [36]. In ceramics, NMs can improve conductivity, but in metal, they can increase electric resistance. The quantum effect takes control at the nanoscale when electron delocalization happens along the axis of nanotubes, nanorods, and nanowires. Discrete energy states take the place of the energy bands due to electron confinement resulting in conducting materials acting as either insulators or semiconductors. This outcome suggests that the metal is transitioning into a semiconductor. For instance, depending on their nanostructure, carbon nanotubes can function as semiconductors or conductors. The number of electron wave modes that contribute to electrical conductivity is decreased in well-defined quantized increments in order to decrease the wire's diameter [1].

Chemical Properties

The chemical properties of this substance dictate its applications. These features include the stability and sensitivity of the NPs to components, including moisture, environment, heat, and light, as well as their reactivity with the target [1]. Applications are determined in part by the NPs' flammability, corrosiveness, anti-corrosiveness, oxidative potential, and reduction potential [37]. NMs exhibit unique catalytic features like reactivity, selectivity, and catalysts [28]. The catalytic performance is significantly improved by catalytic NPs made by 2D sheets of different NMs [38]. In an effort to improve performance, catalysts have been atomically distributed on 2D NM sheets [39].

SYNTHESIS OF NMS

The synthesis of NMs is done primarily using two methods: the top-down approach and the bottom-up approach, as shown in Fig. (**3**).

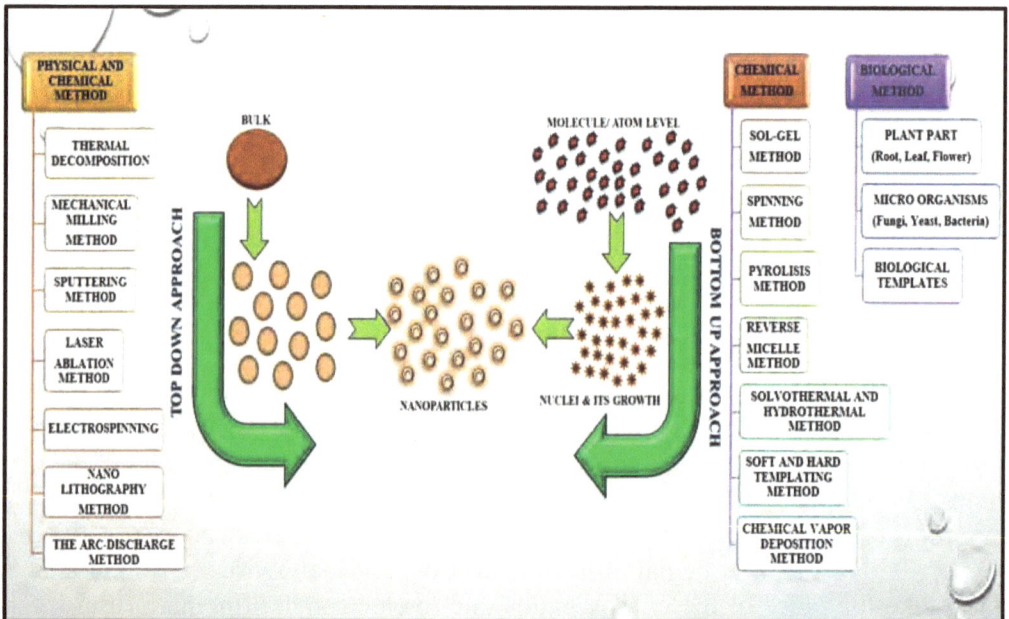

Fig. (3). Top-Down and Bottom-Up Approaches in synthesis of NMs [11].

Top-down Approach

The destructive process, sometimes referred to as the top-down method, breaks down bulk materials into tiny components, which subsequently turn into NMs [1]. Examples of the top-down approach include thermal decomposition, mechanical or ball milling, sputtering method, laser ablation method, electrospinning, nanolithography method, and the arc-discharge method [37 - 41]. During the thermal decomposition process, heat was the cause of the breakdown. It is an endothermic reaction. Heat causes the chemical bonds to break and split into smaller ones. The metal undergoes specific temperature degradation to create the NPs, which are then generated by a chemical reaction [1]. Again, in the top-down approach, the most commonly employed and economical way to produce different types of NPs is mechanical milling. When creating mixes of various phases, mechanical milling works well and is useful for creating nanocomposites [11]. On the other hand, the process of depositing a thin layer of NPs by means of expelled particles striking ions is known as sputtering Fig. (**4**). However, in laser ablation synthesis, the target material is struck by a strong laser beam, which creates NPs. Besides the above-mentioned techniques, one of the most straightforward top-down techniques for creating core-shell ultrathin fibers (nanostructured materials) on a big scale is electrospinning [11]. In this aspect, another useful technique for producing nanoarchitectures using a concentrated electron or light beam is nano lithography. In order to construct the appropriate shape and structure, it is necessary to print the required shape or structure on a light-sensitive material and then selectively remove some of the material. The capacity of nanolithography to create a cluster from a single NP with the required size and form is one of its primary advantages [1]. The production of carbon-based materials such as few-layer graphene (FLG), fullerenes, carbon nanohorns (CNHs), CNTs, and amorphous spherical carbon NPs is its most well-known use [11, 42].

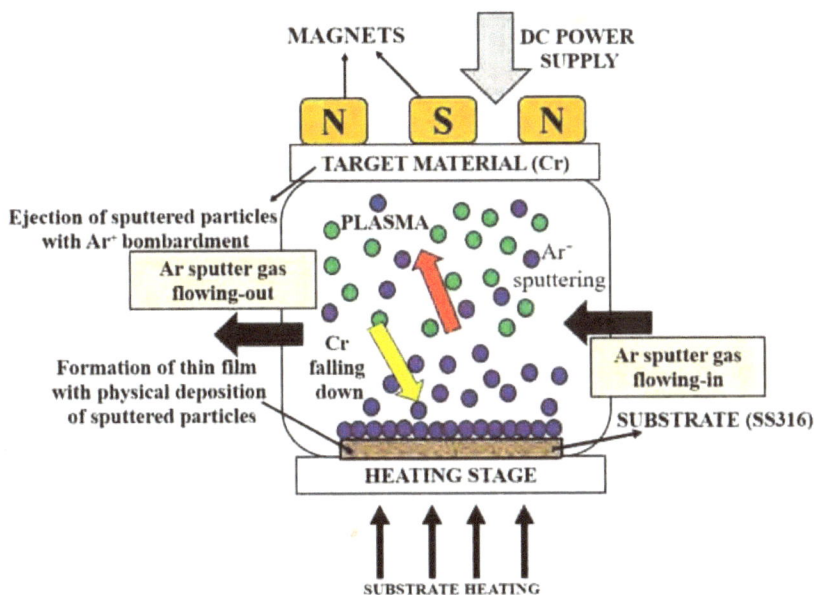

Fig. (4). A schematic diagram of the DC Magnetron Sputtering Process [11].

Bottom-up Approach

The constructive process, sometimes referred to as the bottom-up method, involves the building of material from atoms to clusters to NPs. The bottom-up approach can be further classified into two categories, namely chemical methods and biological or green synthesis [1].

Chemical methods include sol-gel method, pyrolysis method, CVD method, spinning method, reverse micelle method, solvothermal and hydrothermal method, and soft and hard templating method. Synthesized NMs encounter several challenges, such as stability, bioaccumulation, toxic features, high costs, and skilled operations when applied in various applications [43]. To address the aforementioned challenges, it is essential to improve the types, behaviors, and qualities of NMs in practical applications. Nonetheless, these limitations are also opening up exciting new opportunities in this emerging field of research [43]. On the other hand, biological synthesis can build strong, biofunctional NPs by using a variety of species as stable, environmentally friendly precursors.

Biological or Green Synthesis of NMs

The emergence of "green synthesis," as depicted in Fig. (**5**), is gaining significant attention in contemporary materials science and technology research and development as a means to overcome these limitations [19]. Green synthesis aims to enhance the environmental friendliness of materials and NMs by regulating, controlling, and remedying processes. This approach emphasizes waste minimization, pollution control, and the use of non-toxic solvents along with renewable feedstock to elucidate key concepts of green synthesis. Biological or green synthesis involves various elements such as plant parts (leaves, roots, flowers), microorganisms (bacteria, yeast, fungi), and biological templates. These biobased green synthesis techniques rely on different reaction parameters, including solvent type, temperature, pressure, and pH (neutral, basic, or acidic) [43].

Recently, biosynthesis of NPs has been accomplished by employing a range of species, both unicellular and multicellular. It is feasible to think of NP production as a bottom-up process in which the organism produces biomolecules such as proteins, carbohydrates, polysaccharides, and enzymes, which subsequently oxidize or reduce metallic ions to form NPs [44, 45]. The formation of microbial NPs is not completely understood since different types of bacteria interact with metallic ions in different ways. Finally, a microorganism's contacts, pH, temperature, and metabolism all influence the properties of biosynthesized NPs. Depending on the microorganism, NPs might form either inside or outside the cell [45]. Researchers employed cell extracts to manufacture NPs physiologically.

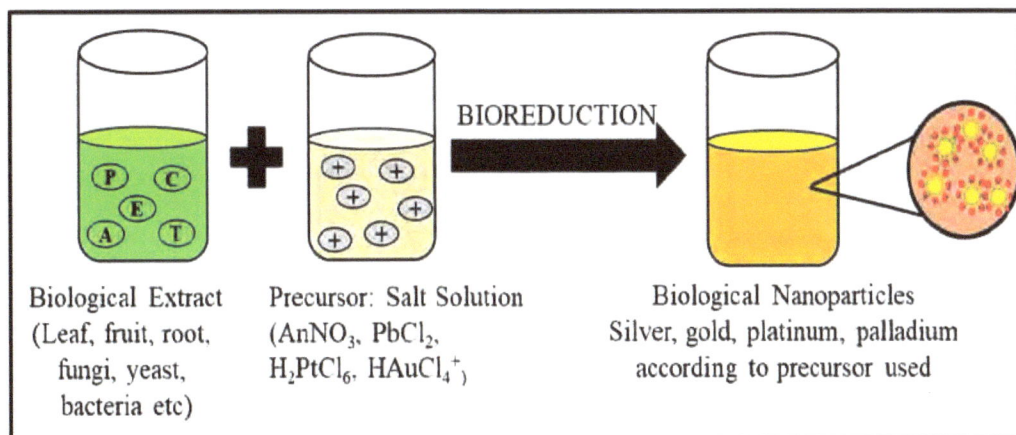

Fig. (5). Biological synthesis of NPs.

Plant-mediated Synthesis of NPs

Different parts of the plant and their metabolites have been effective in the synthesis of NPs [46]. Green synthesis of NPs using plants is a cost-effective, one-step procedure that produces a large number of metabolites while being environmentally benign [47]. The green synthesis of NMs has been achieved using plant extracts from a variety of sources, including *Pinus resinosa, Curcuma longa, Anogeissus latifolia, Musa paradisiaca, Pulicaria glutinosa, Cinnamomum camphora, Diospyros kaki, etc* [45]. Flavones, organic acids, and quinones, naturally occurring in plants, serve as excellent reducing agents for the creation of NPs. Plants have the capability to accumulate specific levels of heavy metals in various parts of their anatomy [43]. Additionally, plants possess a diverse array of active components, including vitamins, polysaccharides, alkaloids, terpenoids, saponins, and tannins. Consequently, biosynthetic methods utilizing plant extracts are increasingly recognized as practical, facile, cost-effective, and efficient means to produce NPs on a large scale, serving as viable alternatives to traditional preparation techniques. In the "one-pot" synthesis process, various plants can be utilized to stabilize and reduce metallic NPs. To explore the diverse applications of metal/metal oxide NPs further, many researchers have employed a green synthesis approach to fabricate these particles using plant leaf extracts. Biomolecules present in plants, including proteins, carbohydrates, and coenzymes, exhibit remarkable capabilities in converting metal salts into NPs. Similar to other biosynthesis methodologies, the synthesis assisted by plant extracts was initially employed to investigate gold and silver metal NPs [43]. Silver and gold NPs have been successfully synthesized using different plants such as Geranium, Aloe Vera, Oat, Alfalfa, Tulsi, Lemon, Neem, Coriander, Mustard, and Lemon Grass. While most studies have focused on the *ex vivo* synthesis of NPs, it has been demonstrated that metallic NPs can also be produced *in vivo* within living plants through the reduction of metal salt ions absorbed as soluble salts [43]. Additionally, *Helianthus annuus, Medicago sativa,* and *Brassica juncea* have been shown to synthesize NPs of zinc, nickel, cobalt, and copper *in vivo.* Various plant leaf extracts, including those from *Coriandrum sativum, Calotropis gigantea, Acalypha indica, Hibiscus rosa-sinensis, Camellia sinensis,* and *Aloe barbadensis Miller,* have also been utilized for the preparation of ZnO NPs [43]. Fig. (**6**) illustrates the synthesis of NPs using plant extracts [48].

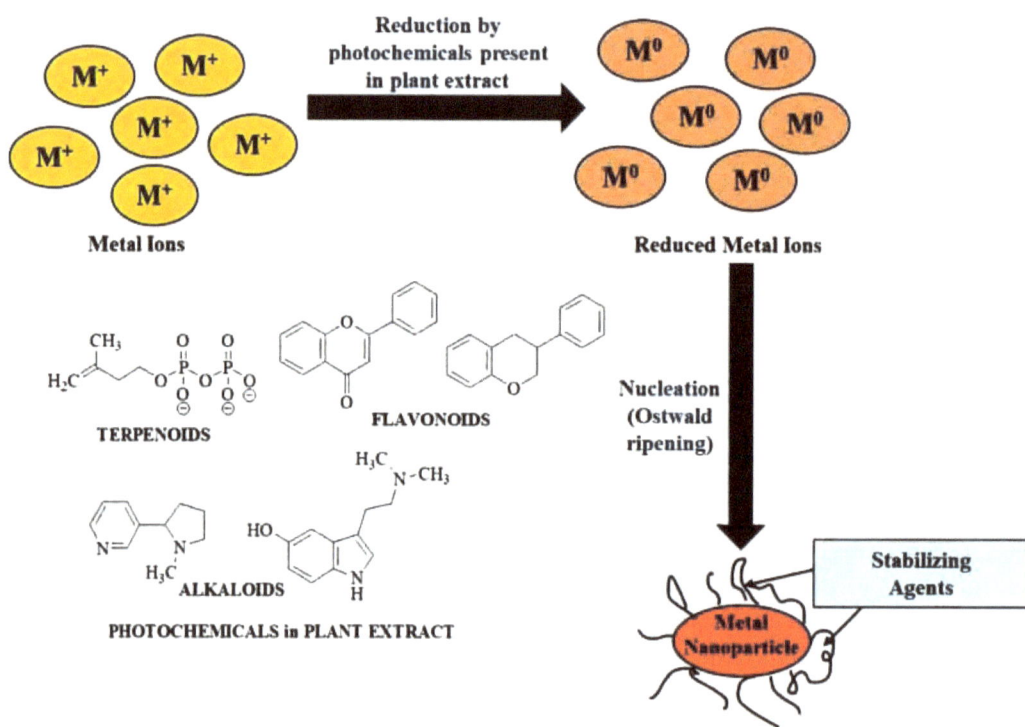

Fig. (6). Plant-mediated biological synthesis of NPs [48].

Microorganism-based Synthesis of NPs

The majority of research has focused on prokaryotes as a means of generating NMs. Because of their environmental ubiquity and capacity for adaptation, bacteria are excellent research subjects. They also grow quickly, are easy to care for, and do not cost much. The growth conditions of temperature, oxygenation, and incubation period can all be readily altered. It is known that bacteria can synthesize inorganic substances both inside and outside of their cells. *Pseudomonas stutzeri* was used to produce Ag-NPs outside of the cells [45].

Various commercial biotechnological applications, including genetic engineering, bioleaching, and bioremediation, have extensively utilized bacterial species [49]. Bacteria are significant candidates for the synthesis of NPs due to their ability to reduce metal ions [50]. Redox reactions *via* intracellular/extracellular pathways commonly occur when metals are reduced into metal ions. Enzymes performing electron shuttle donor activities are involved in the production of silver NPs from *Bacillus licheniformis* [51, 52]. The first bacterial gold NPs were created using *Bacillus subtilis*. Various bacterial strains have been widely employed for the synthesis of bio-reduced silver NPs and gold NPs with distinct sizes and shapes. Magneto tactic bacteria in the ocean's depths produce magnetic particles in anaerobic environments, while photosynthetic bacteria like *Rhodopseudomonas capsulata* produce extracellular gold NPs ranging in size from 10 to 20 nm [43]. Fig. (**7**) illustrates the mechanism of intracellular-cell-bound synthesis of gold NPs using *L. kimchicus (DCY51^T*. Actinomycetes are other microbes that have an important role in the synthesis of NMs. Actinomycetes generate stable, highly polydispersed NPs with a wide range of therapeutic applications. For instance, Streptomyces sp. was found to be effective in producing Zn-NPs, Ag-NPs, and Cu-NPs [45, 54].

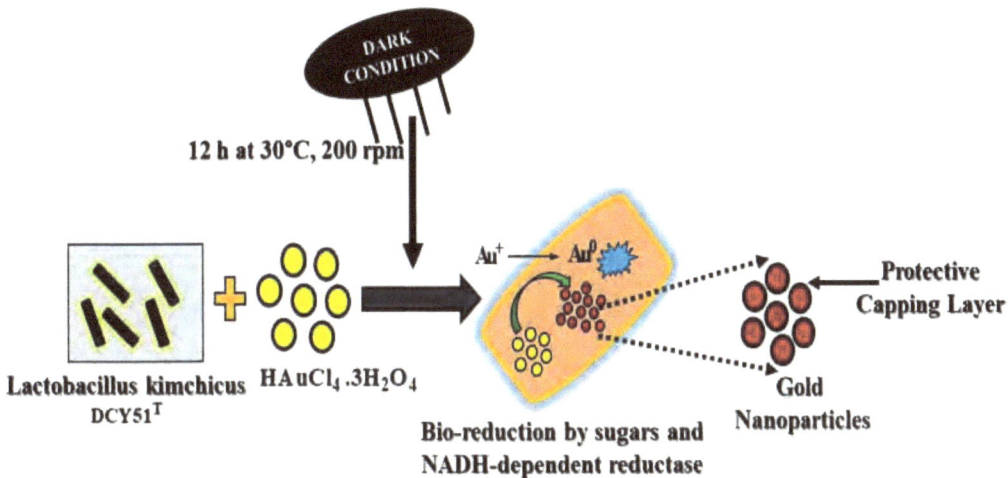

Fig. (7). Gold NPs synthesis using L. *kimchicus (DCY51^T)* [53].

Fungi and Yeast Fabrication of NPs

Another highly efficient method for producing NPs with distinct morphologies involves the production of metal/metal oxide NPs by fungi. Fungi serve as superior biological agents for creating NPs due to their intracellular enzymes. Additionally, enzymes, proteins, and reducing agents present on the cell surfaces of fungi offer many advantages over other organisms. Examples of metal/metal oxide NPs produced using various fungal species include silver, gold, titanium dioxide, and zinc oxide. Nitrate reductase and anthraquinones were identified as the mechanisms by which *Fusarium oxysporum* produced silver NPs [19, 55]. *Sargassum wrightii* algae are utilized in the synthesis of extracellular gold NPs.

Because of their advantages over bacteria in NP synthesis, fungi have drawn great interest in the manufacture of metallic NPs. Significant benefits include the simplicity of downstream processing and scaling up, the economic feasibility, and the presence of mycelia, which offers a higher surface area. Fungal-based NMs are created by a biomineralization technique that comprises the reduction of various metal ions by internal and external enzymes and proteins [56 - 58]. Further research into NM production has been undertaken employing species such as Penicillium, Aspergillus, Fusarium, Verticillium, and Rhizopus. Mohammed *et al.* created hexagonal and nanorod ZnO-NPs and high mono-dispersity components (uniformly distributed) with no agglomeration [57]. This was accomplished by employing *A. niger* and *F. keratoplasticum.* Furthermore, the authors proposed that the fungus-buried protein be bound, lowering orbicular ZnO NPs and preventing the NPs from clumping together. ZnO NPs were also amalgamated utilizing filtrate-cell free (FCF) *Aspergillus terreus* replacement. Following the elucidation of zinc acetate snow with a size range of 10-45 nm, the FCF was patented for producing NP [45, 56 - 58]. Furthermore, FTIR spectra investigation indicated that the produced ZnONPs contained proteins and other biological components. The findings show the various tools utilized to analyze the created ZnO-NPs, including TEM, FTIR, TGA, and XRD studies. At room temperature, *Aspergillus terreus* amalgamated CuO NPs that were separated to remove copper from linked circuits and allowed to persist as nanoforms. CuO NPs were synthesized extracellularly from the biomass of fungal cells isolated from Egyptian soil. The consistency and coating agents around the CuO NPs were discovered to be related to amide groups inside proteins, as revealed by IR analysis. CuO NPs were validated using multiple methods. Outside of cells, NPs are mass-produced and easily processed. The size, monodispersity, particle location, and characteristics of NPs generated by yeast strains from different genera vary greatly. In the majority of the yeast species being studied, these molecules stabilize the complexes and establish the mechanism for NP formation. Resistance refers to a yeast cell's ability to convert ingested metal ions into intricate polymer molecules that are safe for the organism. Yeast production can be easily regulated in laboratory settings for mass synthesis of metal NPs, and there are various advantages to using basic nutrients and lipids to produce yeast strains. Several researchers have documented the successful synthesis of NPs and NMs utilizing yeast. Strains of the yeast *Saccharomyces pombe* and *Candida glabrata* were characterized in order to produce intracellular silver, titanium, cadmium sulfide, selenium, and gold NPs. The extracellular bio-formation of Se

NPs was carried out using yeast extract of *Saccharomyces pombe*. The absorption peak at 300 nm in UV-Vis spectroscopy indicates SeNP synthesis [59 - 61].

Algae Fabrication of NPs

Algae are marine microorganisms that have been shown to manufacture metallic NPs and absorb toxic pollutants such as heavy metals from the environment. If algae processing processes that are both upstream and downstream economically feasible are discovered, algae may have a significant economic impact in the future. The ability of algae to absorb heavy metal ions and restructure them into more flexible forms is widely understood. Because of their ability to reduce metal ions, the synthesis of NMs from a variety of algal materials has emerged as one of the most inventive fields of biochemical approach. Brown, red, blue-green, micro, and macro green algae are the most extensively explored for NP synthesis. For example, reducing tetrachloroaurate ions to make gold NPs was employed to culture dry Chlorella vulgaris microalgae cells for Au-NP production. Brayner *et al*. revealed how cyanobacteria can be used to synthesize palladium, platinum, silver, and gold NPs. *Colpmenia sinusa*, *Ulva facita*, *Pterocladia capillacae*, and *Jania rubins* were the four marine macroalgae that generated NP [45, 62 - 65].

Solvent-based Green Synthesis

'Green synthesis' encompasses two primary approaches: (i) Using water as a solvent and (ii) Using natural sources or extracts as primary components [43]. Ionic and supercritical liquids are famous examples of such burgeoning disciplines. Ionic liquids (ILs) are chosen over other solvents because ILs can efficiently dissolve gases, polar organic molecules, and a variety of metal catalysts, hence aiding biocatalysis, and can function in a wide temperature range due to their constructive thermal stabilities. The majority of them begin to degrade beyond 300 or 400 °C and melt below room temperature, allowing for a wider range of synthesis temperatures than water. ILs do not coordinate, unlike other polar solvents, and their polarities resemble those of alcohol. (f) ILs have dual functionality since they include both cations and anions. ILs can create a wide range of metal NPs, including Au, Ag, Al, Te, Ru, Ir, and Pt [66 - 69]. Since ILs may function as both a protective agent and a reductant, the process of producing NPs is simplified. Bussamara *et al*. performed comparative studies on manganese oxide (Mn_3O_4) NPs utilizing imidazolium ionic liquids and oleylamine. ILs yield smaller NPs with improved dispersity compared to oleylamine solvents (12.1 ± 3.0 nm) [70]. The biodegradability of ionic liquids makes them unsuitable for the production of metallic NPs. Many innovative benign ILs with high biodegradation efficiency are being developed to address these non-biodegradability challenges. Lazarus *et al*. created silver NPs in an ionic liquid ($BmimBF_4$) [71]. NPs of both small, isotropic, spherical, and large, anisotropic hexagonal shapes were created. Supercritical solvents are suitable for "green synthesis" because their properties, such as viscosity, density, and thermal conductivity, alter substantially. Carbon dioxide is the most practical, safe, and inert supercritical fluid. Supercritical water is also an effective solvent system for a variety of procedures. Water's critical pressure is 22.1 MPa, and its temperature is 646 K. Supercritical carbon dioxide can be used to produce copper and silver NPs. According to Sue *et al.*, reduced solubility of metal oxides around the critical point might produce supersaturation and, eventually, the formation of NPs. Kim *et al.* synthesized tungsten oxide (WO_3) NPs using subcritical and supercritical water and methanol [72 - 77]. Table 1 lists some of the biologically produced NPs [22].

Table 1. Examples of NPs that are synthesized by the "green" synthesis method [22].

Nanoparticle	Biogenic Synthesis		Application	Intra/Extra
Gold		*Lactobacillus Kimchicus*	Drug Delivery, Cancer Diagnostics	Intra
Silver		*Bacillus subtilis*	Antibacterial	Extra
Zinc Oxide	Bacteria	*Staphylococcus aureus*	Antibacterial	Intra
Nano selenium		*Bacillus sp.*	Biomedical	Extra
Platinum and Palladium		*Desulfovibrio vulgaris*	Catalyst	Extra
Iron Oxide		*Aspergillus niger* BSC-1	Wastewater treatment	Extra
Copper		*Aspergillus niger*	Antidiabetic and Antibacterial	Extra
Cobalt Oxide	Fungi	*Aspergillus nidulans*	Energy Storage	Extra
Silver		*Aspergillus terreus*	Anticancer, Antibacterial	Extra
Platinum		*Fusarium oxysporum*	Nano Medicine	Extra

FACTORS AFFECTING THE SYNTHESIS OF NPS

The biosynthesis of NPs is influenced by different parameters, including temperature, biomass, precursor concentration, time constant, pH, and the presence of certain enzymes, as shown in Fig. (**8**) [45]. Metal NMs can be made to change in size and shape by either forcing changes in their surroundings or moving their functional molecules. The concentration of metal ions, pH of the reaction mixture, and temperature are critical factors in determining the size and shape of NPs. A few of the planning variables that may have an effect on the rate of intracellular formation of NPs include substrate concentration, pH, temperature, and exposure time. In order to optimize the mycosynthesis of

AgNPs, ZnO NPs, and AuNPs, researchers looked into different parameters like temperature, precursor concentration, pH, and inoculums of biomass [78 - 81].

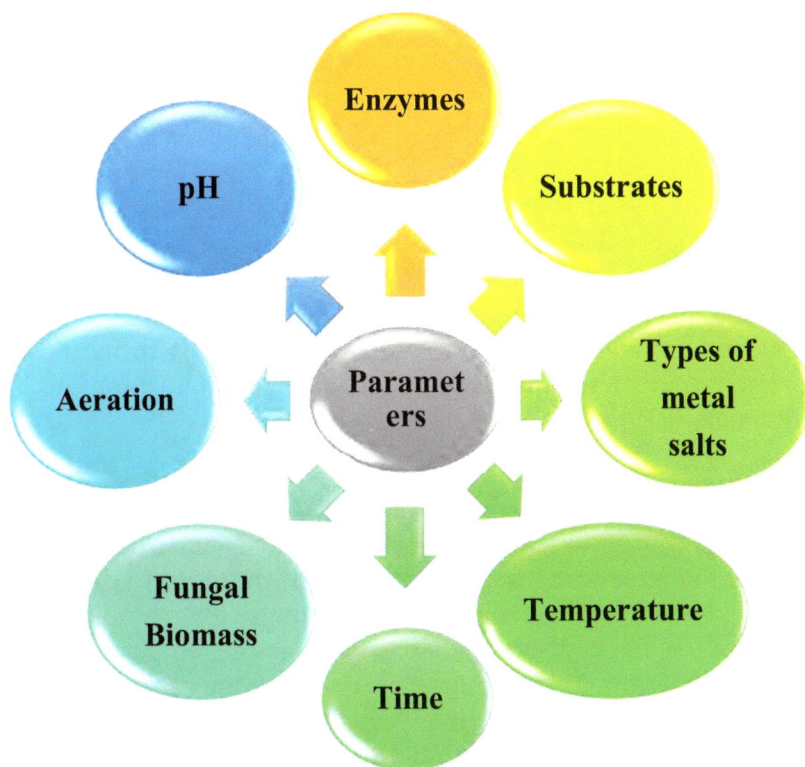

Fig. (8). Factors affecting NPs biosynthesis [45].

Effect of pH

pH influences the formation of NPs, affecting both particle size and shape. Yang and Li demonstrated that products generated at a lower pH exhibited less uniform shape and a tendency to agglomerate. Particles of the proper size and shape can be reliably generated during the synthesis of NPs at different pH levels. During the nucleation and development stages of NPs, pH causes the particles' local surface to protonate and deprotonate molecular atoms. The NPs form a cluster distribution in the colloidal stage, which prevents aggregation at alkaline pH levels. According to Armendari *et al.*, pH is critical in the synthesis of gold NPs. At pH levels of 2, 3, and 4, the resulting NPs are irregular, rod-shaped, icosahedral, hexagonal, and decahedral multi-twinned. Thus, during the nucleation and early development stages, when particle sizes are small, thermodynamically favorable configurations include cubic, truncated octahedron, rhomb-dodecahedron, octahedron, and octagon [19, 82 - 86].

Effect of Concentration of Precursor and Reducing Agents

The concentration-reducing agents and precursors influence the size of the NPs produced. This phenomenon could be induced by an overabundance of reducing chemicals attached to the surface of prepared nuclei, which intensifies the secondary reduction of silver ions. Larger NPs are the result of an increase in the NP growth rate. However, an excess of reducing agents may increase the bridging between produced NPs, leading to NP aggregation. This could be due to an overabundance of metal ions being absorbed on the surface of the prepared nucleus, where larger NPs were produced during the secondary reduction process. According to a few researchers, the proportion of triangles formed in the reaction medium as a function of varied plant extract concentrations indicates that as extract concentration increases, more spherical particles form. To achieve the desired NP size, the reducing agent and precursor concentrations must be at optimal amounts [19, 87, 88].

Effect of Temperature

The rate at which large to small NPs are generated increases with temperature during the NP manufacturing process. High temperatures are often conducive to larger NP nucleation and development. Although low temperatures promote growth, researchers have observed that as the reactive temperature rises, so does the overall reaction rate. Temperature has a wide range of impacts on the size of NPs under both sufficient and insufficient precursor quantities due to its remarkably different influence on the growth kinetics constant k2 and the nucleation kinetics constant k1, respectively. As the reaction temperature rises, the majority of metal ions are consumed in the nucleus synthesis, preventing the secondary reduction process from occurring on the surface of the newly formed nuclei [19]. As a result, there is a greater output of small, widely distributed NPs. The period of incubation in the reaction media has a substantial impact on the characteristics of biologically generated NPs (NPs). The particles' shelf-life may influence their potential; particle aggregation or shrinkage may result in character changes over extended incubation. As a result, there is a greater output of small, widely distributed NPs. The period of incubation in the reaction media has a substantial impact on the characteristics of biologically generated NPs. The particles' self-life may influence their potential; particle aggregation or shrinkage may result in character changes over extended incubation [89 - 93].

MECHANISM OF BIO-SYNTHESIS

Many biological models have distinct mechanisms for the intra- and extracellular production of NMs, as shown in Fig. (9). The intracellular production of NPs is significantly influenced by the cell walls of microorganisms. The enzymes found in the cell wall of the organism bio-reduce the metal ions to NPs, which are subsequently released through the cell wall. We still do not know the precise process for turning biological models into NPs. The enzymes present in the cell wall transform metal ions and metal oxide into NPs which subsequently diffuse out from the cell membrane. A step-by-step method employing *Verticillium sp.* for the intracellular production of NPs explains how bio-reduction, capping, and trapping work in the production of NPs. The cell surface interacts electrostatically and traps the metal ions when it comes into touch with them. Metal ions are transformed into metal and/or metal oxide NPs by the enzymes found in the cell wall. The process involves proteins (enzymes) that are generated by the cell and the intracellular redox value, which reduces Ag^+ to Ag^0. These responses could take place extracellularly or inside cells. Biological compounds function as capping and reductants. They are present in plant extracts and are released by bacteria and fungi. These materials include sugar, carbohydrates, proteins, and enzymes that change metallic ions from (M^+) to (M^0) through an oxidation/reduction process. By aggregating and forming clusters of NMs, the reduced metallic form can be verified by a color change in the reaction mixture [94, 95].

ADVANTAGES OF BIOLOGICAL SYNTHESIS OF NPS

Surprising progress has been made in the field of bioinformed nanostructures and their applications in the past few years. Conventional benefits of bioinformation include ecological creation and maintenance, economic efficiency, and the biocompatibility of mixed NMs. Another benefit of the biogenic process of fusion is that it eliminates the need for an additional stage in physico-chemical mixes where bioactive facilities or microbes are coated or attached to the surface of NPs in order to produce consistent and pharmacologically active atoms [96, 97]. Further, compared to physicochemical procedures, the production of NPs takes a lot less time. Using different microorganisms, a number of intermediates have created quick synthesis pathways with superior NP yield. Despite many advantages that come with the biotic pathway for the formation, there are still important and intriguing issues related to the size and polydispersity of metal oxide NPs. Furthermore, a major effort is made to improve unit size control, feature and amalgamation competency. Consequently, a number of contemporary intelligences have created a unified method for the biosynthesis of monodispersity NPs in terms of size and shape. A highly promising method for efficiently producing microorganisms and NPs is the creation of biofilms. It has been lately acknowledged that the most energetic type of bacterial growth is represented by biosynthesized NPs. Additional advantages of incorporating NPs into microorganisms include their large exterior areas and high mass demands, which can result in increased real and local biosynthesis. Very little is known about the calming mechanism of NPs in biofilms, despite a wealth of study on metal oxide NP fusion in species like bacteria and fungus. The thorough and methodical analysis of the mechanical structure involved in the biofabrication of NPs is one of the main obstacles to bio-mediated synthesis [96 - 99]

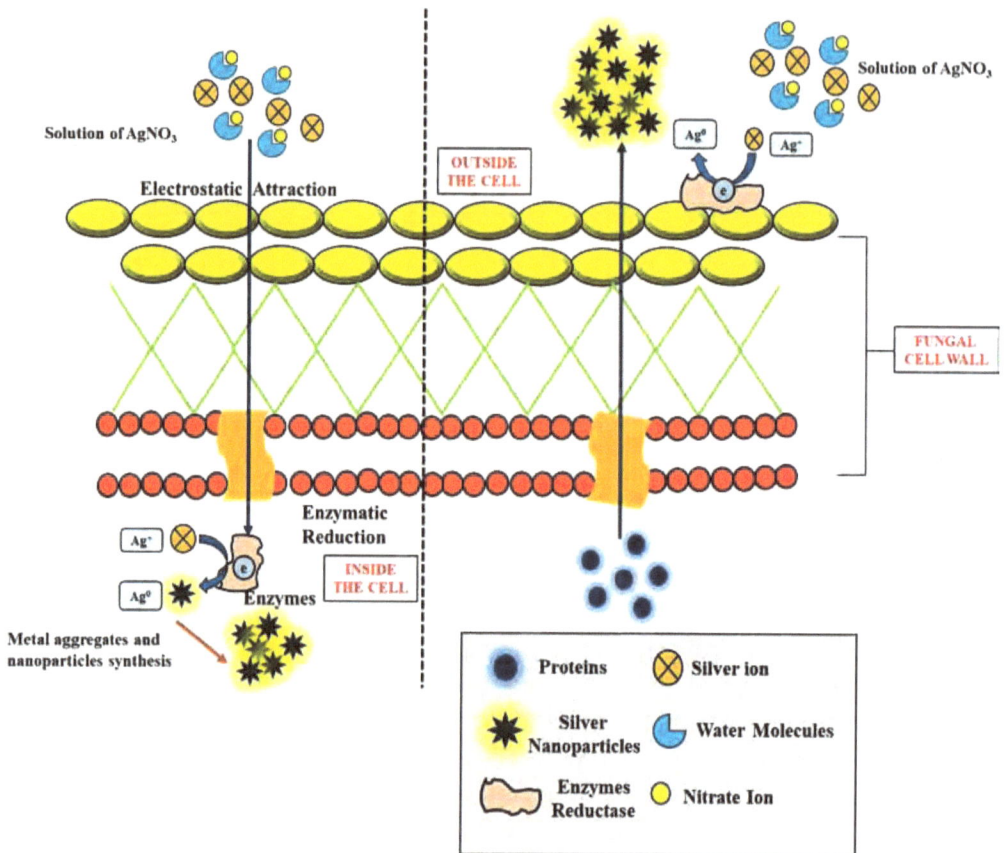

Fig. (9). Intracellular and extracellular mechanism biosynthesis [45].

STABILIZATION AND CHARACTERISATION OF NPS

Numerous studies using extracts from different plant sections have been published on the biogenesis of metal and metal oxide NPs. Hashem and Salem identified the production of selenium NPs (SeNPs) utilizing *Urtica dioica* resulting in a pure, crystalline, spherical shape with sizes ranging from 5 to 43 nm. The UV-Vis spectroscopy absorption peak at 300 nm indicates the production of Se NPs. Additionally, a variety of instruments, including TGA, FTIR, TEM, DLS analysis, and SEM-RDX, were used to characterize the produced SeNP. SeNPs produced through biosynthesis were examined for their antibacterial and antitumor effects [100 - 102]. According to published research, proteins stabilize gold and silver

NPs as they develop. Proteins can attach to gold NPs *via* other protein groups or free amine (NH) groups. These proteins might contain one or more enzymes that cap the gold NPs produced by the reduction process and lower chloroaurate ions. Thus, the stability and capping of the gold NPs are influenced by a different protein. The most frequent method of producing AuNP involves reducing tetrachloroauric acid with different reducing agents. However, because of their high surface energy, gold NPs are very reactive. It is necessary to use a special stabilizer to stop them from building up or precipitating. One method that is frequently used to passivate the surface of AuNP is the biomolecule monolayer. Recently, there has been a lot of promise shown for the application of homopolymers and block polymers that may effectively stabilize AuNP through steric stabilization in novel materials. It is crucial to remember that, particularly when intelligent polymers are used, polymer chains adsorbing on the surface of AuNPs can both functionalize and increase the stability of the gold cores. Furthermore, stabilizing agents that are chemically or adsorbed onto the AuNPs' surface are required. These stabilizing agents, often known as surfactants, are typically charged, which makes the similarly charged NPs resist one another and achieve colloidal stability. AuNPs can be stabilized using a broad variety of stabilizers, including ligands, polymers, surfactants, dendrimers, biomolecules, and others. An essential stage in the biosynthesis of NPs is the physico-chemical characterization of the particles that are generated. Controlling the synthesis of NPs for industrial usage can be made easier by having a better understanding of parameters, including size, shape, surface area, uniformity, and others. It is important to characterize the NMs using a variety of techniques in order to comprehend their varied physio-chemical properties [100 - 104]. The most regularly used tools are divided into the following groups [45]:

  i. X-Ray Diffraction (XRD)
 ii. Scanning Electron Microscopic (SEM)
iii. Transmission Electron Microscope (TEM)
 iv. Ultraviolet-Visible Spectroscopy (UV-Vis)
  v. Fourier Transform-Infrared Spectroscopy (FTIR)
 vi. X-Ray Photoelectron Spectroscopy (XPS)
vii. Atomic Absorption Spectroscopy (AAS)

Examining the structure is essential in understanding the composition and properties of the adhesive compounds. It provides a range of details about the bulk properties of the subject matter. The most popular techniques for investigating the structural properties of NMs include the Zieta size analyzer, FTIR, XPS, XRD, and energy-dispersive X-Ray (EDX). XRD is one of the most important methods for exposing the structural characteristics of NMs. It has enough details about the phases and crystallinity of NMs. Additionally, it provides a ballpark estimate of particle size based on the Debye Scherer computation. XPS is frequently utilized to determine the precise elemental ratio and type of bonding between the elements in NPs materials. UV-Vis is a well-known optical instrument for researching the optical properties of NMs. Since most of the characteristics of NMs are determined by their morphology, morphological features are always of great interest. Although there are several methods for morphological characterization, the most applicable ones are the microscopic approaches like TEM and SEM. SEM is a technology that provides all pertinent information on nanostructures at the nano scale through electron scanning. This method has been applied by many researchers to explore not only the form of their nanostructures but also the dispersion of NMs in bulk or composite materials. Likewise, since TEM is based on the concept of electron transmittance, it might provide details about the bulk materials at low to different magnifications. Investigating the various morphologies of NMs is done using this method [45, 105 - 107].

IMPACT OF ENVIRONMENTAL NANOTECHNOLOGY

Environmental nanotechnology has profoundly influenced all facets of the field, spanning from identification to remediation.

Monitoring and Sensing

Heavy metal analysis is often carried out using techniques such as atomic fluorescence spectroscopy (AFS), atomic absorption spectrometry (AAS), atomic emission spectrometry (AES), and inductively coupled plasma mass spectrometry (ICP-MS). Methods used to detect organic contaminants include high-performance liquid chromatography (HPLC), UV spectroscopy, gas chromatography (GC), GC-MS, and LC-MS. Integrating these analytical techniques enables the comprehensive monitoring of pollutants in soil, air, and water. However, these analyses necessitate costly instrumentation, complex operating methods, and highly skilled personnel. With advances in NMs, photoelectrochemical analysis has developed as a promising tool for chemical and biological monitoring that combines light response and chemical sensing. Furthermore, self-propelled nano/micro motors have recently appeared as an innovative way of environmental sensing. The emergence of nanopollution opens up prospects for the creation and testing of novel monitoring tools [104 - 107].

Environmental Nanocatalysis

Nanocatalysis involves conducting chemical reactions using nanostructured catalysts, which play a crucial role in environmental nanotechnology. For instance, advanced oxidation processes (AOPs) can be significantly improved by catalysis over nanostructured materials. Organic contaminants can be effectively degraded by the Fenton reaction, which uses hydrogen peroxide and ferrous ions to produce hydroxyl radicals. Its drawbacks, such as a limited pH range and sludge formation, can be overcome, though, by using Fenton-like reactions made possible by nanocatalysis. Using sulfate radicals, which have demonstrated exceptional efficacy in breaking down organic contaminants, is another viable strategy. Peroxymonosulfate (PMS) and persulfate (PS) activation techniques have progressed from homogeneous catalysis, heat, and UV light to heterogeneous catalysis using nanostructured catalysts. Notably, the growth of nanocatalysis not only improves catalytic efficiency but also broadens our knowledge of novel mechanisms. Other significant fields that could benefit from the effective use of nanocatalysis include catalytic ozonation, photocatalysis, selective catalytic reduction (SCR), and the elimination of volatile organic compounds (VOCs) and persistent organic pollutants (POPs) [101 - 103].

Adsorption and Nanosorbents

Adsorption, a well-established technology in environmental remediation, has been widely utilized, with activated carbon being among the most commonly used adsorbents. The introduction of NMs has expanded the scope of adsorbents significantly, giving rise to nanosorbents. In the realm of carbon-based materials alone, a variety of nanosorbents have emerged, including nanotubes, quantum dots, nanodiamonds, graphene, nanospheres, fullerene, and mesoporous carbons, each with diverse dimensional properties. A wide range of nanosorbents, many of which are based on graphene, can be generated *via* functionalization processes for a variety of adsorptive applications [105].

Membrane and Nanomembrane

Membrane technology has found extensive use in gas separation, as well as water and wastewater treatment. Traditionally, polymers and ceramics have been employed to fabricate symmetric or asymmetric membranes for these applications. Nonetheless, recent developments in NMs have produced nanoscale-thick membranes or nanomembranes. These innovative membranes offer unique properties that significantly expand the potential applications of membrane technology in environmental processes. The development of nanomembranes and membranes is essential to the advancement of nanofiltration technology [103 - 105].

CONCLUSION AND FUTURE PERSPECTIVES

The advent of NMs and corresponding nanotechnology has presented significant opportunities for traditional environmental technologies, giving rise to the field of environmental nanotechnology. Numerous effective uses have been shown, such as new methods for sensing and monitoring, the use of nanosorbents for selective adsorption, the separation of nanomembranes, and environmental nanocatalysis. Still, there are a lot of obstacles in the way of solving these urgent problems. In environmental processes, it is critical to logically design and apply NMs while maintaining a careful balance between activity, reactivity, and stability.

The remarkable performance observed in environmental applications, whether in terms of catalytic efficiency, is largely attributed to the high reactivity facilitated by the intricate surface and microstructure of NMs. Selective adsorption and degradation are crucial in addressing different environmental challenges. Additionally, the composition and/or structure of the nanosized catalysts would be readily destroyed by severe and dangerous circumstances, which are extremely typical in environmental applications. In environmental applications, stability, which is a crucial criterion of practical viability, is more crucial than in the chemical manufacturing and energy sectors because metal leaching from nanosized metal-based catalysts and/or newly generated intermediates might result in secondary contaminants. All of the activity, selectivity, and stability should be taken into consideration in the rational design of NMs and the exact control of the applications. Mechanistic research is required to comprehend environmental processes and nanotechnology in a better way. Furthermore, by examining processes at the nano/bio interfaces, it is necessary to fully explore the potential environmental risks associated with NMs. The involvement and deliberate attempts in environmental nanotechnology would support overall sustainability in the future. Despite the vast research that has been done and published on NMs, it is not always possible to produce desired NMs at a mature enough scale or to use them safely. Considerable efforts must be devoted to monitoring and remedying environmental pollution. Even the application of theoretical findings to practical environmental solutions may necessitate the involvement of chemical engineering efforts.

ACKNOWLEDGEMENT

The author would like to acknowledge the financial assistance provided by the Department of Science & Technology and Biotechnology, Government of West Bengal, India (Memo No. 860(Sanc.)/STBT-11012(25)/5/2019-ST SEC dated 03/11/2023).

REFERENCES

[1]   Mekuye B, Abera B. Nanomaterials: An overview of synthesis, classification, characterization, and applications. Nano Select 2023; 4(8): 486-501.
[http://dx.doi.org/10.1002/nano.202300038]

[2]   Pal SL, Jana U, Manna PK, Mohanta GP, Manavalan R. J Anim Plant Sci 2011; 228: 234.

[3]   Buzea C, Pacheco II, Robbie K. NMs and NPs: sources and toxicity. Biointerphases 2007; 2(4): MR17-71.
[http://dx.doi.org/10.1116/1.2815690] [PMID: 20419892]

[4]   Findik F. Nanomaterials and their applications. Periodicals of Engineering and Natural Sciences (PEN) 2021; 9(3): 62-75.
[http://dx.doi.org/10.21533/pen.v9i3.1837]

[5]   Pacheco-Torgal F, Jalali S. Nanotechnology: Advantages and drawbacks in the field of construction and building materials. Constr Build Mater 2011; 25(2): 582-90.
[http://dx.doi.org/10.1016/j.conbuildmat.2010.07.009]

[6]   Afolalu SA, Soetan SB, Ongbali SO, Abioye AA, Oni AS. Morphological characterization and physio-chemical properties of nanoparticle-review. IOP Conference Series: Materials Science and Engineering. 2019; 640: pp. (1)012065-5.
[http://dx.doi.org/10.1088/1757-899X/640/1/012065]

[7]   Gaffet E. 2011.NMs: a review of the definitions, applications, health effects. How to implement secure development Nanomat\'eriaux: une revue des d\'efinitions, des applications, des effets sanitaires et des moyens\a mettre en oeuvre pour un d\'eveloppement s\'ecuris\'e. preprint arXiv:11062206

[8]   Abdalkreem TM. Optical properties of gold and silver nanoparticles. 2018.

[9]   De M, Ghosh PS, Rotello VM. Applications of NPs in biology. Adv Mater 2008; 20(22): 4225-41.
[http://dx.doi.org/10.1002/adma.200703183]

[10]   Franco-Luján VA, Montejo-Alvaro F, Ramírez-Arellanes S, Cruz-Martínez H, Medina DI. Nanomaterial-Reinforced Portland-Cement-Based Materials: A Review. Nanomaterials (Basel) 2023; 13(8): 1383.
[http://dx.doi.org/10.3390/nano13081383] [PMID: 37110968]

[11]   Baig N, Kammakakam I, Falath W. Nanomaterials: a review of synthesis methods, properties, recent progress, and challenges. Materials Advances 2021; 2(6): 1821-71.
[http://dx.doi.org/10.1039/D0MA00807A]

[12]   Heiligtag FJ, Niederberger M. The fascinating world of nanoparticle research. Mater Today 2013; 16(7-8): 262-71.
[http://dx.doi.org/10.1016/j.mattod.2013.07.004]

[13]   Walter P, Welcomme E, Hallégot P, et al. Early use of PbS nanotechnology for an ancient hair dyeing formula. Nano Lett 2006; 6(10): 2215-9.
[http://dx.doi.org/10.1021/nl061493u] [PMID: 17034086]

[14]   Jeevanandam J, Barhoum A, Chan YS, Dufresne A, Danquah MK. Review on nanoparticles and nanostructured materials: history, sources, toxicity and regulations. Beilstein J Nanotechnol 2018; 9(1): 1050-74.
[http://dx.doi.org/10.3762/bjnano.9.98] [PMID: 29719757]

[15]   Freestone I, Meeks N, Sax M, Higgitt C. The Lycurgus cup—a roman nanotechnology. Gold Bull 2007; 40(4): 270-7.
[http://dx.doi.org/10.1007/BF03215599]

[16]   Santamaria A. Historical overview of nanotechnology and nanotoxicology Nanotoxicity: methods and protocols 2012; 1-12.

[17] Nasrollahzadeh M, Sajadi SM, Sajjadi M, Issaabadi Z. An introduction to nanotechnology. Interface science and technology. Elsevier 2019; 28: pp. 1-27.

[18] Taniguchi N. On the Basic Concept of Nanotechnology. Proc Int Conf Prod Eng. 1974;Part II:18–23.

[19] Patil S, Chandrasekaran R. Biogenic nanoparticles: a comprehensive perspective in synthesis, characterization, application and its challenges. J Genet Eng Biotechnol 2020; 18(1): 67.
[http://dx.doi.org/10.1186/s43141-020-00081-3] [PMID: 33104931]

[20] Ramsden J. Nanotechnology: an introduction. William Andrew 2016.
[http://dx.doi.org/10.1016/B978-0-323-39311-9.00007-8]

[21] Osman AI, Zhang Y, Farghali M, Rashwan AK, Eltaweil AS, El-Monaem A, *et al.* Synthesis of green NPs for energy, biomedical, environmental, agricultural, and food applications: A review. Environ Chem Lett 2024; 1-47.
[http://dx.doi.org/10.1007/s10311-023-01648-5]

[22] Mughal B, Zaidi SZJ, Zhang X, Hassan SU. Biogenic NPs: Synthesis, characterisation and applications. Appl Sci (Basel) 2021; 11(6): 2598.
[http://dx.doi.org/10.3390/app11062598]

[23] Iravani S. Green synthesis of metal nanoparticles using plants. Green Chem 2011; 13(10): 2638-50.
[http://dx.doi.org/10.1039/c1gc15386b]

[24] Wu Q, Miao W, Zhang Y, Gao H, Hui D. Mechanical properties of nanomaterials: A review. Nanotechnol Rev 2020; 9(1): 259-73.
[http://dx.doi.org/10.1515/ntrev-2020-0021]

[25] Nam NH, Luong NH. NPs: Synthesis and applications. Materials for biomedical engineering. Elsevier 2019; pp. 211-40.
[http://dx.doi.org/10.1016/B978-0-08-102814-8.00008-1]

[26] Peng P, Hu A, Gerlich AP, Zou G, Liu L, Zhou YN. Joining of silver NMs at low temperatures: processes, properties, and applications. ACS Appl Mater Interfaces 2015; 7(23): 12597-618.
[http://dx.doi.org/10.1021/acsami.5b02134] [PMID: 26005792]

[27] Bokov D, Turki Jalil A, Chupradit S, *et al.* NM by sol-gel method: synthesis and application. Adv Mater Sci Eng 2021; 2021(1): 5102014.
[http://dx.doi.org/10.1155/2021/5102014]

[28] Khalid K, Tan X, Mohd Zaid HF, *et al.* Advanced in developmental organic and inorganic nanomaterial: a review. Bioengineered 2020; 11(1): 328-55.
[http://dx.doi.org/10.1080/21655979.2020.1736240] [PMID: 32138595]

[29] Fang J, Chen YC. Nanomaterials for photohyperthermia: a review. Curr Pharm Des 2013; 19(37): 6622-34.
[http://dx.doi.org/10.2174/1381612811319370006] [PMID: 23621537]

[30] Flores-Rojas GG, López-Saucedo F, Vera-Graziano R, Mendizabal E, Bucio E. Magnetic NPs for medical applications: Updated review. Macromol 2022; 2(3): 374-90.
[http://dx.doi.org/10.3390/macromol2030024]

[31] Huynh KH, Pham XH, Kim J, *et al.* Synthesis, properties, and biological applications of metallic alloy NPs. Int J Mol Sci 2020; 21(14): 5174.
[http://dx.doi.org/10.3390/ijms21145174] [PMID: 32708351]

[32] Horikoshi S, Serpone N, Eds. Microwaves in NPsynthesis: Fundamentals and Applications. Germany: John Wiley & Sons 2013.
[http://dx.doi.org/10.1002/9783527648122.ch1]

[33] Dolez PI, Ed. Nanoengineering: global approaches to health and safety issues. Elsevier 2015.

[34] Cho G, Park Y, Hong YK, Ha DH. Ion exchange: an advanced synthetic method for complex

nanoparticles. Nano Converg 2019; 6(1): 17.
[http://dx.doi.org/10.1186/s40580-019-0187-0] [PMID: 31155686]

[35] Khan Y, Sadia H, Ali Shah SZ, *et al.* Classification, synthetic, and characterization approaches to NPs, and their applications in various fields of nanotechnology: A review. Catalysts 2022; 12(11): 1386.
[http://dx.doi.org/10.3390/catal12111386]

[36] Krishnan SK, Singh E, Singh P, Meyyappan M, Nalwa HS. A review on graphene-based nanocomposites for electrochemical and fluorescent biosensors. RSC Advances 2019; 9(16): 8778-881.
[http://dx.doi.org/10.1039/C8RA09577A] [PMID: 35517682]

[37] Ijaz I, Gilani E, Nazir A, Bukhari A. Detail review on chemical, physical and green synthesis, classification, characterizations and applications of nanoparticles. Green Chem Lett Rev 2020; 13(3): 223-45.
[http://dx.doi.org/10.1080/17518253.2020.1802517]

[38] Zhu W, Guo Y, Ma B, *et al.* Fabrication of highly dispersed Pd nanoparticles supported on reduced graphene oxide for solid phase catalytic hydrogenation of 1,4-bis(phenylethynyl) benzene. Int J Hydrogen Energy 2020; 45(15): 8385-95.
[http://dx.doi.org/10.1016/j.ijhydene.2019.10.094]

[39] Liu X, Xu M, Wan L, *et al.* Superior catalytic performance of atomically dispersed palladium on graphene in CO oxidation. ACS Catal 2020; 10(5): 3084-93.
[http://dx.doi.org/10.1021/acscatal.9b04840]

[40] Albeladi A B. Synthesis of silver NPs and their sensing applications

[41] Nouailhat A. An introduction to nanoscience and nanotechnology. John Wiley & Sons 2010; 10.

[42] Zhang D, Ye K, Yao Y, *et al.* Controllable synthesis of carbon nanomaterials by direct current arc discharge from the inner wall of the chamber. Carbon 2019; 142: 278-84.
[http://dx.doi.org/10.1016/j.carbon.2018.10.062]

[43] Singh J, Dutta T, Kim KH, Rawat M, Samddar P, Kumar P. 'Green' synthesis of metals and their oxide nanoparticles: applications for environmental remediation. J Nanobiotechnology 2018; 16(1): 84.
[http://dx.doi.org/10.1186/s12951-018-0408-4]

[44] Alsharif SM, Salem SS, Abdel-Rahman MA, *et al.* Multifunctional properties of spherical silver nanoparticles fabricated by different microbial taxa. Heliyon 2020; 6(5): e03943.
[http://dx.doi.org/10.1016/j.heliyon.2020.e03943] [PMID: 32518846]

[45] Salem SS. A mini review on green nanotechnology and its development in biological effects. Arch Microbiol 2023; 205(4): 128.
[http://dx.doi.org/10.1007/s00203-023-03467-2] [PMID: 36944830]

[46] Mittal AK, Chisti Y, Banerjee UC. Synthesis of metallic nanoparticles using plant extracts. Biotechnol Adv 2013; 31(2): 346-56.
[http://dx.doi.org/10.1016/j.biotechadv.2013.01.003] [PMID: 23318667]

[47] Ahn EY, Jin H, Park Y. Assessing the antioxidant, cytotoxic, apoptotic and wound healing properties of silver nanoparticles green-synthesized by plant extracts. Mater Sci Eng C 2019; 101: 204-16.
[http://dx.doi.org/10.1016/j.msec.2019.03.095] [PMID: 31029313]

[48] Shah M, Fawcett D, Sharma S, Tripathy S, Poinern G. Green synthesis of metallic NPs *via* biological entities. Materials (Basel) 2015; 8(11): 7278-308.
[http://dx.doi.org/10.3390/ma8115377] [PMID: 28793638]

[49] Gericke M, Pinches A. Microbial production of gold nanoparticles. Gold Bull 2006; 39(1): 22-8.
[http://dx.doi.org/10.1007/BF03215529]

[50] Iravani S. Bacteria in nanoparticle synthesis: current status and future prospects. Int Sch Res Notices

2014; 2014: 1-18.
[http://dx.doi.org/10.1155/2014/359316] [PMID: 27355054]

[51]   Eming SA, Martin P, Tomic-Canic M. Wound repair and regeneration: mechanisms, signaling, and translation Science translational medicine 2014; 6(265): 265sr6-6.https://doi.org/https://doi.org/10.1126/scitranslmed.3009337.Wound

[52]   Kalishwaralal K, Deepak V, Ramkumarpandian S, Nellaiah H, Sangiliyandi G. Extracellular biosynthesis of silver nanoparticles by the culture supernatant of Bacillus licheniformis. Mater Lett 2008; 62(29): 4411-3.
[http://dx.doi.org/10.1016/j.matlet.2008.06.051]

[53]   Mathivanan K, Selva R, Chandirika JU, *et al.* Biologically synthesized silver nanoparticles against pathogenic bacteria: Synthesis, calcination and characterization. Biocatal Agric Biotechnol 2019; 22: 101373.
[http://dx.doi.org/10.1016/j.bcab.2019.101373]

[54]   Manivasagan P, Venkatesan J, Sivakumar K, Kim SK. Actinobacteria mediated synthesis of nanoparticles and their biological properties: A review. Crit Rev Microbiol 2016; 42(2): 209-21.
[PMID: 25430521]

[55]   Durán N, Marcato PD, Alves OL, De Souza GIH, Esposito E. Mechanistic aspects of biosynthesis of silver nanoparticles by several Fusarium oxysporum strains. J Nanobiotechnology 2005; 3(1): 8.
[http://dx.doi.org/10.1186/1477-3155-3-8]

[56]   Ahmad A, Mukherjee P, Mandal D, *et al.* Enzyme mediated extracellular synthesis of CdS nanoparticles by the fungus, Fusarium oxysporum. J Am Chem Soc 2002; 124(41): 12108-9.
[http://dx.doi.org/10.1021/ja027296o] [PMID: 12371846]

[57]   Mohamed AA, Abu-Elghait M, Ahmed NE, Salem SS. Eco-friendly mycogenic synthesis of ZnO and CuO NPs for *in vitro* antibacterial, antibiofilm, and antifungal applications. Biol Trace Elem Res 2021; 199(7): 2788-99.
[http://dx.doi.org/10.1007/s12011-020-02369-4] [PMID: 32895893]

[58]   Spagnoletti FN, Spedalieri C, Kronberg F, Giacometti R. Extracellular biosynthesis of bactcricidal Ag/AgCl nanoparticles for crop protection using the fungus Macrophomina phaseolina. J Environ Manage 2019; 231: 457-66.
[http://dx.doi.org/10.1016/j.jenvman.2018.10.081] [PMID: 30388644]

[59]   Lian S, Diko C S, Yan Y, *et al.* Characterization of biogenic selenium NPs derived from cell-free extracts of a novel yeast Magnusiomyces ingens 3 Biotech 9: 1-8.2019;

[60]   Fouda A, EL-Din Hassan S, Salem SS, Shaheen TI. In-Vitro cytotoxicity, antibacterial, and UV protection properties of the biosynthesized Zinc oxide nanoparticles for medical textile applications. Microb Pathog 2018; 125: 252-61.
[http://dx.doi.org/10.1016/j.micpath.2018.09.030] [PMID: 30240818]

[61]   Boroumand Moghaddam A, Namvar F, Moniri M, Md Tahir P, Azizi S, Mohamad R. NPs biosynthesized by fungi and yeast: a review of their preparation, properties, and medical applications. Molecules 2015; 20(9): 16540-65.
[http://dx.doi.org/10.3390/molecules200916540] [PMID: 26378513]

[62]   Uzair B, Liaqat A, Iqbal H, *et al.* Green and cost-effective synthesis of metallic NPs by algae: Safe methods for translational medicine. Bioengineering (Basel) 2020; 7(4): 129.
[http://dx.doi.org/10.3390/bioengineering7040129] [PMID: 33081248]

[63]   Nowicka B. Heavy metal–induced stress in eukaryotic algae—mechanisms of heavy metal toxicity and tolerance with particular emphasis on oxidative stress in exposed cells and the role of antioxidant response. Environ Sci Pollut Res Int 2022; 29(12): 16860-911.
[http://dx.doi.org/10.1007/s11356-021-18419-w] [PMID: 35006558]

[64]   El-Refaey AA, Salem SS. Algae materials for bionanopesticides: NPs and composites. Algae

Materials. Academic Press 2023; pp. 219-30.
[http://dx.doi.org/10.1016/B978-0-443-18816-9.00004-6]

[65]   El-Rafie HM, El-Rafie MH, Zahran MK. Green synthesis of silver nanoparticles using polysaccharides extracted from marine macro algae. Carbohydr Polym 2013; 96(2): 403-10.
[http://dx.doi.org/10.1016/j.carbpol.2013.03.071] [PMID: 23768580]

[66]   Bouquillon S, Courant T, Dean D, *et al.* Biodegradable ionic liquids: selected synthetic applications. Aust J Chem 2007; 60(11): 843-7.
[http://dx.doi.org/10.1071/CH07257]

[67]   Carter EB, Culver SL, Fox PA, *et al.* Sweet success: ionic liquids derived from non-nutritive sweetenersElectronic supplementary information (ESI) available: experimental details; IR spectra. See http://www.rsc.org/suppdata/cc/b3/b313068a/. Chem Commun (Camb) 2004; (6): 630-1.
[http://dx.doi.org/10.1039/b313068a] [PMID: 15010753]

[68]   Harjani JR, Singer RD, Garcia MT, Scammells PJ. Biodegradable pyridinium ionic liquids: design, synthesis and evaluation. Green Chem 2009; 11(1): 83-90.
[http://dx.doi.org/10.1039/B811814K]

[69]   Imperato G, König B, Chiappe C. Ionic green solvents from renewable resources. Eur J Org Chem 2007; 2007(7): 1049-58.
[http://dx.doi.org/10.1002/ejoc.200600435]

[70]   Bussamara R, Melo WWM, Scholten JD, *et al.* Controlled synthesis of Mn_3O_4 nanoparticles in ionic liquids. Dalton Trans 2013; 42(40): 14473-9.
[http://dx.doi.org/10.1039/c3dt32348j] [PMID: 23970370]

[71]   Lazarus LL, Riche CT, Malmstadt N, Brutchey RL. Effect of ionic liquid impurities on the synthesis of silver nanoparticles. Langmuir 2012; 28(45): 15987-93.
[http://dx.doi.org/10.1021/la303617f] [PMID: 23092200]

[72]   Fürstner A, Ackermann L, Beck K, *et al.* Olefin metathesis in supercritical carbon dioxide. J Am Chem Soc 2001; 123(37): 9000-6.
[http://dx.doi.org/10.1021/ja010952k] [PMID: 11552807]

[73]   Wittmann K, Wisniewski W, Mynott R, *et al.* Supercritical carbon dioxide as solvent and temporary protecting group for rhodium-catalyzed hydroaminomethylation. Chemistry 2001; 7(21): 4584-9.
[http://dx.doi.org/10.1002/1521-3765(20011105)7:21<4584::AID-CHEM4584>3.0.CO;2-P]   [PMID: 11757649]

[74]   Pollet P, Eckert CA, Liotta CL. Solvents for sustainable chemical processes. WIT Trans Ecol Environ 2011; 154: 21-31.
[http://dx.doi.org/10.2495/CHEM110031]

[75]   Ohde H, Hunt F, Wai CM. Synthesis of silver and copper NPs in a water-in-supercritical-carbon dioxide microemulsion. Chem Mater 2001; 13(11): 4130-5.
[http://dx.doi.org/10.1021/cm010030g]

[76]   Sue K, Adschiri T, Arai K. Predictive model for equilibrium constants of aqueous inorganic species at subcritical and supercritical conditions. Ind Eng Chem Res 2002; 41(13): 3298-306.
[http://dx.doi.org/10.1021/ie010956y]

[77]   Kim M, Lee BY, Ham HC, *et al.* Facile one-pot synthesis of tungsten oxide (WO_3-x) nanoparticles using sub and supercritical fluids. J Supercrit Fluids 2016; 111: 8-13.
[http://dx.doi.org/10.1016/j.supflu.2016.01.011]

[78]   Shaheen T I, Salem S S, Fouda A. Current advances in fungal nanobiotechnology: Mycofabrication and applications Microbial nanobiotechnology: Principles and applications 113-43.2021;

[79]   Al-Kordy HMH, Sabry SA, Mabrouk MEM. Statistical optimization of experimental parameters for extracellular synthesis of zinc oxide nanoparticles by a novel haloalaliphilic Alkalibacillus sp.W7. Sci Rep 2021; 11(1): 10924.

[http://dx.doi.org/10.1038/s41598-021-90408-y] [PMID: 34035407]

[80]   Desai MP, Patil RV, Harke SS, Pawar KD. Bacterium mediated facile and green method for optimized biosynthesis of gold NPs for simple and visual detection of two metal ions. J Cluster Sci 2021; 32(2): 341-50.
[http://dx.doi.org/10.1007/s10876-020-01793-9]

[81]   Koçer AT, Özçimen D. Eco-friendly synthesis of silver NPs from macroalgae: optimization, characterization and antimicrobial activity. Biomass Convers Biorefin. 2022:1–12.
[http://dx.doi.org/10.1007/s13399-022-02506-0]

[82]   Ndikau M, Noah NM, Andala DM, Masika E. Green synthesis and characterization of silver NPs using Citrullus lanatus fruit rind extract. Int J Anal Chem 2017; 1-9.
[http://dx.doi.org/10.1155/2017/8108504]

[83]   Velgosová O, Mražíková A, Marcinčáková R. Influence of pH on green synthesis of Ag nanoparticles. Mater Lett 2016; 180: 336-9.
[http://dx.doi.org/10.1016/j.matlet.2016.04.045]

[84]   Chitra K, Annadurai G. Antibacterial activity of pH-dependent biosynthesized silver NPs against clinical pathogen. BioMed Res Int 2014; 1-6.
[http://dx.doi.org/10.1155/2014/725165]

[85]   Barisik M, Atalay S, Beskok A, Qian S. Size dependent surface charge properties of silica NPs. J Phys Chem C 2014; 118(4): 1836-42.
[http://dx.doi.org/10.1021/jp410536n]

[86]   Anigol L B, Charantimath J S, Gurubasavaraj P M. Effect of concentration and pH on the size of silver NPs synthesized by green chemistry Org Med Chem Int J 2017; 3(5): 1-5.

[87]   Rokade S S, Joshi K A, Mahajan K, *et al.* Novel anticancer platinum and palladium NPs from Barleria prionitis Global journal of nanomedicine 2017; 2(5): 555-600.

[88]   Shankar SS, Rai A, Ahmad A, Sastry M. Controlling the optical properties of lemongrass extract synthesized gold nanotriangles and potential application in infrared-absorbing optical coatings. Chem Mater 2005; 17(3): 566-72.
[http://dx.doi.org/10.1021/cm048292g]

[89]   Kaviya S, Santhanalakshmi J, Viswanathan B. Green synthesis of silver NPs using Polyalthia longifolia leaf extract along with D-sorbitol: study of antibacterial activity. J Nanotechnol 2011; 2011: 1-5.
[http://dx.doi.org/10.1155/2011/152970]

[90]   Kredy HM. The effect of pH, temperature on the green synthesis and biochemical activities of silver NPs from Lawsonia inermis extract. Journal of Pharmaceutical Sciences and Research 2018; 10(8): 2022-6.

[91]   Liu H, Zhang H, Wang J, Wei J. Effect of temperature on the size of biosynthesized silver nanoparticle: deep insight into microscopic kinetics analysis Arabian Journal of Chemistry 2020; 13(1): 1011-9.https://doi.org/https://doi.org/10.1016/j.arabjc

[92]   Patra JK, Baek KH. Green nanobiotechnology: factors affecting synthesis and characterization techniques. Journal of NMs 2015; 2014: 219-9.https://doi.org/https://doi.org/10.1155/(2014)/417305

[93]   Yang N, Li WH. Mango peel extract mediated novel route for synthesis of silver NPs and antibacterial application of silver NPs loaded onto non-woven fabrics. Ind Crops Prod 2013; 48: 81-8.https://doi.org/https://doi.org/10.1016/j.indcrop.2013.04.001
[http://dx.doi.org/10.1016/j.indcrop.2013.04.001]

[94]   Mukherjee P, Ahmad A, Mandal D, *et al.* Fungus-mediated synthesis of silver NPs and their immobilization in the mycelial matrix: a novel biological approach to nanoparticle synthesis. Nano Lett 2001; 1(10): 515-9.
[http://dx.doi.org/10.1021/nl0155274]

[95]   Qamar SUR, Ahmad JN. Nanoparticles: Mechanism of biosynthesis using plant extracts, bacteria, fungi, and their applications. J Mol Liq 2021; 334: 116040.
       [http://dx.doi.org/10.1016/j.molliq.2021.116040]

[96]   Salem SS. Bio-fabrication of selenium NPs using Baker's yeast extract and its antimicrobial efficacy on food borne pathogens. Appl Biochem Biotechnol 2022; 194(5): 1898-910.
       [http://dx.doi.org/10.1007/s12010-022-03809-8] [PMID: 34994951]

[97]   Neha Agarwal, Vijendra Singh Solanki, Brijesh Pare, Neetu Singh, Sreekantha B. Jonnalagadda, Current trends in nanocatalysis for green chemistry and its applications- a mini-review, Current, Opinion in Green and Sustainable Chemistry, 41, 2023, 100788, ISSN 2452-2236.
       [http://dx.doi.org/10.1016/j.cogsc.2023.100788]

[98]   Tanzil AH, Sultana ST, Saunders SR, Shi L, Marsili E, Beyenal H. Biological synthesis of nanoparticles in biofilms. Enzyme Microb Technol 2016; 95: 4-12.
       [http://dx.doi.org/10.1016/j.enzmictec.2016.07.015] [PMID: 27866625]

[99]   Iravani S, Varma RS. Bacteria in heavy metal remediation and nanoparticle biosynthesis. ACS Sustain Chem& Eng 2020; 8(14): 5395-409.
       [http://dx.doi.org/10.1021/acssuschemeng.0c00292]

[100] Pourali P, Yahyaei B, Afsharnezhad S. Bio-synthesis of gold NPs by Fusarium oxysporum and assessment of their conjugation possibility with two types of β-lactam antibiotics without any additional linkers. Microbiology 2018; 87(2): 229-37.
       [http://dx.doi.org/10.1134/S0026261718020108]

[101] Bhambure R, Bule M, Shaligram N, Kamat M, Singhal R. Extracellular biosynthesis of gold NPs using Aspergillus niger–its characterization and stability. Chem Eng Technol 2009; 32(7): 1036-41.
       [http://dx.doi.org/10.1002/ceat.200800647]

[102] Doghish AS, Hashem AH, Shehabeldine AM, Sallam AAM, El-Sayyad GS, Salem SS. Nanocomposite based on gold nanoparticles and carboxymethyl cellulose: Synthesis, characterization, antimicrobial, and anticancer activities. J Drug Deliv Sci Technol 2022; 77: 103874.
       [http://dx.doi.org/10.1016/j.jddst.2022.103874]

[103] Hashem AH, Salem SS. Green and ecofriendly biosynthesis of selenium nanoparticles using *Urtica dioica* (stinging nettle) leaf extract: Antimicrobial and anticancer activity. Biotechnol J 2022; 17(2): 2100432.
       [http://dx.doi.org/10.1002/biot.202100432] [PMID: 34747563]

[104] Ibrahim S, Ahmad Z, Manzoor MZ, Mujahid M, Faheem Z, Adnan A. Optimization for biogenic microbial synthesis of silver nanoparticles through response surface methodology, characterization, their antimicrobial, antioxidant, and catalytic potential. Sci Rep 2021; 11(1): 770.
       [http://dx.doi.org/10.1038/s41598-020-80805-0] [PMID: 33436966]

[105] Das C, Paul SS, Saha A, *et al.* Silver-based NMs as therapeutic agents against coronaviruses: a review. Int J Nanomedicine 2020; 15: 9301-15.
       [http://dx.doi.org/10.2147/IJN.S280976] [PMID: 33262589]

[106] Fouda A, Salem SS, Wassel AR, Hamza MF, Shaheen TI. Optimization of green biosynthesized visible light active CuO/ZnO nano-photocatalysts for the degradation of organic methylene blue dye. Heliyon 2020; 6(9): e04896.
       [http://dx.doi.org/10.1016/j.heliyon.2020.e04896] [PMID: 32995606]

[107] Salem SS, Fouda A. Green synthesis of metallic NPs and their prospective biotechnological applications: an overview. Biol Trace Elem Res 2021; 199(1): 344-70.
       [http://dx.doi.org/10.1007/s12011-020-02138-3] [PMID: 32377944]

CHAPTER 2

Aquatic Milieu and Nanotechnology: A Critical Review on Remediation of Pollutants

Mishu Singh[1,*]

[1] *Department of Chemistry, Pt. D. D. U. Govt. Girls P.G College, Lucknow, India*

Abstract: Access to sources of clean water is necessary for the existence of all living beings on Earth. In freshwater environments, a wide variety of species, from minuscule to mega, can be found. However, constant contamination of water bodies has changed the freshwater ecosystems. Every year, the problem of water pollution gets worse, which eventually affects the limited supply of freshwater resources. All the anthropogenic activities have caused long-term negative effects on the delicate structure of freshwater ecosystems. Wastewater can be treated in several ways before being released into recipient water bodies. However, these conventional techniques are unable to meet the necessary standards for wastewater treatment for a variety of reasons. Furthermore, there is reason for concern regarding the efficacy of these currently available conventional treatments. For environmental remediation, different skillful technologies such as physicochemical reactions, filtration, adsorption, and photocatalysis are employed to eliminate impurities from various environmental matrices. Materials based on nanotechnology have superior qualities and are especially useful for these kinds of procedures because of their low volume-to-surface area ratio, which frequently leads to increased reactivity. Based on the information presented in this chapter, it appears that using nanotechnology to treat wastewater could be advantageous, efficient, and environmentally friendly. It is selective but effective to clean only organic-based pollution. Additionally, due to their extraordinary adsorption behavior, Nanomaterials (NMs) are verified as disinfectants, pathogen identifiers, and antibacterial agents in environmental remediation.

Keywords: Disinfectant, Heavy metals, Industrial effluents, Nanoparticles, Nanotechnology, Remediation, Water pollution.

INTRODUCTION

Undoubtedly, environmental pollution is one of the biggest issues faced by the modern society. Continuous research endeavors are dedicated to developing innovative technologies for removing contaminants from air, water, and soil. The numerous dangerous contaminants include heavy metals, organic compounds, oil

---

* **Corresponding author Mishu Singh:** Department of Chemistry, Pt. D. D. U. Govt. Girls P.G College, Lucknow, India; E-mail: mishusingh17@gmail.com

Neha Agarwal, Vijendra Singh Solanki, Neetu Singh & Maulin P. Shah (Eds.)

spills, pesticides, herbicides, fertilizers, hazardous gases, industrial effluents, sewage, and particulate matter. Many approaches can be used for environmental remediation because different kinds of materials are used in this process. Recent studies have emphasized the utilization of NMs in the advancement of environmental remediation technologies due to the challenges posed by the complex mixture of pollutants and their high volatility, making their clean-up and degradation challenging.

Water stress affects half of the world's population. Roughly 70% of the industrial effluent is released into waterways without being adequately treated. Incorporating NMs into current water and wastewater treatment processes could be a promising improvement. Owing to their lower cost, NMs offer tremendous potential for wastewater purification. The direct application of nanoparticles (NPs) in wastewater and water decontamination techniques is not without drawbacks. The activity of these NPs is greatly reduced as a result of these particles' propensity to aggregate in liquefied systems or beds. The impact of NPs on human health and the aquatic environment remains uncertain. It is, therefore, advisable to develop such a device or method that, during wastewater treatment, can reduce the release of NPs.

Freshwater makes up only 3% of the water on Earth, with the remaining 97% being salt water. Jose *et al.* reported that only 1 percent of the 3 percent is in the quantifiable form, and the other 2/3 is in the frozen form, and this scarcity is a terrible situation for developing nations. Pressure on the water bodies is increasing due to the unchecked population growth. Conventional techniques for treating wastewater are less efficient and not cost-effective [1]. According to Sharma and Sharma, 2013 [2], novel techniques such as nanotechnology are being continuously researched to enhance wastewater treatment techniques. NPs, thus, are defined as materials and particles with dimensions ranging from 1 to 100 nm.

After the scanning tunneling microscope (STM) was developed in the 1980s, nanotechnology saw a significant upsurge. Building block-sized particles are arranged in nanostructures and NMs, which have a nm size range. Materials at the nanoscale possess distinct qualities such as stability, morphological traits, and adsorption capacities that set them apart from macro-scale counterparts with similar structures, and these qualities increase their efficacy as wastewater treatment agents [3]. Nanotechnology offers significant advantages for the environment, notably in the cost-effective and efficient elimination of contaminants from wastewater [4]. Certain NMs have distinct modes of operation; some eliminate pollutants, while others isolate and segregate them. The numerous advantages of this technology, such as its high efficiency and cost-effective wastewater treatment, increase its dependability in addressing a wide range of global issues.

There are parallel endeavors beneath each investigative course in nanotechnology, and the gigantic sum of cash is streaming beneath its investigations and advancement plans. Thus, advancements may happen earlier than anticipated. The main objective of this review paper is to provide an overview of the interest and fascination that researchers and analysts have shown in utilizing nanotechnology for advanced water treatment. Additionally, it seeks to acknowledge the significant obstacles and potential directions for future advancements in the field of nanotechnology and its applications in water treatment.

NMS FOR WATER DISINFECTION

When it comes to pollution remediation, the materials used must not contribute to further environmental pollution. This is why the use of biodegradable materials is highly compelling in this field. By utilizing biodegradable materials, there is no waste left behind after the treatment process, which not only enhances consumer confidence and acceptance of the technology but also provides a safer and more environmentally friendly option for pollutant remediation. Additionally, these materials offer a solution to the problem of low efficiencies caused by off-targeting, making them particularly attractive in the development of new technologies that aim to capture contaminants. Therefore, much research is focused on applying nanotechnology principles and merging them with modification of material surfaces to create engineered materials capable of overcoming many of the challenges associated with the removal of contaminants. Some of the most important factors to reflect on while creating NMs are target-specific capture, cost-effectiveness, creation within the lab with green chemistry, nontoxic, ease of biodegradability, recyclability, and post-use recovery (possibility of regeneration). Despite the possible benefits of the previously listed NMs, some of them are inherently unstable and, therefore, require preparation [5].

Several NPs, like Ag, *etc.*, have pronounced adsorption capability [6, 7]. Being passive oxidants, these NPs produce less harmful by-products. There are certain drawbacks to using nanotechnology to decontaminate wastewater [8]. For example, to remove contaminants, an NP must come in direct contact with the target organism, which lowers the particle's activity. Furthermore, the lack of NMs following the treatment process prevents wastewater from subsequently exhibiting antimicrobial activity [9]. Various environmental pollutant remedial approaches are shown in Fig. (**1**).

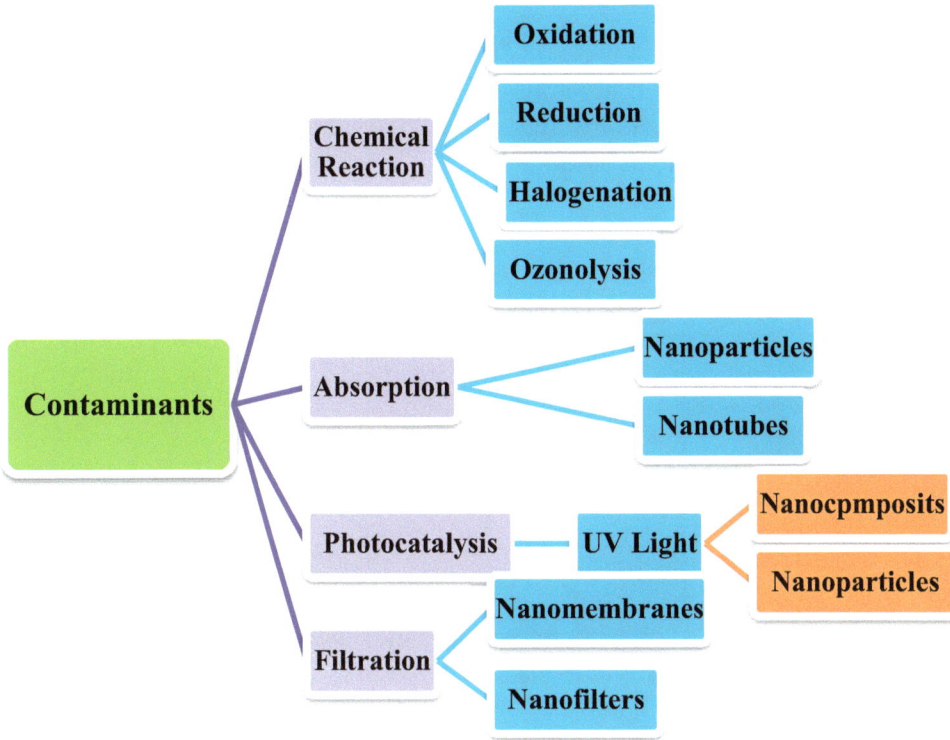

Fig. (1). Environmental pollutant's remediation approaches.

APPLICATIONS OF NANOTECHNOLOGY IN WATER TREATMENT

NMs are the carters of the nanotechnology insurgency and a key obstacle mediator for the submissions of nanotechnology to tackle this worldwide water crisis. These particles have several crucial physio-biochemical properties, which make them extraordinarily appealing for water filtration. Moreover, NPs can be incorporated with different chemical groups to outspread their partiality toward a given set of compounds. They can also assist as selective and recyclable agents for harmful metallic particles, radioactive nuclei, and natural and inorganic solvents/solutes in watery arrangements. These particles give exceptional openings to create more proficient water-purification reagents due to their expansive properties [10]. The NPs can be used in the distinctive two areas:

  i. Spotting and detection
 ii. Handling and removal

Reactive materials are applied in nanoremediation techniques to help detoxify and transform pollutants. These substances start the harmful pollutants' chemical reduction and catalysis. The range of strategies beyond nanotechnology is evident in remediation. Comparing adsorptive to reactive and *in situ* to ex situ are two key approaches between conventional and advanced technologies in this regard. Reactive remediation techniques influence the degradation of contaminants, while absorptive remediation techniques remove contaminants (particularly metals) by sequestration. Ex situ approach for treatment occurs once the pollutant has been moved to a more suitable site; *in situ* treatment deals with contaminants while they are still present in the material [11].

TYPES OF NPS AND THEIR POTENTIAL APPLICATIONS IN WATER TREATMENT

Polymeric Nanoadsorbents

Polymeric nanosorbents such as dendrimers (bulky branched atoms) can be used to remove organics and many metals. Most metals will be adsorbed from the customized outer layer, while natural compounds will be adsorbed from the inner hydrophobic shell. Using this combined dendrimer-filtration framework, almost all copper can be effectively purified. The adsorbent is effectively initiated by changes in pH [12]. Adsorbents for the removal of dyes and other anionic compounds from waste materials are possible by forming chitosan-dendrimer nanostructures. The sorbent is biodegradable, biocompatible, and non-toxic, and they can remove up to 99% of colors [13].

Water containing organic contaminants and heavy metals can be removed by dendrimers, a type of polymeric adsorbent. The adsorption of metal and organic contaminants onto adsorbents relies on the formation of complex and diverse electrostatic interactions [14]. The dendrimers containing metals are retrieved by ultrafiltration at an acidic pH of 4. Nanoadsorbents are readily incorporated into the treatment processes that are currently in use due to their many advantages. Nano-adsorbents work efficiently for the removal of arsenic; economically [15]. ADSORBSIA, a nano-crystalline titanium dioxide, is found very useful for the elimination of arsenic from the water used for drinking purposes. Owing to their advantages, nano-adsorbents are simple to incorporate into current treatment procedures. For optimal system efficiency, their powdered form can be applied within slurry- reactors, where they can work with more efficacies; however, an additional system is required for the recovery of the NPs [16]. Although fixed-bed reactors often encounter mass transfer limitations, they eliminate the need for an additional separation process. Nanoadsorbents have been widely used for the removal of arsenic because they exhibit good performance and are also more cost-effective than other adsorbents. Nano-adsorbents can be employed to load beads or granules within the fixed or fluidized absorbers [17].

Nanofilters

Nanofiltration membranes exhibit pore sizes ranging from 1 to 10 nm, falling between those used in microfiltration and ultrafiltration yet larger than those employed in reverse osmosis. Typically composed of polymeric thin films, commonly utilized materials include polyethene terephthalate and aluminum metal. The control of pore dimensions is influenced by factors such as pH, temperature, and duration of refinement. Polyethylene terephthalate-based films and similar materials are commonly known as "track-etch" layers [18]. On the other hand, films composed of metals, such as alumina layers, are generated through the electrochemical deposition of a thin layer of Al_2O_3 from aluminum metal in a low pH acidic environment [19].

Nanofilters play a crucial role in wastewater treatment processes due to their high permeation capabilities, particularly in eliminating monovalent and bivalent ions that conventional filters struggle to permeate. They serve as a barrier between reverse osmosis and ultrafiltration, aiding in the treatment of hard water. Nanofiltration membranes effectively remove various contaminants from water, meeting stringent water purification standards and allowing the treated water to meet strict criteria for reuse. The incorporation of various NMs, including carbon nanotubes, titanium dioxide, chitosan, silver NPs, and fullerene NPs, further enhances the adsorption capacity of Nanofilters [20, 21]. These NMs are recognized for their antibacterial properties, posing minimal risk as they do not generate hazardous by-products in water. Different NMs can be applied in water treatment, either by directly interacting with bacterial cells or by oxidizing cellular components. While nanotechnology presents promising solutions for disinfecting wastewater, challenges exist, such as the requirement for direct contact between NPs and microbial cell membranes, as highlighted by previous studies [22].

Nanofiltration membranes possess pore sizes ranging from 1 to 10 nm, which fall between those employed in microfiltration and ultrafiltration but are larger than those used in reverse osmosis [23]. These membranes consist of polymeric thin films with commonly used materials being polyethene terephthalate and aluminum metal. The control of pore dimensions is influenced by factors such as pH, temperature, and refinement duration, resulting in pore densities ranging from 1 to 10^6 pores per cm^2. Films crafted from polyethene terephthalate and similar substances are commonly known as "track-etch" layers. Conversely, films composed of metals, like alumina layers, are created through the electrochemical induction of a thin Al_2O_3 layer from aluminum metal in a low pH acidic environment [24].

Nanofilters play a crucial role in wastewater treatment through membrane filtration, enabling high permeation. Their effectiveness lies in the removal of monovalent and bivalent ions that conventional filters cannot permeate, making nanofilters valuable in treating hard water and serving as a boundary between reverse osmosis and ultrafiltration [2]. Nanofilters excel in eliminating various contaminants from water, meeting stringent water purification standards. The water that undergoes nanofiltration meets extremely strict criteria for reuse, as these filters effectively remove many pollutants, minimizing the need for further water decontamination [25].

Various NMs, such as carbon nanotubes, titanium dioxide, chitosan, silver NPs, and fullerene NPs, are recognized for their notable adsorption capacity and well-documented antibacterial properties. These inert and mildly oxidizing NPs exhibit a lack of hazardous by-products when introduced into water. NMs offer versatile applications in water treatment, encompassing direct attack on bacterial cells or oxidation of cellular components. However, the implementation of nanotechnology for wastewater disinfection comes with certain limitations. Feng *et al.* [7] emphasized that the effective removal of microorganisms from wastewater using NPs necessitates direct contact with their cell membranes.

Nanocomposite Membranes

Nanocomposite films represent an emerging class of filtration materials comprising a combination of blended matrix films and surface-functionalized films. Blended matrix films incorporate nanofillers within a structural framework, typically utilizing inorganic materials embedded in a polymeric or inorganic oxide matrix. These inorganic nanofillers, such as metal oxide NPs like Al_2O_3 and TiO_2, offer advantages in terms of enhanced mechanical and thermal stability, as well as increased surface-to-mass ratio [26]. This results from their larger specific surface area. The incorporation of zeolites into these films contributes to heightened hydrophilicity, leading to improved water permeability. Thin film nanocomposite films are semi-permeable membranes that have a distinctive layer on the upper surface commonly employed in reverse osmosis applications. The hydrophobic surface of ordered mesoporous carbons can be treated, and only a small percentage of it is needed to enhance the hydrophilicity of the film surface, resulting in significantly increased pure water permeability. Innovative thin film nanocomposite films have been developed by incorporating super hydrophilic

NPs into a polyamide thin film, enhancing penetration efficiency and reducing fouling potential. These NPs are designed to attract water, exhibiting high permeability by absorbing water and repelling contaminants. Furthermore, the hydrophilic NPs embedded in the film act as a deterrent to organic compounds and bacteria, which commonly lead to the clogging of conventional films over time. An additional strategy to prevent film clogging involves the functionalization of surfaces with chemical substances that can oxidize organic contaminants, thereby preventing the accumulation of fouling layers [27].

NMs for Detection of Trace Contaminants

Trace element concentration and detection in wastewater are two applications for NMs. The detection of organic pollutants and trace metal ions is a promising application for carbon nano-tubules [28]. Other NMs, such as modified nano-Au particles, have demonstrated superior capacity for high sensitivity and selectivity Hg^{2+} detection [29].

Nano-clay

Montmorillonite and allophane are the predominant types of nano clays, and they occur naturally in the clay fraction of soil. Montmorillonite is a crystalline hydrous phyllosilicate, representing a type of layer silicate. Recently, layered minerals known as "organo-clays" or "organically-modified montmorillonites" have been developed as polymer-clay nanocomposite materials. These spaces allow the minerals to adsorb water molecules and both positive and negative ions. Adsorbed ions in clays exchange interactions with the external environment. Clays have one major drawback despite their great utility in a variety of applications. To get around this issue, scientists have been searching for a method to use molecular pillars to support and prop up the clay layers [30]. Most clays can expand, creating additional space between their layers to accommodate ionic species and adsorbed water. The petroleum industry has expressed a keen interest in utilizing nanoclays to extract hydrocarbons from the refinery process. Additionally, nanoclays have undergone testing for the treatment of surface and groundwater, as well as the removal of hazardous organic chemicals originating from the pesticide and pharmaceutical industries. Clays possess a property known as swelling, wherein the gaps between their layers can enlarge, allowing them to absorb ionic species and water. In the context of the petroleum industry, nanoclays are being explored as a potential solution for extracting hydrocarbons from water used in refinery processes [31]. Furthermore, these nanoclays have been tested for their efficacy in treating both surface and groundwater, addressing environmental concerns related to various industries, including pesticides and pharmaceuticals.

Thin Film Nanocomposite (TFN) Membranes

The main objective of the material's development is to incorporate zeolites, nano-silver (nano-Ag), nano-titanium dioxide (nano-TiO_2), and carbon nanotubes (CNTs) into TFN through a doping process. This addition of NPs significantly alters the properties of the membrane [32]. The hypothesis suggests that the small and hydrophilic pores of nanozeolites establish preferential pathways for water flow. TFN membranes, specifically doped with 250 nm nano-zeolites at 0.2 wt percent, demonstrated moderately higher permeability and superior salt rejection (>99.4 percent) compared to commercially available reverse osmosis (RO) membranes [32]. Nanozeolites are frequently used as dopants, contributing to increased permeability and imparting a negative charge to the membrane. The enhanced water permeability through zeolite-filled pores may be attributed to interactions at the zeolite-polymer interface. Furthermore, the incorporation of antimicrobial agents through nanozeolites imparts anti-fouling characteristics to the membrane, as demonstrated in studies [34, 35]. This development holds promise for improving the performance and functionality of membranes in various applications.

Nanotubes

In recent years, the water industry has seen the introduction of various NMs thanks to advancements in nanotechnology, which have yielded promising outcomes. Among these materials, CNTs have garnered significant attention due to their unique properties. CNTs come in two main types: single-walled carbon nanotubes (SWNTs) and multi-walled carbon nanotubes (MWNTs), both of which are rolled into tube-like structures. These CNTs possess exceptional characteristics that make them highly desirable for a wide range of applications. They exhibit strong antimicrobial properties, allow for higher water flux compared to other porous materials of similar size, offer tunable pore size and surface chemistry, display high electrical and thermal conductivity, possess special adsorption capabilities, and can be functionalized to increase their affinity towards specific target molecules [36]. However, it is important to note that unfunctionalized CNTs have been found to be generally toxic and insoluble in water in some studies. To address this issue and enhance their dispersion in water while improving their reusability, CNTs are functionalized with various functional groups such as amines, carboxyl, and hydroxyl, among others. This functionalization process enhances their biocompatibility and water solubility. Moreover, nitric acid is commonly used to oxidize multi-walled carbon nanotubes, thereby increasing their absorption capacity. This oxidation process creates reactive sites either on the tips of the nanotubes or as defects on the sidewalls, which enhances the level of adsorption, according to a research work [37].

Super Sand Nano Channel

A groundbreaking filtration material known as "super sand" has been created through the application of graphite oxide, an NM, onto conventional sand. The components required for producing super sand, including graphite and regular sand, are both affordable and readily obtainable. Additionally, the manufacturing process of super sand can be conducted at or close to room temperature. These favorable aspects have sparked optimism among experts, who foresee the potential for super sand to become a cost-efficient and highly effective solution for water filtration in the coming years [37].

Nano-sensors

Another potential avenue in the future of nanotechnology lies in the development of nano-sensors. Nano-sensors serve the purpose of detecting Nano molecules and communicating their data to other devices. Their fundamental reason is to develop restorative nano items, silicon computer chips, and nano-robot manufacturers. Nano-sensors moreover diminished the discussed contaminations and found the solid discussed poisons. In the future, nanosensors will move forward and will show the world of innovation [38].

NPs of coinage metals: Gold and Silver

The principal motivation for utilizing noble metals stems from their inherent property of exhibiting minimal reactivity when examined on a larger scale. Recent advancements in the field of noble metal NP synthesis have provided valuable insights into their behavior and reactivity at the nanoscale level. When it comes to the application of noble metal NP-based chemistry in drinking water filtration, it proves to be particularly effective in addressing three key types of contaminants: halogenated organics, which include pesticides, heavy metals, and microorganisms [39].

The innovative chemistry of NPs of gold was found during the oxidation of CO [40]. It was built up that gold supported on different metal oxides may be a valuable candidate for alkene hydrogenation. Many curious disclosures emerged from this research. Gold NP upheld on alumina is a great framework for the expulsion of Mercury from water. Adsorption capacity was examined employing a column test and was observed utilizing UV spectroscopy. It was affirmed by control tests that unadulterated alumina alone is incapable of evacuating mercury from water [41].

Silver has been extensively utilized in households since ancient times, with silver vessels serving to preserve perishable items and purify water. In the early 20th century, a porous metallic silver mesh, known as "Katadyn silver", was developed for use as an antibacterial water filter. Efforts were also made to immobilize silver in a zero-valent form on activated carbon for water purification purposes. A modern avenue of research focused on the biosynthesis of noble metal NPs emerged, leading to increased scientific interest in exploring microorganisms as bio-factories [42]. Various microorganisms, including plants, plant extracts, bacteria, fungi, and human cells, have been investigated for their ability to synthesize noble metal NPs. Silver is particularly well-studied as an oligo-dynamic material due to its broad spectrum of antimicrobial effectiveness, low toxicity, and ease of incorporation into various substrates for active sanitization applications. The most recognized adverse health effect associated with silver is argyria, an irreversible darkening of the skin and mucous membranes resulting from excessive exposure to ionic silver (Ag(I), Ag^+) [43].

Small-scale Water Treatment Systems

In regions with scarce water resources, a nano wipe fabric has emerged as a highly efficient solution for collecting water droplets from the air, outperforming traditional polypropylene nets. This innovation in NMs has the potential to increase water capture by up to ten times. Moreover, these NMs have been modified not only to collect but also to filter the collected water from contaminants. Another significant application of nanotechnology in addressing clean water issues is individual water treatment. To make a global impact on the availability of clean drinking water, new technologies are needed to treat water directly at the point of use [44]. This concept involves providing individuals with devices that enable them to filter water at their taps, wells, or homes, thereby decentralizing the process of water treatment. This approach is particularly beneficial in developing countries, where large-scale treatment plants are scarce, and there is a significant need for cost-effective methods to provide clean drinking water to communities. Additionally, small-scale water treatment systems are less vulnerable to bioterrorism compared to large plants. Community-based water treatment would be most effective if tailored to remove specific contaminants found in local water sources, a task that may require the use of nanotechnology [39]. Nanosorbents, nanocatalysts, smart membranes, nanosensors, and other nanotechnology-based solutions can form the basis of innovative small-scale water treatment systems [45]. Achieving the goal of individual water treatment may prove to be more feasible than integrating nanotechnology into existing centralized water treatment plants operated by public utilities.

NANOTECHNOLOGY AS A RISK TO HUMAN HEALTH: A CHALLENGE CONCEALED

Despite the widespread promotion of the benefits of nanotechnology, discussions regarding the potential impacts of its extensive use are insufficient. Both proponents and critics of nanotechnology face challenges in presenting their arguments due to the limited data available to support either side. Given the rapid pace of development in this field and the significant attention it is receiving, safety concerns have also been raised about the use of NMs in various products. Some have drawn comparisons between high-aspect-ratio NPs and asbestos fibers. To address this issue, several organizations are taking initiatives like the European Nano-safe Consortium, which aims to assess the potential risks posed by NMs. NMs have the ability to enter the human body through various routes; accidental or incidental exposure during production or use is likely to occur through the lungs, leading to rapid translocation to other vital organs *via* the bloodstream. However, residual chemicals remaining during wastewater treatment generate a variety of by-products through reactions with pollutants. Continuous exposure to these by-products through drinking water, inhalation, and dermal contact during routine indoor activities may pose serious risks to human health. For instance, residual aluminum salts in treated water may contribute to Alzheimer's disease. Concerning CNTs, existing information on the toxicity of CNTs in drinking water is limited with many unanswered questions. Furthermore, the impacts of traditional water treatment chemicals on human health have been studied extensively but there has been no comparative analysis. Thus, a qualitative comparison of the human health impacts of both residual CNTs and conventional water treatment chemicals is necessary. Regarding the eco-toxicity of nanoscale Titanium dioxide, Silicon dioxide, and Zinc oxide water suspensions, comparative assessments have shown that the potential eco-toxicity of these nano-sized dioxides varies, with antibacterial activity increasing with particle concentration. Advertised NP sizes did not correspond to actual particle sizes. Aggregation also produced additional-sized particles with similar antibacterial activity at a given concentration. These findings underscore the need for caution in the use and disposal of such engineered NMs to prevent unintended environmental impacts, as well as the importance of further research to enhance risk management [46].

The effects of downsizing the oxide of zinc to micro and nanoscale on biological parameters were investigated. The zinc oxide NPs exhibited toxicity at lower doses, in this way disturbing future nanotoxicology inquiries that should be centered on the significance of dosage measurements. The ability of NPs to penetrate the outer layer of the skin was the focus of several discussions. However, there are questionable issues accessible for molecule infiltration due to diverse exploratory setups. Despite the inherent risks associated with the use of this technology, it is crucial to consider its potential advantages as well. In another articulation, "we must see at nanotechnology in both green and ruddy or dark sides".

CONCLUSION AND FUTURE DIRECTIONS

Despite extensive research exploring nanotechnology's potential for environmental remediation, unresolved concerns persist. Further investigation is crucial to fully grasp how nanotechnology can effectively tackle environmental pollutants in real-world settings, especially concerning contaminated air, water, and soil stemming from industrial activities. While the methodologies for employing diverse nanotechnologies are well-established, limited knowledge exists about the destiny of these substances post-degradation of contaminants. Despite claims of recyclability for some materials, their diminishing efficacy over time emphasizes the necessity of understanding their fate to avert environmental harm.

Additional inquiries should address obstacles before NMs can be successfully deployed in water treatment. Thorough evaluations of NMs' safety, encompassing toxicity, transport, and environmental behavior, are imperative. While nanotechnology shows promise in addressing pollution challenges, its primary hurdle lies in preventing contamination to foster clean and sustainable practices. Despite potential risks linked with NMs, researchers strive to harness nanotechnology's advantages while mitigating adverse effects. Nanotechnology, akin to any other field of knowledge, presents both opportunities and risks, highlighting the imperative need for responsible and cautious application.

REFERENCES

[1]   Zekić E, Vuković Z, Halkijević I. Application of nanotechnology in wastewater treatment. Gradevinar 2018; 70(4): 315-23.
[http://dx.doi.org/10.14256/JCE.2165.2017]

[2]   Sharma V, Sharma A. Nanotechnology: An emerging future trend in wastewater treatment with its innovative products and processes International Journal of Enhanced Research in Science Technology & Engineering 2013; 1: 2.

[3]   Kanchi S. Nanotechnology for water treatment. Int J Environ Anal Chem 2014; 1(2)

[4]   Theron J, Walker JA, Cloete TE. Nanotechnology and water treatment: applications and emerging opportunities. Crit Rev Microbiol 2008; 34(1): 43-69.
[http://dx.doi.org/10.1080/10408410701710442] [PMID: 18259980]

[5]   Singh M. Remediation of Aquatic Milieu and Nanotechnology: A Review International Journal of All Research Education and Scientific Methods (IJARESM) 2023; 11(10): 546.

[6]   Duhan JS, Kumar R, Kumar N, Kaur P, Nehra K, Duhan S. Nanotechnology: The new perspective in precision agriculture. Biotechnol Rep (Amst) 2017; 15: 11-23.
[http://dx.doi.org/10.1016/j.btre.2017.03.002] [PMID: 28603692]

[7]   Feng QL, Wu J, Chen GQ, Cui FZ, Kim TN, Kim JO. A mechanistic study of the antibacterial effect

of silver ions onEscherichia coli andStaphylococcus aureus. J Biomed Mater Res 2000; 52(4): 662-8.
[http://dx.doi.org/10.1002/1097-4636(20001215)52:4<662::AID-JBM10>3.0.CO;2-3]   [PMID: 11033548]

[8]   Chorawala KK, Mehta MJ. Applications of nanotechnology in wastewater treatment. International Journal of Innovative and Emerging Research in Engineering 2015; 2(1): 21-6.

[9]   Lens PNL, Virkutye J, Jegatheesan V, Kim SH, Al-Abed S. Nanotechnology for water and wastewater treatment. IWA Publishing 2013.
[http://dx.doi.org/10.2166/9781780404592]

[10]   Inoue Y, Hoshino M, Takahashi H, *et al.* Bactericidal activity of Ag–zeolite mediated by reactive oxygen species under aerated conditions. J Inorg Biochem 2002; 92(1): 37-42.
[http://dx.doi.org/10.1016/S0162-0134(02)00489-0] [PMID: 12230986]

[11]   Cai Y, Jiang G, Liu J, Zhou Q. Multiwalled carbon nanotubes as a solid-phase extraction adsorbent for the determination of bisphenol A, 4-n-nonylphenol, and 4-tert-octylphenol. Anal Chem 2003; 75(10): 2517-21.
[http://dx.doi.org/10.1021/ac0263566] [PMID: 12919000]

[12]   Amin MT, Alazba AA, Manzoor U. A review of removal of pollutants from water/wastewater using different types of nanomaterials. Adv Mater Sci Eng 2014; 2014: 1-24.
[http://dx.doi.org/10.1155/2014/825910]

[13]   Bottino A, Capannelli G, D'Asti V, Piaggio P. Preparation and properties of novel organic–inorganic porous membranes. Separ Purif Tech 2001; 22-23(1-2): 269-75.
[http://dx.doi.org/10.1016/S1383-5866(00)00127-1]

[14]   Crooks RM, Zhao M, Sun L, Chechik V, Yeung LK. Dendrimer-encapsulated metal nanoparticles: synthesis, characterization, and applications to catalysis. Acc Chem Res 2001; 34(3): 181-90.
[http://dx.doi.org/10.1021/ar000110a] [PMID: 11263876]

[15]   Li J, Liu H, Paul Chen J. Microplastics in freshwater systems: A review on occurrence, environmental effects, and methods for microplastics detection. Water Res 2018; 137(15): 362-74.
[http://dx.doi.org/10.1016/j.watres.2017.12.056] [PMID: 29580559]

[16]   Sylvester DJ, Bowman DM. Navigating the Patent Landscapes for Nanotechnology: English Gardens or Tangled Grounds? Biomedical Nanotechnology: Methods and Protocols. Sarah J Hurst (ed), Springer, New York 2011; pp. 359-78.

[17]   Vijayageetha V A, Annamalai V, Pandiarajan A. A study on the nanotechnology in water and wastewater treatment IOSR Journal of Applied Physics (IOSR-JAP) 2018; 10(4): 28-31.

[18]   Roco MC. Nanotechnology Research Directions for Societal Needs in 2020', Science Policy Reports The Long View of Nanotechnological Development: The National Nanotechnology Initiative at 10 Years 2011; 1: 1-28.
[http://dx.doi.org/10.1007/978-94-007-1168-6]

[19]   Pendergast MTM, Nygaard JM, Ghosh AK, Hoek EMV. Using nanocomposite materials technology to understand and control reverse osmosis membrane compaction. Desalination 2010; 261(3): 255-63.
[http://dx.doi.org/10.1016/j.desal.2010.06.008]

[20]   Jose AJ, Jacob AM, Manjush KC, Kappen J. Chitosan in water purification technology. In: Ahmad S, Hussain CM, Eds. Green and sustainable advanced materials: Applications. 2018.
[http://dx.doi.org/10.1002/9781119528463.ch5]

[21]   Duran A, Tuzen M, Soylak M. Preconcentration of some trace elements via using multiwalled carbon nanotubes as solid phase extraction adsorbent 2009.
[http://dx.doi.org/10.1016/j.jhazmat.2009.03.119]

[22]   Yamanaka M, Hara K, Kudo J. Bactericidal actions of a silver ion solution on Escherichia coli, studied by energy-filtering transmission electron microscopy and proteomic analysis. Appl Environ Microbiol 2005; 71(11): 7589-93.

[http://dx.doi.org/10.1128/AEM.71.11.7589-7593.2005] [PMID: 16269810]

[23] Fuwad A, Ryu H, Malmstadt N, Kim SM, Jeon TJ. Biomimetic membranes as potential tools for water purification: Preceding and future avenues. Desalination 2019; 458: 97-115.
[http://dx.doi.org/10.1016/j.desal.2019.02.003]

[24] Hogen-Esch T, Pirbazari M, Ravindran V, Yurdacan HM, Kim W. High-Performance Membranes for Water Reclamation Using Polymeric and Nanomaterials US Patent No 20160038885A 2019.

[25] Li Q, Mahendra S, Lyon DY, Brunet L, Liga MV, Li D, Alvarez PJJ. Antimicrobial nanomaterials for water disinfection and microbial control: Potential applications and implications. Environ Sci Technol. 2008;42(12): 4606–4615.

[26] Yadav VB, Gadi R, Kalra S. Clay based nanocomposites for removal of heavy metals from water: A review. J Environ Manage 2019; 232: 803-17.
[http://dx.doi.org/10.1016/j.jenvman.2018.11.120] [PMID: 30529868]

[27] Unuabonah EI, Taubert A. Clay–polymer nanocomposites (CPNs): Adsorbents of the future for water treatment. Appl Clay Sci 2014; 99: 83-92.
[http://dx.doi.org/10.1016/j.clay.2014.06.016]

[28] Wang X, Cai W, Lin Y, Wang G, Liang C. Mass production of micro/nanostructured porous ZnO plates and their strong structurally enhanced and selective adsorption performance for environmental remediation. J Mater Chem 2010; 20(39): 8582-90.
[http://dx.doi.org/10.1039/c0jm01024c]

[29] Lin YH, Tseng WL. Ultrasensitive sensing of Hg(2+) and CH(3)Hg(+) based on the fluorescence quenching of lysozyme type VI-stabilized gold nanoclusters. Anal Chem 2010; 82(22): 9194-200.
[http://dx.doi.org/10.1021/ac101427y] [PMID: 20954728]

[30] Zhu HY, Li JY, Zhao J-C, Churchman GJ. Photocatalysts prepared from layered clays and titanium hydrate for degradation of organic pollutants in water. Appl Clay Sci 2005; 28(1-4): 79-88.
[http://dx.doi.org/10.1016/j.clay.2004.05.001]

[31] Tahoon, Mohamed A., Saifeldin M. Siddeeg, Norah Salem Alsaiari, Wissem Mnif, and Faouzi Ben Rebah, Effective Heavy Metals Removal from Water Using Nanomaterials: A Review. Processes 8, 2020; 6: 645.
[http://dx.doi.org/10.3390/pr8060645]

[32] Yin J, Yang Y, Hu Z, Deng B. Attachment of silver nanoparticles (AgNPs) onto thin-film composite (TFC) membranes through covalent bonding to reduce membrane biofouling. J Membr Sci 2013; 441: 73-82.
[http://dx.doi.org/10.1016/j.memsci.2013.03.060]

[33] Sung JH, Ji JH, Yoon JU, *et al.* Lung function changes in Sprague-Dawley rats after prolonged inhalation exposure to silver nanoparticles. Inhal Toxicol 2008; 20(6): 567-74.
[http://dx.doi.org/10.1080/08958370701874671] [PMID: 18444009]

[34] Lind ML, Ghosh AK, Jawor A, *et al.* Influence of zeolite crystal size on zeolite-polyamide thin film nanocomposite membranes. Langmuir 2009; 25(17): 10139-45. a
[http://dx.doi.org/10.1021/la900938x] [PMID: 19527039]

[35] Lind ML, Jeong BH, Subramani A, Huang X, Hoek EMV. Effect of mobile cation on zeolite-polyamide thin film nanocomposite membranes. J Mater Res 2009; 24(5): 1624-31. b
[http://dx.doi.org/10.1557/jmr.2009.0189]

[36] Montemagno C, Schmidt J, Tozzi S. Biomimetic Membranes US Patent No 20040049230 2004.

[37] Singh M. Nano-adsorbents for the Effective Removal of Heavy Metal Ions from Domestic and Industrial Wastewater: A Review of Current Scenario. Research Communications 2023; 1(1): 112-24.

[38] Khajeh M, Laurent S, Dastafkan K. Nanoadsorbents: classification, preparation, and applications (with emphasis on aqueous media). Chem Rev 2013; 113(10): 7728-68.

[http://dx.doi.org/10.1021/cr400086v] [PMID: 23869773]

[39]   Singh M. A Comprehensive Review: Nanomembranes and Nanosorbents for Water Treatment International Journal of Research and Innovation in Applied Science (IJRIAS) s 2022; 7(9): 01-6.

[40]   Bond GC, Sermon PA, Webb G, Buchanan DA, Wells PB. Hydrogenation over supported gold catalysts. J Chem Soc Chem Commun 1973; 13(13): 444b.
[http://dx.doi.org/10.1039/c3973000444b]

[41]   Lisha KP, Anshup , Pradeep T. Towards a practical solution for removing inorganic mercury from drinking water using gold nanoparticles. Gold Bull 2009; 42(2): 144-52.
[http://dx.doi.org/10.1007/BF03214924]

[42]   Kalhapure RS, Sonawane SJ, Sikwal DR, *et al.* Solid lipid nanoparticles of clotrimazole silver complex: An efficient nano antibacterial against Staphylococcus aureus and MRSA. Colloids Surf B Biointerfaces 2015; 136: 651-8.
[http://dx.doi.org/10.1016/j.colsurfb.2015.10.003] [PMID: 26492156]

[43]   Taurozzi JS, Arul H, Bosak VZ, *et al.* Effect of filler incorporation route on the properties of polysulfone–silver nanocomposite membranes of different porosities. J Membr Sci 2008; 325(1): 58-68.
[http://dx.doi.org/10.1016/j.memsci.2008.07.010]

[44]   Aragon M, Kottenstette R, Dwyer B, *et al.* Arsenic pilot plant operation and results Anthony. Sandia National Laboratories 2007.

[45]   Kumar VS, Nagaraja BM, Shashikala V, *et al.* Highly efficient Ag/C catalyst prepared by electro-chemical deposition method in controlling microorganisms in water. J Mol Catal Chem 2004; 223(1-2): 313-9.
[http://dx.doi.org/10.1016/j.molcata.2003.09.047]

[46]   Buzea C, Pacheco II, Robbie K. Nanomaterials and nanoparticles: Sources and toxicity. Biointerphases 2007; 2(4): MR17-71.
[http://dx.doi.org/10.1116/1.2815690] [PMID: 20419892]

CHAPTER 3

Contemporary Execution of Nanomaterials in Wastewater Treatment

Sheerin Masroor[1,*] and **Ajay Kumar Tiwari**[2]

¹ Department of Chemistry, A.N. College, Patliputra University, Patna 800013, Bihar, India

² Department of Chemistry, School of Applied Sciences, Uttaranchal University, Dehradun, Uttarakhand 248007, India

Abstract: Water of high quality must be readily available to all living beings on the planet. With water resources becoming scarce, the primary need of the modern period is wastewater treatment. While there are other methods, such as adsorption, flocculation, and filtration, they are only employed in the initial stages of wastewater treatment. Nanomaterials (NMs) and recent advances in technology have garnered interest in wastewater treatment. The ability of nanoparticles (NPs) to catalyze reactions, adsorb substances, reactivity, and larger surface area makes them highly valuable in wastewater treatment. Diverse varieties of NMs are employed to eliminate distinct pollutants from wastewater to make it eco-friendly in use. Activated carbon, graphene, carbon nanotubes, metal oxide NPs (*e.g.*, TiO_2, ZnO, and iron oxides), zero-valent metal NPs (*e.g.*, Ag, Fe, and Zn), and nanocomposites, titanium oxide, and magnesium oxide are a few types of nanoadsorbents that are utilized in wastewater treatment to extract heavy metals. Both organic and inorganic contaminants can be eliminated from water using nanocatalysts, such as electrocatalysts and photocatalysts. Special destruction or removal of some organic contaminants with the use of semiconducting NPs, either alone or in conjunction with ozonation, the Fenton process, or sonolysis is being done. Additionally, the topic of how well nanotechnology works against different parameters to provide pure water in an environmentally responsible manner is covered. The advances gained in wastewater treatment through the use of NPs are the main concerns of this chapter.

Keywords: Contaminants, Metal NPs, Metal oxide NPs, Nanocatalysts, Nanocomposites, Nanotechnology, Wastewater treatment.

INTRODUCTION

Water pollution due to anthropogenic activities is a global problem in present times. Even though environmental factors are also responsible for the deteriorated quality of water, the term "pollution" typically suggests that human activities are

---

* **Corresponding author Sheerin Masroor:** Department of Chemistry, A.N. College, Patliputra University, Patna 800013, Bihar, India; E-mail: masroor.sheerin@gmail.com

Neha Agarwal, Vijendra Singh Solanki, Neetu Singh & Maulin P. Shah (Eds.)

the major cause of water pollution. The main cause of water pollution is the discharge of contaminated wastewater into the surface or groundwater, consequently affecting the environment and ecosystem badly. In this view, water pollution control and the treatment of wastewater is a matter of great concern. Unfortunately, only approximately 1% of the available water is fit for consumption by human beings. According to an estimation from the World Health Organization (WHO) in 2015, more than 1.1 billion people are facing potable water scarcity worldwide. This global scarcity of pure water is a consequence of rapid industrialization and a growing population [1]. Another major challenge is the purification of water and wastewater by the removal of organic and inorganic pollutants to bring it to a permissible level [2]. It becomes more challenging because the majority of the existing technologies have certain drawbacks, such as high energy requirements, incomplete removal of pollutants, and the creation of toxic sludge during the wastewater purification process [3, 4]. Nanotechnology is an emerging technique that works against different parameters to provide pure water in an environmentally friendly manner. Therefore, the authors have made an effort to cover the advances and potential of NMs in wastewater treatment in a comprehensive manner, along with other basic techniques.

BASIC TECHNIQUES FOR WASTEWATER TREATMENT

Wastewater treatment is an essential procedure to safeguard the environment and the public's health. Physical, chemical, biological, tertiary, and disinfection are the five fundamental principles of wastewater treatment [5].

Primary/Physical Treatment

Using physical procedures like filtration, sedimentation, and screening, solid particles are removed from wastewater as a part of physical treatment. To remove big solids, including plastics, sticks, rags, *etc.*, from the wastewater, screens are used, followed by sedimentation, where the wastewater is allowed to settle in a basin so that smaller particles (such as sand and silt) settle down. The wastewater's tiny residual particles are subsequently eliminated by filtering.

Chemical Treatment

Utilizing chemicals to extract impurities from wastewater is known as chemical treatment. In order to eliminate contaminants by sedimentation or filtration, chemicals are added to the wastewater that causes them to be coagulated or precipitated during the process. For good results, physical and chemical treatments are frequently combined.

Biological Treatment

Biological treatment is done to remove organic matter in wastewater utilizing microorganisms like bacteria, algae, and fungi. Biological treatment mineralizes organic matter into simpler and more stable compounds. Although it is a natural process that occurs in an aqueous environment, in wastewater treatment plants, the process is artificially accelerated and controlled by regulating the growth conditions of microorganisms, including temperature and pH levels.

Tertiary Treatment

At the last phase of the treatment process, known as tertiary treatment, the water is given a thorough cleaning before being allowed to return to the environment. Dissolved solids and other contaminants that were not eliminated during the earlier treatment stages must be removed during this process. Reverse osmosis, activated carbon adsorption, and membrane filtration are a few examples of tertiary treatment procedures.

Disinfection

In order to guarantee that the wastewater is safe to be released into the environment, any pathogens, bacteria, or viruses must be eliminated during the final stage of wastewater treatment, known as disinfection. Ozone, chlorine, and UV light are a few examples of disinfection techniques.

The biological methods currently in use for wastewater purification have the disadvantage of being extremely slow processes and can occasionally make drinking water more toxic by releasing microorganisms [6]. Applying these biological systems to wastewater containing nonbiodegradable contaminants is also ineffective. Physical filtration techniques also produce toxic sludge that poses significant disposal challenges, and it is challenging to attain 100% purification using these techniques. The potential applications of NMs in wastewater treatment have been documented in many studies [7, 8]. The field of NMs plays a wide range of roles in the treatment of industrial and municipal wastewater. These include the use of nanoadsorbents [9, 10], semiconducting nanocatalysts [11, 12], electrocatalysts [13], antimicrobial materials [14], and nanomembranes [15]. To effectively degrade or remove organic pollutants, semiconducting NPs can be used alone or in conjunction with ozonation, the Fenton process, or sonolysis. Therefore, applications of NMs in wastewater treatment are listed and critically reviewed in this chapter. Additionally, the efficiency of nanotechnology in antimicrobial activity to generate pure water through a sustainable method is also discussed. The functions of NMs in adsorption methods—more especially, carbon-based nanoadsorbents—to eliminate heavy metal pollution from industrial wastewater are also thoroughly covered.

DIFFERENT TYPES OF CONTAMINANTS IN WASTEWATER

These are divided into inorganic [16 - 21], biological [22 - 24], and organic categories based on their nature. An overview of the main contaminants is provided in Table **1**.

Table 1. Overview of the main contaminants present in water treatment.

Nature of Contaminants	Types	Examples	Transmission Route	Negative Effects on Living Beings
Biological	Bacteria, fungi, viruses, and protozoa.	Salmonella Klebsiella Escherichia coli Poliovirus Rotavirus Enterovirus Cryptosporidium ssp. Giardia ssp. Entamoeba histolytica	Excreta	Colitis, Renal insufficiency, Diarrhoea, Homolytic uremic kidney, Salmonellosis, and typhoid fever. Legionnaires' illness.
Inorganic	Heavy Metals, Bioaccumulative NMs	Cadmium (Cd) Lead (Pb) Chromium (Cr) Zinc (Zn) Mercury (Hg), Oxides of different metals.	Industries	Different body infections. Modifications to the body's production and use of neurotransmitters. Production of reactive oxygen species (ROS), which have the potential to induce oxidative stress.
Organic	Pesticides such as fungicides, insecticides, and herbicides, Bioaccumulative and Persistent.		Soil, water	

DIFFERENT TYPES OF NMS USED IN WASTEWATER TREATMENT

Numerous NPs with diverse physical, chemical, and biological properties can be found in organic, inorganic, and composite materials [25]. Metal/metal oxide NPs oxides, carbon nanotubes (CNTs), metal-organic frameworks (MOFs), and graphene-based and polymer-based NMs have been among the most studied NMs in wastewater treatment. Carbon-based NPs make up the majority of organic NMs

[26]. Metals and metal oxides, including copper (Cu), iron (Fe), zinc (Zn), aluminum (Al), iron oxide (FeO), titanium oxide (TiO2), and zinc oxide (ZnO), form the basis of NMs. These materials are categorized as inorganic [27], organic-inorganic, organic-organic, and inorganic-inorganic and composite materials [28].

Given that nanotechnology cannot be used for water and wastewater treatment, it makes the most sense to include NMs in industrial water treatment processes. According to the types of materials they employ, NMs can be divided into three major categories: membranes, catalysts, and adsorbents at the nanoscale [29].

Nanoadsorbents

Nanosized particles with a great affinity for adsorbing substances are known as adsorptive NPs, and they can be organic or inorganic in nature. Thanks to their key attributes, which include their tiny size, strong reactivity, greater surface energy, and catalytic potential, these NPs have the potential applications in removing many types of contaminants. Four types of NPs can be used in adsorption processes: metallic, magnetic, mixed oxide, and metal oxide.

Nanocatalysts

High photocatalytic activity is produced in nano-catalysis when light interacts with metallic NPs, making it more and more popular. Hydroxyl radicals that are produced in a photocatalytic process kill organic materials and bacteria. To be classified as a nanocatalyst, a nano photocatalyst needs to fulfill specific requirements. These include being inert and having a concentration in air and water that is below the maximum allowed limit. Additionally, it must precipitate and form ordinary particles and agglomerates.

Nano-membranes

Particles of wastewater are separated by a nano-membrane. Heavy metals, dyes, and other pollutants can be effectively removed using these filters.

NMs such as nanotubes, nanoribbons, and nanofibers are employed as nano-membranes.

COMMONLY USED NPS IN WATER TREATMENT

Science that studies things at the nanoscale level is called nanoscience. Materials containing at least one component with a dimension smaller than 100 nm are used in nanotechnology [6]. Since these materials are so small compared to the typical materials, they have very different mechanical, electrical, optical, and magnetic properties. NMs are highly adsorbent and reactive and have a vast surface area due to their small size. There have also been reports of NPs being extremely mobile in solution. There have been reports of the removal of microorganisms, inorganic anions, organic contaminants, and heavy metals using different kinds of NMs. For their possible uses in wastewater treatment, a wide range of NMs, including carbon nanotubes (CNTs), metal/metal oxide NPs, nanocomposites, zero-valent (ZV) metal NPs, nanoporous-activated carbon, graphene, fullerenes, and cellulose NMs, have been thoroughly studied.

Metal NPs

Zinc NPs (ZnNPs)

Zinc has been explored as an alternative to iron in the degradation of pollutants in wastewater treatment using ZV metal NPs [30]. Zinc has a higher standard reduction potential than iron (Zinc −0.762V, Fe −0.440 V), making it a stronger reductant. Consequently, zinc NPs may degrade contaminants faster than nano-zero-valent iron (nZVI) particles. A majority of research has been done on the dehalogenation reaction in relation to the use of nano-zero-valent zinc (nZVZ) [31]. Several studies have also shown that nZVZ may be effective in reducing contaminants, but its use is mostly restricted to the breakdown of halogenated organic compounds, particularly CCl_4. There have not been many reports of nZVZ treating other types of pollutants. As a result, nZVZ applications at contaminated field locations have not yet been accomplished on a pilot or full-scale basis [32].

Silver NPs (AgNPs)

Strong antibacterial activities against a variety of pathogens, such as viruses [33], bacteria [34], and fungi [35], are attributed to the high toxicity of AgNPs to microorganisms. AgNPs are excellent antibacterial agents and are frequently used to clean water. There is an ongoing discussion over the precise mechanism underlying AgNPs' antibacterial activities. It has been observed that AgNPs can attach and permeate the bacterial cell wall and alter its structures, making it more permeable [36]. Furthermore, free radicals may be produced when AgNPs come into contact with bacteria. They are thought to be the cause of cell death since they can harm the cell membrane [37]. AgNPs, if applied directly, may cause certain problems, such as their propensity to congregate in aqueous solutions, which gradually lowers their efficiency [38]. Owing to their strong antibacterial activity and affordability, AgNPs affixed to the filter materials have been deemed promising for wastewater remediation [39].

Iron NPs (FeNPs)

There is an increasing interest in the use of different ZV metal NPs, such as Fe, Zn, Al, and Ni, in the treatment of wastewater. In Table **1**, the standard reduction potentials for these metals are presented [40]. In comparison to many redox-labile pollutants, nZVI or nZVZ has a fair possibility of acting as reducing agents due to their moderate standard reduction potential. Fe has some notable advantages over Zn in wastewater remediation, even though Fe has a poorer reduction capacity. These advantages include low cost, strong adsorption capabilities, precipitation, and oxidation. Consequently, the most researched ZV metal NPs are ZVI NPs. These NPs have excellent reducing ability and good adsorption capabilities due to their enormous specific surface area and incredibly small size [41]. The majority of its exceptional effectiveness in pollutant removal can be attributed to these features.

Metal Oxide NPs

Zinc Oxide NPs (ZnONPs)

Owing to its special qualities, which include a high oxidation capacity and good photocatalytic activity, ZnONPs have become a promising option in water and wastewater treatment. ZnONPs are environmentally friendly in nature because they are perfect for treating sewage since they are not only gentle on the environment but also get along with living things [42 - 45]. Furthermore, because of their almost identical band gap energies, ZnO and TiO_2 NPs have comparable photocatalytic capacities. On the other hand, ZnONPs are less expensive than TiO_2NPs [46]. Additionally, compared to other semiconducting metal oxides, ZnONPs may absorb more quanta and a larger variety of solar spectra [47].

Additionally, photo corrosion prevents ZnONPs from being applied, which leads to limited photocatalytic efficacy [48]. Metal doping with anionic, cationic, rare-earth, and metal dopants is a frequent way to increase ZnONPs' photodegradation effectiveness [49]. Furthermore, many studies have confirmed that combining ZnONPs with other semiconductors, such as CdO [50], TiO_2 [51], SnO_2 [52], CeO_2 [53], graphene oxide (GO) [54], and reduced graphene oxide (RGO) [55] is a workable strategy to increase photodegradation efficiency of ZnONPs.

Titanium Dioxide NPs (TiO₂-NPs)

The TiO_2 semiconductor electrode was detected by Fujishima and Honda in the year 1972 [56] by a photocatalytic technique, which later gained significant interest. The photocatalytic degradation method has demonstrated efficacy in wastewater treatment by oxidizing pollutants and decomposing them into CO_2,

H_2O, and reusable anions like NO_3^-, PO_3^-, and Cl^-. Recently, TiO_2 has been the subject of great interest in research because of its strong photocatalytic activity, affordable price, photostability, non-toxicity, easy accessibility, and chemical and biological stability [57]. The three natural states of titanium dioxide are anatase, rutile, and brookite. Anatase is still regarded as a useful substance for nano-photocatalysis [58].

Iron Oxide NPs (FeONPs)

FeONPs are becoming more and more popular as adsorbents for heavy metals remediation from wastewater since they are easily obtainable and simple to use. Due to their accessibility and ease of usage, FeONPs have gained great interest as a means of eliminating heavy metals. Magnetic magnetite (Fe_3O_4), magnetic maghemite (Fe_2O_4), and nonmagnetic hematite (Fe_2O_3) are frequently utilized as nano-adsorbents. It is difficult to recover nanoadsorbents from contaminated water due to their small size. With the use of an external magnetic field, these can be readily extracted and recovered from the system. As a result, they have been potentially used in water systems to adsorb different heavy metals [59 - 61]. The adsorption properties of FeONPs have been enhanced by adding different ligands, such as ethylenediamine tetraacetic acid (EDTA) and meso-2,--dimercaptosuccinic acid (DMSA)) [62] or polymers (*e.g.*, copolymers of acrylic acid and crotonic acid) [63].

NMS BASED ON CARBON

The non-metallic element carbon is the second most plentiful element in the human body after oxygen and the sixth most abundant element. Because of their non-toxicity, high abundance, high surface area, stable structure, porosity, and high sorption capacity, carbon-based NMs have been extensively used in the removal of heavy metals and dyes from wastewater [64]. The electrical, thermal, and mechanical properties of carbon NMs are exceptional. They have excellent stability, good conductivity, minimal toxicity, and environmental friendliness because they are made entirely of carbon. Carbon is widely regarded as a biocompatible substance because it makes up a sizable portion of the human body. Their vast surface area, linear geometry, and superior electrical conductivity make their surface very accessible to the electrolyte. Anisotropic heat conductivity is also a strong characteristic of NMs based on carbon. Because of this characteristic, carbon-based NMs are utilized in high-end computer electronics, where uncooled chip temperatures can approach $100°C$. Their significant capacity to adsorb a wide range of pollutants due to large surface area and selectivity towards aromatics are the reasons behind their potential advantages in wastewater treatment [65].

Graphene

Graphene is a two-dimensional (2D) layer of sp^2 hybridized carbon atoms that form a hexagonal honeycomb-like crystal lattice (Fig. **1**). This substance is the building block of all graphitic forms. Graphite sheets (3D) are split into graphene, rolled graphene layers produce carbon nanotubes (1D), and graphene becomes fullerene (0D) when folded [66].

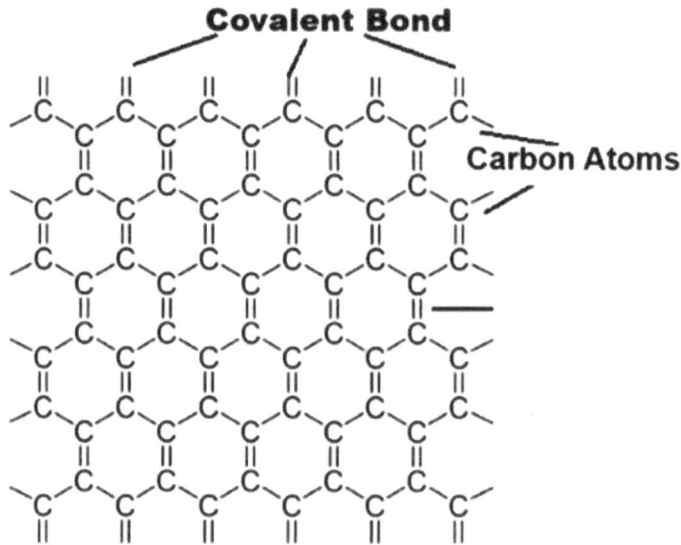

Fig. (1). Hexagonal honeycomb-like crystal lattice of graphene.

Graphene and its derivative structures have emerged as innovative and effective materials in water and wastewater treatment and purification [67].

Fullerenes

Carbon allotropes made up of fused rings of five to seven atoms that are joined by single and double bonds to form a closed or partially closed mesh are known as fullerenes (Fig. **2**). The molecules might have other shapes, such as tubes, ellipsoids, or hollow spheres. The empirical formula for fullerenes with a closed mesh topology is C_n, which is commonly abbreviated as C_n. In this formula, n represents the number of carbon atoms in the fullerene. There might be more than one isomer for particular values of n. Buckminsterfullerene (C_{60}), the most well-known member of the family, is named after Buckminster Fuller. Due to their informal similarity to the association football ("soccer") standard ball, the closed fullerenes, particularly C_{60}, are often known as "buckyballs." Bucky onions are the term given to nested closed fullerenes. Fullerite is the term for the bulk solid form of either pure or mixed fullerenes [68]. According to Thilgen and Diederich [69], they are a type of carbon allotropes whose molecules resemble hollow spheres, tubes, or ellipsoids and are made up of an even number of coordinated carbon atoms to form convex closed polyhedral [70]. According to Manawi and co-workers [71], fullerenes are conventionally made using chemical vapor deposition and arc discharge vaporization of graphite. Many investigations have been done on the use of fullerene as an adsorbent for organic compounds in contaminated waterways [72]. The presence of fullerenes enhances the hydrophobicity and enhances the ability of adsorption of heavy metals from aquatic environments [73].

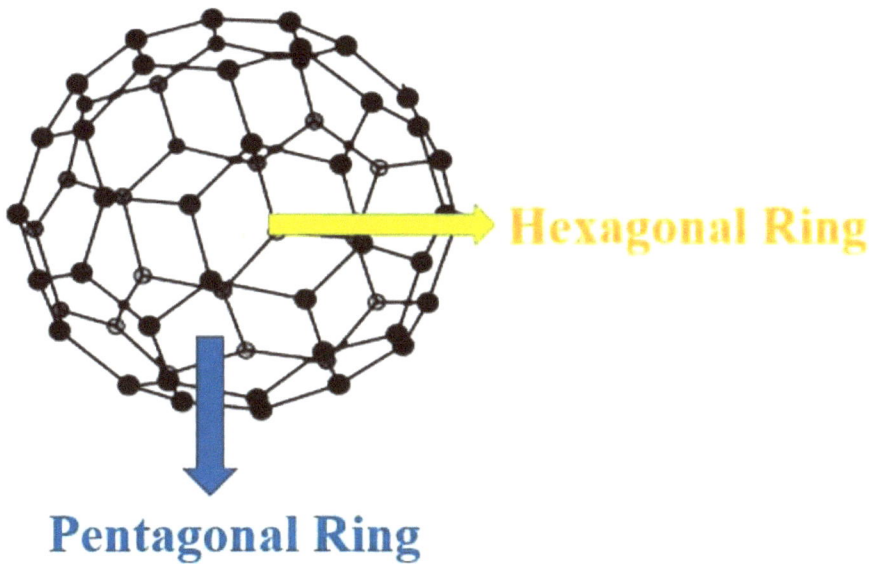

Fig. (2). Structure of Fullerenes.

Carbon Nanotubes (CNTs)

Graphene sheets are rolled up into cylinders as thin as 1 nm in diameter to create CNTs [74]. Because of their special qualities, CNTs have gained a lot of attention due to remarkable adsorption efficiencies for various types of contaminants, including ethyl benzene [75], Pb^{2+}, Cu^{2+}, and Cd^{2+} [76], dichlorobenzene [77] and Zn^{2+} [78], as well as dyes [79]. These are attributed to their exceptionally great surface area and highly porous structures [80, 81].

The structure of CNTs allows for the presentation of two different forms: single-walled carbon nanotubes (SWNTs) (Fig. **3a**) and multi-walled carbon nanotubes

(MWCNTs) (Fig. **3b**) [82, 83]. Both MWCNTs [84 - 86] and SWCNTs [87] have been used in wastewater remediation in recent years. CNTs may effectively adsorb a wide range of substances because of several contaminant-CNT interactions, such as the hydrophobic effect, π-π interactions, covalent bonds, H-bonds, and electrostatic forces [88]. Additionally, organic molecules with carboxylic, hydroxyl, and amino groups attach themselves to the surface of carbon nanotubes (CNTs) to create H-bonds that supply electrons [89]. Van der Waals forces also cause individual carbon nanotubes (10–100) to cluster into pores. According to Upadhyayula and his coworkers [90], these pores function as mesopores or give the CNT structure a greater surface area, which can concentrate microorganisms (viruses, bacteria, and parasites) with a high affinity.

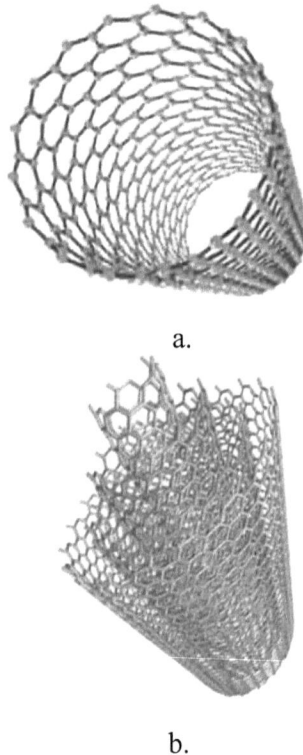

a.

b.

Fig. (3). The structure of carbon nanotubes (CNTs), **3a**. SWNTs, **3b**. MWNTs.

Notwithstanding their remarkable qualities, carbon nanotubes' restricted production volume and high cost have a major impact on their development and potential uses. To construct structural components, CNTs cannot be employed in isolation without a matrix or supporting media [91].

NMS MADE FROM CELLULOSE

The customizable form, size, and surface chemistry of cellulose nanocrystals (CNs) give rise to a variety of intriguing features, including high mechanical strength, reinforcing abilities, biodegradability, environmental friendliness, and self-assembly in aqueous conditions [92 - 100]. CNs are one kind of natural polysaccharides that are frequently employed as drug delivery agents in the pharmaceutical industry [101] or as nanocarriers to efficiently remove dyes [102]. In recent decades, there has been a great focus on developing novel technologies and applications that make use of cellulose NPs, such as antimicrobial materials, heavy metals, dyes, and adsorbents [103]. All plant fibers are mostly made of cellulose, which is among the most plentiful materials in the world [104]. Its availability, biodegradability, renewability, and affordability make it the most promising biopolymer [105]. With lengthy chains, cellulose is a linear polymer. $(C_6H_{10}O_5)$n is its molecular formula, which is made up of β-Dglucopyranose units connected by β-1,4-glycosidic linkages to create cellobiose, the core unit of cellulose, a dimer. Fibers made of cellulose have highly crystalline, compressed domains and loosely packed, amorphous chains [106].

NANOCOMPOSITES

A composite that contains a nanoscale morphology, such as NPs, nanotubes, or lamellar nanostructure, is called a nanocomposite. They, too, are composed of many phases; the diameters of at least one of the phases ought to fall between 10 and 100 nm. Nanocomposites, as an emerging technology, offer advantageous alternatives to many engineering materials that have limitations. Dispersed matrix and dispersed phase components are the basis for classifying nanocomposites [107]. It has been feasible to create a variety of fascinating new materials with unique features using creative synthetic methods. Because of their exact binding activity (chelation, absorption, and ion exchange), nanocomposites can easily remove bacteria, viruses, and contaminants from wastewater.

The application of NPs in water management has been linked to a number of real-world issues, including buildup, difficult separation, drainage into contact water, and adverse effects on human and environmental health [108]. The choice of hosts for nanocomposites is crucial, and it even affects how well the finished products function. The most likely path to advance water nanotechnology from the lab to large-scale applications was thought to be *via* nanocomposites up until this point [109].

Nanocomposites Made from Metal

Some of the examples relating to this class are given as under:

a. The antibacterial characteristics of nanosilver-fiber composites have been acknowledged for polymer-supported nanosilver, and these composites exhibit good inhibitory activity against both gram-positive and negative bacteria [110].
b. Beads made of silver-alginate composite show the ability to effectively disinfect water for portability [111].
c. The unique nanocomposite that was manufactured and contained AgNPs and mesoporous alumina has been used to remove toxic dyes from synthetic waste, such as reactive yellow, methyl orange, and bromothymol blue [112].

Nanocomposites made from Metal Oxide

To combat environmental pollution problems, metal oxide nanocomposite (MONC) devices are frequently employed as photocatalysts and adsorbents. For the purpose of eliminating different contaminants, MONCs are combined with graphene, silica, other oxides, CNT, and polymers [113]. MONC offers a modern method for enhancing charge separation and promoting charge transfer processes to change the characteristics of semiconductor metal oxide photocatalysts [114]. Some of the examples relating to this class are given here:

a. CNT-reinforced alumina composite was created by elevating CNT over energetic alumina doped with Fe and Ni. Many elements that might start high-capacity synthesis, such as CNTs, amorphous carbon, and different surface functional groups like carboxyl, carbonyl, and hydroxyl groups, had an impact on the composite [115].
b. The newest and most efficient type of metal oxide-based composites is called TiO_2. Due to its nontoxicity and capacity for photo-oxidative degradation of contaminants [116], benzene derivatives [117], and carbamazepine [118], the TiO_2 nanocomposite has drawn increased attention in the field of water purification. CNT/TiO_2 composites also demonstrated a strong ability to photodegrade these contaminants.
c. TiO_2-zeolite NCs for industrial dye water treatment: a new approach [119].

Nanocomposites made from Carbon

Comprising two or more elements, carbon nanocomposites (CNCs) are manufactured to make a composite combination with CNTs. Some of the examples are given below:

a. MWCNTs-COOH hybrid material was created by the primary host that worked as an active sorbent for the separation of Cd^{+2} and Pb^{+2} at trace levels [120].
b. Novel NCs made of biomass are utilized to extract specific heavy metals (As, Cr, Cu, Pd, and Zn) from wastewater [121].
c. According to Tian *et al.* [122], by utilizing FeONPs and AgNPs on the surface of graphene oxide (GO), scientists were able to create a unique nanocomposite that is environmentally friendly, effective, and synergistic. The compound, when combined with pure AgNPs, provides specifically enhanced bactericidal activity against both Gram-positive and Gram-negative bacteria. Gram-positive bacteria can be killed at low agent concentrations by the GO-ION--Ag composite, which is one of GO's wonderful benefits.

CONCLUSION AND FUTURE RECOMMENDATIONS

Wastewater-containing deposits and water dumps have had a detrimental worldwide impact, lowering the quality of natural water bodies for both humans and wildlife. Wastewater bodies are becoming more difficult to treat due to the variety of resistant compounds as a result of industrial development and material consumption. Medications, colorants, insecticides, heavy metals, and polycyclic aromatic hydrocarbons are some of the more prevalent and identified instances. Taking all of the above into consideration, the scientific community has focused heavily on creating alternatives that can lessen the damage and are affordable, readily available, and sustainable. This review has demonstrated a broad range of prospective nanotechnologies that have been tried to treat and enhance the organoleptic properties of drinking water and wastewater, supplying a safe and harmless liquid to society and the environment. Relevant contributions have been made by nanotechnology through a variety of functional materials, including metal oxides, polymers, catalysts, nanotubes, graphene, and cellulose materials. These materials indirectly aid in the breakdown of the aforementioned pollutants. NMs have been seen as an effective substitute, but they also have several drawbacks. One of these is that industrial implementation has not allowed for a significant advancement in their use for treating large volumes of wastewater because there are not many studies conducted at this scale. To rule out NPs as toxic, teratogenic, and, in extreme circumstances, carcinogenic agents, toxicological tests are also advised. However, there have not been any documented instances of intoxication as a result of their use or disposal. These drawbacks, however, do not take away from their promise as a strong wastewater treatment platform. With the main goal of addressing the possible hazards of water pollution, NMs have advantageous properties that make them environmentally friendly based on the concepts of green chemistry and circular economy.

REFERENCES

[1]   Adeleye AS, Conway JR, Garner K, Huang Y, Su Y, Keller AA. Engineered nanomaterials for water treatment and remediation: Costs, benefits, and applicability. Chem Eng J 2016; 286: 640-62.
[http://dx.doi.org/10.1016/j.cej.2015.10.105]

[2]   Naushad M, Sharma G, Kumar A, *et al.* Efficient removal of toxic phosphate anions from aqueous environment using pectin based quaternary amino anion exchanger. Int J Biol Macromol 2018; 106: 1-10.
[http://dx.doi.org/10.1016/j.ijbiomac.2017.07.169] [PMID: 28774808]

[3]   Ferroudj N, Nzimoto J, Davidson A, *et al.* Maghemite NPs and Maghemite/Silica Nanocomposite Microspheres as Magnetic Fenton Catalysts for the Removal of Water Pollutants Appl Catal B Environ 2013; 136: 9-18.

[4]   Qu X, Brame J, Li Q, Alvarez PJJ. Nanotechnology for a safe and sustainable water supply: enabling integrated water treatment and reuse. Acc Chem Res 2013; 46(3): 834-43.
[http://dx.doi.org/10.1021/ar300029v] [PMID: 22738389]

[5]   What are the 5 basic principles of wastewater treatment? [Internet]. Trity Enviro. [cited 2025 Jun 20]. Available from: https://trityenviro.com/bd/what-are-the-5-basic-principles-of-wastewater-treatment/

[6]   Naushad M, Ahamad T, Al-Maswari BM, Abdullah Alqadami A, Alshehri SM. Nickel ferrite bearing nitrogen-doped mesoporous carbon as efficient adsorbent for the removal of highly toxic metal ion from aqueous medium. Chem Eng J 2017; 330: 1351-60.
[http://dx.doi.org/10.1016/j.cej.2017.08.079]

[7]   Zhang Q, Xu R, Xu P, *et al.* Performance study of ZrO_2 ceramic micro-filtration membranes used in pretreatment of DMF wastewater. Desalination 2014; 346: 1-8.
[http://dx.doi.org/10.1016/j.desal.2014.05.006]

[8]   Kyzas GZ, Matis KA. Nanoadsorbents for pollutants removal: A review. J Mol Liq 2015; 203: 159-68.
[http://dx.doi.org/10.1016/j.molliq.2015.01.004]

[9]   Tang X, Zhang Q, Liu Z, Pan K, Dong Y, Li Y. Removal of Cu(II) by loofah fibers as a natural and low-cost adsorbent from aqueous solutions. J Mol Liq 2014; 199: 401-7.
[http://dx.doi.org/10.1016/j.molliq.2014.09.033]

[10]   El-Sayed MEA. Nanoadsorbents for water and wastewater remediation. Sci Total Environ. 2020 Oct 15;739:139903.Epub 2020 Jun 3.
[http://dx.doi.org/doi: 10.1016/j.scitotenv.2020.139903] [PMID: 32544683]

[11]   Babu SG, Vijayan AS, Neppolian B, Ashokkumar M. SnS2/rGO: an efficient photocatalyst for the complete degradation of organic contaminants. Materials Focus 2015; 4(4): 272-6. a
[http://dx.doi.org/10.1166/mat.2015.1247]

[12]   Kumar PS, Selvakumar M, Babu SG, Karuthapandian S. Veteran cupric oxide with new morphology and modified bandgap for superior photocatalytic activity against different kinds of organic contaminants (acidic, azo and triphenylmethane dyes). Mater Res Bull 2016; 83: 522-33.
[http://dx.doi.org/10.1016/j.materresbull.2016.06.043]

[13]   Dutta AK, Maji SK, Adhikary B. γ-Fe_2O_3 nanoparticles: An easily recoverable effective photo-catalyst for the degradation of rose bengal and methylene blue dyes in the waste-water treatment plant. Mater Res Bull 2014; 49: 28-34.
[http://dx.doi.org/10.1016/j.materresbull.2013.08.024]

[14]   Chaturvedi S, Dave PN, Shah NK. Applications of nano-catalyst in new era. J Saudi Chem Soc 2012; 16(3): 307-25.
[http://dx.doi.org/10.1016/j.jscs.2011.01.015]

[15]   Ouyang X, Li W, Xie S, *et al.* Hierarchical CeO_2 nanospheres as highly-efficient adsorbents for dye removal. New J Chem 2013; 37(3): 585-8.
[http://dx.doi.org/10.1039/c3nj41095a]

[16]   Fu Z, Xi S. The effects of heavy metals on human metabolism. Toxicol Mech Methods 2020; 30(3): 167-76.
       [http://dx.doi.org/10.1080/15376516.2019.1701594] [PMID: 31818169]

[17]   Rajasurya V, Surani S. Legionnaires disease in immunocompromised host. In: Surani S, Varon J, Eds. Hospital Acquired Infection and Legionnaires' Disease. London, UK: IntechOpen 2019.

[18]   Fowler CC, Galán JE. Decoding a Salmonella Typhi regulatory network that controls typhoid toxin expression within human cells. Cell Host Microbe 2018; 23(1): 65-76.e6.
       [http://dx.doi.org/10.1016/j.chom.2017.12.001] [PMID: 29324231]

[19]   Kumar B L, Gopal D S. Effective Role of Indigenous Microorganisms for Sustainable Environment 3 Biotech 2015; 5(6): 867-76.

[20]   Akpor OB, Ogundeji MD, Olaolu DT, Aderiye BI. Microbial roles and dynamics in wastewater treatment systems: an overview. International Journal of Pure & Applied Bioscience 2014; 2(1): 156-68.

[21]   Verma R, Dwivedi P. Heavy metal water pollution-A case study. Recent Research in Science and Technology 2013; 5(5): 98-9.

[22]   Kawser Ahmed M, Baki MA, Kundu GK, Saiful Islam M, Monirul Islam M, Muzammel Hossain M. Human health risks from heavy metals in fish of Buriganga river, Bangladesh. Springerplus 2016; 5(1): 1697.
       [http://dx.doi.org/10.1186/s40064-016-3357-0] [PMID: 27757369]

[23]   El-sayed MEA. Nanoadsorbents for water and wastewater remediation. Sci Total Environ 2020; 739: 139903.
       [http://dx.doi.org/10.1016/j.scitotenv.2020.139903] [PMID: 32544683]

[24]   Fatima D. Microorganism-based biological agents in wastewater treatment: potential use and benefits in agriculture. Field Practices for Wastewater Use in Agriculture. In: V. K. Tripathi & M. R. Goyal (eds), 1st edn (Apple Academic Press, ed.). Taylor & Francis Group, New York, United States, 2021; pp. 69-80.
       [http://dx.doi.org/10.1201/9781003034506-6]

[25]   Barik B, Nayak PS, Dash P. Nanomaterials in wastewater treatments. Nanotechnology in the Beverage Industry 2020; 20: 185-206.
       [http://dx.doi.org/10.1016/B978-0-12-819941-1.00007-9]

[26]   Grimsdale AC, Müllen K. The chemistry of organic nanomaterials. Angew Chem Int Ed 2005; 44(35): 5592-629.
       [http://dx.doi.org/10.1002/anie.200500805] [PMID: 16136610]

[27]   Landsiedel R, Ma-Hock L, Kroll A, *et al.* Testing metal-oxide nanomaterials for human safety. Adv Mater 2010; 22(24): 2601-27.
       [http://dx.doi.org/10.1002/adma.200902658] [PMID: 20512811]

[28]   Taylor-Pashow KML, Della Rocca J, Huxford RC, Lin W. Hybrid nanomaterials for biomedical applications. Chem Commun (Camb) 2010; 46(32): 5832-49.
       [http://dx.doi.org/10.1039/c002073g] [PMID: 20623072]

[29]   Schodek DL, Ferreira P, Ashby MF. NMs, nanotechnologies, and design: an introduction for engineers and architects. Butterworth- Heinemann 2009.

[30]   Bokare V, Jung J, Chang YY, Chang YS. Reductive dechlorination of octachlorodibenzo-p-dioxin by nanosized zero-valent zinc: Modeling of rate kinetics and congener profile. J Hazard Mater 2013; 250-251: 397-402.
       [http://dx.doi.org/10.1016/j.jhazmat.2013.02.020] [PMID: 23500419]

[31]   Tratnyek PG, Salter AJ, Nurmi JT, Sarathy V. Environmental applications of zerovalent metals: iron *vs.* zinc

[http://dx.doi.org/10.1021/bk-2010-1045.ch009]

[32]   Pare, Brijesh & Joshi, Roshni & Mehta, Sanika & Solanki, Vijendra & Gupta, Rupesh & Agarwal, Neha & Yadav, Virendra. (2024). Preparation and characterisation of BiOCl nano photocatalyst for the remediation of wastewater under LED light. International Journal of Environmental Analytical Chemistry. 1-25.
[http://dx.doi.org/10.1080/03067319.2024.2442086]

[33]   Borrego B, Lorenzo G, Mota-Morales JD, *et al.* Potential application of silver nanoparticles to control the infectivity of Rift Valley fever virus *in vitro* and *in vivo*. Nanomedicine 2016; 12(5): 1185-92.
[http://dx.doi.org/10.1016/j.nano.2016.01.021] [PMID: 26970026]

[34]   Kalhapure RS, Sonawane SJ, Sikwal DR, *et al.* Solid lipid nanoparticles of clotrimazole silver complex: An efficient nano antibacterial against Staphylococcus aureus and MRSA. Colloids Surf B Biointerfaces 2015; 136: 651-8.
[http://dx.doi.org/10.1016/j.colsurfb.2015.10.003] [PMID: 26492156]

[35]   Krishnaraj C, Ramachandran R, Mohan K. andP. T. Kalaichelvan, "Optimization for rapid synthesis of silver NPs and its effect on phytopathogenic fungi,". Spectrochim Acta A Mol Biomol Spectrosc 2012; 93: 95-9.
[http://dx.doi.org/10.1016/j.saa.2012.03.002] [PMID: 22465774]

[36]   Sondi I, Salopek-Sondi B. Silver nanoparticles as antimicrobial agent: a case study on E. coli as a model for Gram-negative bacteria. J Colloid Interface Sci 2004; 275(1): 177-82.
[http://dx.doi.org/10.1016/j.jcis.2004.02.012] [PMID: 15158396]

[37]   Danilczuk M, Lund A, Sadlo J, Yamada H, Michalik J. Conduction electron spin resonance of small silver particles. Spectrochim Acta A Mol Biomol Spectrosc 2006; 63(1): 189-91.
[http://dx.doi.org/10.1016/j.saa.2005.05.002] [PMID: 15978868]

[38]   Li X, Lenhart JJ, Walker HW. Aggregation kinetics and dissolution of coated silver nanoparticles. Langmuir 2012; 28(2): 1095-104.
[http://dx.doi.org/10.1021/la202328n] [PMID: 22149007]

[39]   Quang DV, Sarawade PB, Jeon SJ, *et al.* Effective water disinfection using silver nanoparticle containing silica beads. Appl Surf Sci 2013; 266: 280-7.
[http://dx.doi.org/10.1016/j.apsusc.2012.11.168]

[40]   Rivero-Huguet M, Marshall WD. Reduction of hexavalent chromium mediated by micron- and nano-scale zero-valent metallic particles. J Environ Monit 2009; 11(5): 1072-9.
[http://dx.doi.org/10.1039/b819279k] [PMID: 19436867]

[41]   Matheson LJ, Tratnyek PG. Reductive dehalogenation of chlorinated methanes by iron metal. Environ Sci Technol 1994; 28(12): 2045-53.
[http://dx.doi.org/10.1021/es00061a012] [PMID: 22191743]

[42]   Reynolds DC, Look DC, Jogai B, Litton CW, Cantwell G, Harsch WC. Valence-Band Ordering in ZnO Physical Review B—Condensed Matter and Materials Physics. 1999; 60: pp. (4)2340-4.
[http://dx.doi.org/10.1103/PhysRevB.60.2340]

[43]   Janotti A, Van de Walle CG. Fundamentals of zinc oxide as a semiconductor. Rep Prog Phys 2009; 72(12): 126501.
[http://dx.doi.org/10.1088/0034-4885/72/12/126501]

[44]   Chen Y, Bagnall DM, Koh H, *et al.* Plasma assisted molecular beam epitaxy of ZnO on c -plane sapphire: Growth and characterization. J Appl Phys 1998; 84(7): 3912-8.
[http://dx.doi.org/10.1063/1.368595]

[45]   Schmidt-Mende L, MacManus-Driscoll JL. ZnO – nanostructures, defects, and devices. Mater Today 2007; 10(5): 40-8.
[http://dx.doi.org/10.1016/S1369-7021(07)70078-0]

[46]   Daneshvar N, Salari D, Khataee AR. Photocatalytic degradation of azo dye acid red 14 in water on

ZnO as an alternative catalyst to TiO2. J Photochem Photobiol Chem 2004; 162(2-3): 317-22.
[http://dx.doi.org/10.1016/S1010-6030(03)00378-2]

[47]   Behnajady M, Modirshahla N, Hamzavi R. Kinetic study on photocatalytic degradation of C.I. Acid Yellow 23 by ZnO photocatalyst. J Hazard Mater 2006; 133(1-3): 226-32.
[http://dx.doi.org/10.1016/j.jhazmat.2005.10.022] [PMID: 16310945]

[48]   Gomez-Solís C, Ballesteros JC, Torres-Martínez LM, *et al.* Rapid synthesis of ZnO nano-corncobs from Nital solution and its application in the photodegradation of methyl orange. J Photochem Photobiol Chem 2015; 298: 49-54.
[http://dx.doi.org/10.1016/j.jphotochem.2014.10.012]

[49]   Lee KM, Lai CW, Ngai KS, Juan JC. Recent developments of zinc oxide based photocatalyst in water treatment technology: A review. Water Res 2016; 88: 428-48.
[http://dx.doi.org/10.1016/j.watres.2015.09.045] [PMID: 26519627]

[50]   Samadi M, Pourjavadi A, Moshfegh AZ. Role of CdO addition on the growth and photocatalytic activity of electrospun ZnO nanofibers: UV *vs.* visible light. Appl Surf Sci 2014; 298: 147-54.
[http://dx.doi.org/10.1016/j.apsusc.2014.01.146]

[51]   Pant HR, Park CH, Pant B, Tijing LD, Kim HY, Kim CS. Synthesis, characterization, and photocatalytic properties of ZnO nano-flower containing TiO$_2$ NPs. Ceram Int 2012; 38(4): 2943-50.
[http://dx.doi.org/10.1016/j.ceramint.2011.11.071]

[52]   Uddin MT, Nicolas Y, Olivier C, *et al.* Nanostructured SnO$_2$-ZnO heterojunction photocatalysts showing enhanced photocatalytic activity for the degradation of organic dyes. Inorg Chem 2012; 51(14): 7764-73.
[http://dx.doi.org/10.1021/ic300794j] [PMID: 22734686]

[53]   Liu IT, Hon MH, Teoh LG. The preparation, characterization and photocatalytic activity of radical-shaped CeO$_2$/ZnO microstructures. Ceram Int 2014; 40(3): 4019-24.
[http://dx.doi.org/10.1016/j.ceramint.2013.08.053]

[54]   Dai K, Lu L, Liang C, *et al.* Graphene oxide modified ZnO nanorods hybrid with high reusable photocatalytic activity under UV-LED irradiation. Mater Chem Phys 2014; 143(3): 1410-6.
[http://dx.doi.org/10.1016/j.matchemphys.2013.11.055]

[55]   Zhou X, Shi T, Zhou H. Hydrothermal preparation of ZnO-reduced graphene oxide hybrid with high performance in photocatalytic degradation. Appl Surf Sci 2012; 258(17): 6204-11.
[http://dx.doi.org/10.1016/j.apsusc.2012.02.131]

[56]   Fujishima A, Honda K. Electrochemical photolysis of water at a semiconductor electrode. Nature 1972; 238(5358): 37-8.
[http://dx.doi.org/10.1038/238037a0] [PMID: 12635268]

[57]   Guesh K, Mayoral Á, Márquez-Álvarez C, Chebude Y, Díaz I. Enhanced photocatalytic activity of TiO$_2$ supported on zeolites tested in real wastewaters from the textile industry of Ethiopia. Microporous Mesoporous Mater 2016; 225: 88-97.
[http://dx.doi.org/10.1016/j.micromeso.2015.12.001]

[58]   Yamakata A, Vequizo JJM. Curious behaviors of photogenerated electrons and holes at the defects on anatase, rutile, and brookite TiO2 powders: A review. J Photochem Photobiol Photochem Rev 2019; 40: 234-43.
[http://dx.doi.org/10.1016/j.jphotochemrev.2018.12.001]

[59]   Tan L, Xu J, Xue X, *et al.* Multifunctional nanocomposite Fe$_3$O$_4$@SiO$_2$–mPD/SP for selective removal of Pb(II) and Cr(VI) from aqueous solutions. RSC Advances 2014; 4(86): 45920-9.
[http://dx.doi.org/10.1039/C4RA08040H]

[60]   Lei Y, Chen F, Luo Y, Zhang L. Three-dimensional magnetic graphene oxide foam/Fe3O4 nanocomposite as an efficient absorbent for Cr(VI) removal. J Mater Sci 2014; 49(12): 4236-45.
[http://dx.doi.org/10.1007/s10853-014-8118-2]

[61]   Ngomsik AF, Bee A, Talbot D, Cote G. Magnetic solid–liquid extraction of Eu(III), La(III), Ni(II) and Co(II) with maghemite nanoparticles. Separ Purif Tech 2012; 86: 1-8.
[http://dx.doi.org/10.1016/j.seppur.2011.10.013]

[62]   Warner CL, Addleman RS, Cinson AD, *et al.* High-performance, superparamagnetic, nanoparticle-based heavy metal sorbents for removal of contaminants from natural waters. ChemSusChem 2010; 3(6): 749-57.
[http://dx.doi.org/10.1002/cssc.201000027] [PMID: 20468024]

[63]   Ge F, Li MM, Ye H, Zhao BX. Effective removal of heavy metal ions Cd2+, Zn2+, Pb2+, Cu2+ from aqueous solution by polymer-modified magnetic nanoparticles. J Hazard Mater 2012; 211-212: 366-72.
[http://dx.doi.org/10.1016/j.jhazmat.2011.12.013] [PMID: 22209322]

[64]   Apul OG, Wang Q, Zhou Y, Karanfil T. Adsorption of aromatic organic contaminants by graphene nanosheets: Comparison with carbon nanotubes and activated carbon. Water Res 2013; 47(4): 1648-54.
[http://dx.doi.org/10.1016/j.watres.2012.12.031] [PMID: 23313232]

[65]   Khin MM, Nair AS, Babu VJ, Murugan R, Ramakrishna S. A review on nanomaterials for environmental remediation. Energy Environ Sci 2012; 5(8): 8075-109.
[http://dx.doi.org/10.1039/c2ee21818f]

[66]   Shan SJ, Zhao Y, Tang H, Cui FY. A mini-review of carbonaceous NMs for removal of contaminants from wastewater. IOP Conf Ser Earth Environ Sci 2017; 68: 012003.
[http://dx.doi.org/10.1088/1755-1315/68/1/012003]

[67]   Ali I, Basheer AA, Mbianda XY, *et al.* Graphene based adsorbents for remediation of noxious pollutants from wastewater. Environ Int 2019; 127: 160-80.
[http://dx.doi.org/10.1016/j.envint.2019.03.029] [PMID: 30921668]

[68]   Available from: https://en.wikipedia.org/wiki/Fullerene

[69]   Thilgen C, Diederich F. Structural aspects of fullerene chemistry--a journey through fullerene chirality. Chem Rev 2006; 106(12): 5049-135.
[http://dx.doi.org/10.1021/cr0505371] [PMID: 17165683]

[70]   Burakov AE, Galunin EV, Burakova IV, *et al.* Adsorption of heavy metals on conventional and nanostructured materials for wastewater treatment purposes: A review. Ecotoxicol Environ Saf 2018; 148: 702-12.
[http://dx.doi.org/10.1016/j.ecoenv.2017.11.034] [PMID: 29174989]

[71]   Manawi YM, Ihsanullah , Samara A, Al-Ansari T, Atieh MA. A review of carbon NMs' synthesis *via* the chemical vapor deposition (CVD) method. Materials (Basel) 2018; 11(5): 822.
[http://dx.doi.org/10.3390/ma11050822] [PMID: 29772760]

[72]   Lucena R, Simonet BM, Cárdenas S, Valcárcel M. Potential of nanoparticles in sample preparation. J Chromatogr A 2011; 1218(4): 620-37.
[http://dx.doi.org/10.1016/j.chroma.2010.10.069] [PMID: 21071033]

[73]   Umair Azhar, Huma Ahmad, Hafsa Shafqat, Muhammad Babar, Hafiz Muhammad Shahzad Munir, Muhammad Sagir, Muhammad Arif, Afaq Hassan, Nova Rachmadona, Saravanan Rajendran, Muhammad Mubashir, Kuan Shiong Khoo, Remediation techniques for elimination of heavy metal pollutants from soil: A review, Environmental Research, Volume 214, Part 4, 2022, 113918, ISSN 0013-9351.
[http://dx.doi.org/10.1016/j.envres.2022.113918]

[74]   Chatterjee A, Deopura BL. Carbon nanotubes and nanofibre: An overview. Fibers Polym 2002; 3(4): 134-9.
[http://dx.doi.org/10.1007/BF02912657]

[75]   Lu C, Su F, Hu S. Surface modification of carbon nanotubes for enhancing BTEX adsorption from

aqueous solutions. Appl Surf Sci 2008; 254(21): 7035-41.
[http://dx.doi.org/10.1016/j.apsusc.2008.05.282]

[76]   Li YH, Ding J, Luan Z, *et al.* Competitive adsorption of Pb^{2+}, Cu^{2+} and Cd^{2+} ions from aqueous
       solutions by multiwalled carbon nanotubes. Carbon 2003; 41(14): 2787-92.
       [http://dx.doi.org/10.1016/S0008-6223(03)00392-0]

[77]   Peng X, Li Y, Luan Z, *et al.* Adsorption of 1,2-dichlorobenzene from water to carbon nanotubes.
       Chem Phys Lett 2003; 376(1-2): 154-8.
       [http://dx.doi.org/10.1016/S0009-2614(03)00960-6]

[78]   Cho HH, Wepasnick K, Smith BA, Bangash FK, Fairbrother DH, Ball WP. Sorption of aqueous Zn[II]
       and Cd[II] by multiwall carbon nanotubes: the relative roles of oxygen-containing functional groups
       and graphenic carbon. Langmuir 2010; 26(2): 967-81.
       [http://dx.doi.org/10.1021/la902440u] [PMID: 19894751]

[79]   Madrakian T, Afkhami A, Ahmadi M, Bagheri H. Removal of some cationic dyes from aqueous
       solutions using magnetic-modified multi-walled carbon nanotubes. J Hazard Mater 2011; 196: 109-14.
       [http://dx.doi.org/10.1016/j.jhazmat.2011.08.078] [PMID: 21930344]

[80]   Chen W, Duan L, Zhu D. Adsorption of polar and nonpolar organic chemicals to carbon nanotubes.
       Environ Sci Technol 2007; 41(24): 8295-300.
       [http://dx.doi.org/10.1021/es071230h] [PMID: 18200854]

[81]   Lin D, Xing B. Adsorption of phenolic compounds by carbon nanotubes: role of aromaticity and
       substitution of hydroxyl groups. Environ Sci Technol 2008; 42(19): 7254-9.
       [http://dx.doi.org/10.1021/es801297u] [PMID: 18939555]

[82]   Müller M, Maultzsch J, Wunderlich D, Hirsch A, Thomsen C. Raman spectroscopy on chemically
       functionalized carbon nanotubes. Phys Status Solidi, B Basic Res 2007; 244(11): 4056-9.
       [http://dx.doi.org/10.1002/pssb.200776119]

[83]   Ali ME, Das R, Maamor A, Hamid SBA. Multifunctional carbon nanotubes (CNTs): a new dimension
       in environmental remediation. Adv Mat Res 2014; 832: 328-32.

[84]   Bingbing Qiu, Xuedong Tao, Hao Wang, Wenke Li, Xiang Ding, Huaqiang Chu, Biochar as a low-
       cost adsorbent for aqueous heavy metal removal: A review, Journal of Analytical and Applied
       Pyrolysis,Volume 155, 2021, 105081, ISSN 0165-2370
       [http://dx.doi.org/10.1016/j.jaap.2021.105081.]

[85]   Samuel Sunday Ogunsola, Mayowa Ezekiel Oladipo, Peter Olusakin Oladoye, Mohammed Kadhom,
       Carbon nanotubes for sustainable environmental remediation: A critical and comprehensive review,
       Nano-Structures & Nano-Objects, Volume 37, 2024, 101099, ISSN 2352-507X.
       [http://dx.doi.org/10.1016/j.nanoso.2024.101099]

[86]   Baby, R., Saifullah, B. & Hussein, M.Z. Carbon Nanomaterials for the Treatment of Heavy Metal-
       Contaminated Water and Environmental Remediation. Nanoscale Res Lett 14, 341 (2019).
       [http://dx.doi.org/10.1186/s11671-019-3167-8]

[87]   Yang CM, Park JS, An KH, *et al.* Selective removal of metallic single-walled carbon nanotubes with
       small diameters by using nitric and sulfuric acids. J Phys Chem B 2005; 109(41): 19242-8.
       [http://dx.doi.org/10.1021/jp053245c] [PMID: 16853485]

[88]   Muhammad Sajid, Mohammad Asif, Nadeem Baig, Muhamed Kabeer, Ihsanullah Ihsanullah, Abdul
       Wahab Mohammad, Carbon nanotubes-based adsorbents: Properties, functionalization, interaction
       mechanisms, and applications in water purification, Journal of Water Process Engineering, Volume 47,
       2022, 102815, ISSN 2214-7144.
       [http://dx.doi.org/10.1016/j.jwpe.2022.102815]

[89]   Abdelbasir SM, Shalan AE. An overview of nanomaterials for industrial wastewater treatment. Korean
       J Chem Eng 2019; 36(8): 1209-25.
       [http://dx.doi.org/10.1007/s11814-019-0306-y]

[90]   Upadhyayula VKK, Deng S, Mitchell MC, Smith GB. Application of carbon nanotube technology for removal of contaminants in drinking water: A review. Sci Total Environ 2009; 408(1): 1-13.
[http://dx.doi.org/10.1016/j.scitotenv.2009.09.027] [PMID: 19819525]

[91]   Nigar Anzar, Rahil Hasan, Manshi Tyagi, Neelam Yadav, Jagriti Narang, Carbon nanotube - A review on Synthesis, Properties and plethora of applications in the field of biomedical science, Sensors International, Volume 1, 2020, 100003, ISSN 2666-3511.
[http://dx.doi.org/10.1016/j.sintl.2020.100003]

[92]   Li T, Chen C, Brozena AH, *et al.* Developing fibrillated cellulose as a sustainable technological material. Nature 2021; 590(7844): 47-56.
[http://dx.doi.org/10.1038/s41586-020-03167-7] [PMID: 33536649]

[93]   Chen W, Yu H, Lee SY, Wei T, Li J, Fan Z. Nanocellulose: a promising nanomaterial for advanced electrochemical energy storage. Chem Soc Rev 2018; 47(8): 2837-72.
[http://dx.doi.org/10.1039/C7CS00790F] [PMID: 29561005]

[94]   Zhu H, Luo W, Ciesielski PN, *et al.* Wood-Derived Materials for Green Electronics, Biological Devices, and Energy Applications. Chem Rev 2016; 116(16): 9305-74.
[http://dx.doi.org/10.1021/acs.chemrev.6b00225] [PMID: 27459699]

[95]   Jorfi M, Foster EJ. Recent advances in nanocellulose for biomedical applications. J Appl Polym Sci 2015; 132(14): app.41719.
[http://dx.doi.org/10.1002/app.41719]

[96]   De France K, Zeng Z, Wu T, Nyström G. Functional Materials from Nanocellulose: Utilizing Structure–Property Relationships in Bottom☐Up Fabrication. Adv Mater 2021; 33(28): 2000657.
[http://dx.doi.org/10.1002/adma.202000657] [PMID: 32267033]

[97]   Mautner A. Nanocellulose water treatment membranes and filters: a review. Polym Int 2020; 69(9): 741-51.
[http://dx.doi.org/10.1002/pi.5993]

[98]   Yang X, Biswas SK, Han J, *et al.* Surface and Interface Engineering for Nanocellulosic Advanced Materials. Adv Mater 2021; 33(28): 2002264.
[http://dx.doi.org/10.1002/adma.202002264] [PMID: 32902018]

[99]   Heise K, Kontturi E, Allahverdiyeva Y, *et al.* Nanocellulose: Recent Fundamental Advances and Emerging Biological and Biomimicking Applications. Adv Mater 2021; 33(3): 2004349.
[http://dx.doi.org/10.1002/adma.202004349] [PMID: 33289188]

[100] Habibi Y, Lucia LA, Rojas OJ. Cellulose nanocrystals: chemistry, self-assembly, and applications. Chem Rev 2010; 110(6): 3479-500.
[http://dx.doi.org/10.1021/cr900339w] [PMID: 20201500]

[101] Raj V, Lee JH, Shim JJ, Lee J. Recent findings and future directions of grafted gum karaya polysaccharides and their various applications: A review. Carbohydr Polym 2021; 258: 117687.
[http://dx.doi.org/10.1016/j.carbpol.2021.117687] [PMID: 33593560]

[102] Raj V, Raorane CJ, Lee JH, Lee J. Appraisal of Chitosan-Gum Arabic-Coated Bipolymeric Nanocarriers for Efficient Dye Removal and Eradication of the Plant Pathogen *Botrytis cinerea*. ACS Appl Mater Interfaces 2021; 13(40): 47354-70.
[http://dx.doi.org/10.1021/acsami.1c12617] [PMID: 34596375]

[103] Choudhury RR, Sahoo SK, Gohil JM. Potential of bioinspired cellulose nanomaterials and nanocomposite membranes thereof for water treatment and fuel cell applications. Cellulose 2020; 27(12): 6719-46.
[http://dx.doi.org/10.1007/s10570-020-03253-z]

[104] Olivera S, Muralidhara HB, Venkatesh K, Guna VK, Gopalakrishna K, Kumar K Y. Potential applications of cellulose and chitosan nanoparticles/composites in wastewater treatment: A review. Carbohydr Polym 2016; 153: 600-18.

[http://dx.doi.org/10.1016/j.carbpol.2016.08.017] [PMID: 27561533]

[105] Carpenter AW, de Lannoy CF, Wiesner MR. Cellulose nanomaterials in water treatment technologies. Environ Sci Technol 2015; 49(9): 5277-87.
[http://dx.doi.org/10.1021/es506351r] [PMID: 25837659]

[106] Abouzeid RE, Khiari R, El-Wakil N, Dufresne A. Current state and new trends in the use of cellulose NMs for wastewater treatment. Biomacromolecules 2019; 20(2): 573-97.
[http://dx.doi.org/10.1021/acs.biomac.8b00839] [PMID: 30020778]

[107] Wang J, Kaskel S. KOH activation of carbon-based materials for energy storage. J Mater Chem 2012; 22(45): 23710.
[http://dx.doi.org/10.1039/c2jm34066f]

[108] Gehrke I, Geiser A, Somborn-Schulz A. Innovations in nanotechnology for water treatment. Nanotechnol Sci Appl 2015; 8: 1-17.
[http://dx.doi.org/10.2147/NSA.S43773] [PMID: 25609931]

[109] Zhang Y, Wu B, Xu H, *et al.* Nanomaterials-enabled water and wastewater treatment. NanoImpact 2016; 3-4: 22-39.
[http://dx.doi.org/10.1016/j.impact.2016.09.004]

[110] Botes M, Eugene Cloete T. The potential of nanofibers and nanobiocides in water purification. Crit Rev Microbiol 2010; 36(1): 68-81.
[http://dx.doi.org/10.3109/10408410903397332] [PMID: 20088684]

[111] Lin S, Huang R, Cheng Y, Liu J, Lau BLT, Wiesner MR. Silver nanoparticle-alginate composite beads for point-of-use drinking water disinfection. Water Res 2013; 47(12): 3959-65.
[http://dx.doi.org/10.1016/j.watres.2012.09.005] [PMID: 23036278]

[112] Yahyaei B, Azizian S, Mohammadzadeh A, Pajohi-Alamoti M. Chemical and biological treatment of waste water with a novel silver/ordered mesoporous alumina nanocomposite. J Indian Chem Soc 2015; 12(1): 167-74.
[http://dx.doi.org/10.1007/s13738-014-0470-2]

[113] Lateef A, Nazir R. Metal nanocomposites: synthesis, characterization and their applications In: P DS, (ed) Science and applications of tailored nanostructures. 1st edn. One central press, Italy 2017; pp. 239-40.

[114] Ray C, Pal T. Recent advances of metal-metal oxide nanocomposites and their tailored nanostructures in numerous catalytic applications. J Mater Chem A 5(20):9465–9487. [115] Sankararamakrishnan N, Jaiswal M, Verma N (2014) Composite nanofloral clusters of carbon nanotubes and activated alumina: an efficient sorbent for heavy metal removal. Chem Eng J 2017; 235: 1-9.

[116] Ge M-Z, Cao C-Y, Huang J-Y, *et al.* Synthesis, modification, and photo/photoelectrocatalytic degradation applications of TiO2 nanotube arrays: a review. Nanotechnol Rev 2016; 5(1)
[http://dx.doi.org/10.1515/ntrev-2015-0049]

[117] Silva CG, Faria JL. Photocatalytic oxidation of benzene derivatives in aqueous suspensions: Synergic effect induced by the introduction of carbon nanotubes in a TiO_2 matrix. Appl Catal B 2010; 101(1-2): 81-9.
[http://dx.doi.org/10.1016/j.apcatb.2010.09.010]

[118] Martínez C, Canle L M, Fernández MI, Santaballa JA, Faria J. Kinetics and mechanism of aqueous degradation of carbamazepine by heterogeneous photocatalysis using nanocrystalline TiO_2, ZnO and multi-walled carbon nanotubes–anatase composites. Appl Catal B 2011; 102(3-4): 563-71.
[http://dx.doi.org/10.1016/j.apcatb.2010.12.039]

[119] Senusi F, Shahadat M, Ismail S, Hamid SA. Recent advancement in membrane technology for water purification. In: Oves M, Ed. Modern age environmental problems and their remediation, Recent Advancement. 1st ed. Springer International Publishing AG 2018; pp. 1-237.
[http://dx.doi.org/10.1007/978-3-319-64501-8_9]

[120] Ma, Ruirui, Juan Li, Ping Zeng, Liang Duan, Jimin Dong, Yunxia Ma, and Lingkong Yang. 2024. "The Application of Membrane Separation Technology in the Pharmaceutical Industry" Membranes 14, no. 1: 24.
[http://dx.doi.org/10.3390/membranes14010024]

[121] Muneeb M, Zahoor M, Muhammad B, AliKhan F, Ullah R, AbdEl-Salam NM. Removal of heavy metals from drinking water by magnetic carbon nanostructures prepared from biomass. J Nanomater. 2017;2017:10.

[122] Tian T, Shi X, Cheng L, *et al.* Graphene-based nanocomposite as an effective, multifunctional, and recyclable antibacterial agent. ACS Appl Mater Interfaces 2014; 6(11): 8542-8.
[http://dx.doi.org/10.1021/am5022914] [PMID: 24806506]

CHAPTER 4

Carbon NMs in Environmental Remediation

Alisha[1] and **Sudesh Kumar**[1,*]

[1] Department of Chemistry, National Council of Educational Research and Training (NCERT), New Delhi, Delhi 110016, India

Abstract: Ecological concerns like polluted drinking water have impacted every facet of our existence. Ecological restoration hinges predominantly on employing diverse methods,` such as absorption, adsorption, chemical processes, light-induced catalysis, and purification of water, for the elimination of pollutants from distinct ecological mediums like terrain, aqua, and atmosphere. Nanoscience is a cutting-edge scientific discipline possessing the capacity to address numerous ecological hurdles through the manipulation of the dimensions and configuration of substances at a nanoscopic level. Carbon nanomaterials (NMs) are exceptional due to their harmless characteristics, large area, simplified decomposition, and notably beneficial ecological restoration. In this context, this chapter discusses the mechanistic pathways and uses of carbon materials for the light-catalyzed and adsorption elimination of contaminants present in polluted water. Carbon materials enable improved adsorption owing to robust bonding between contaminants and binding regions. In light-induced chemical reactions, increased efficacy is credited to the enhanced capture of radiance and diminished reassembly of light-activated charge carriers. The recent advancements achieved in the elimination of contaminants from contaminated water utilizing diverse forms of carbon NMs as adsorptive agents, including graphene, carbon nanotubes, activated carbon, and fullerenes, are examined.

Keywords: Adsorption, Carbon NMs, Environmental remediation, Pollutants, Water.

INTRODUCTION

Air contamination is defined as the existence of unwanted chemical species that obstruct the environment or induce harmful impacts on human species and other living forms. Rapid commercialization and the rise in human population have contributed to extensive urban development, resulting in a rapid escalation of environmental pollution [1]. Enhancing the quality of soil, water, and atmosphere is a significant hurdle in the contemporary world. Recognizing and mitigating air contaminants, along with their proactive prohibition, are crucial measures for

* **Corresponding author Sudesh Kumar:** Department of Chemistry, National Council of Educational Research and Training (NCERT), New Delhi, Delhi 110016, India; E-mail: sudeshneyol@gmail.com

Neha Agarwal, Vijendra Singh Solanki, Neetu Singh & Maulin P. Shah (Eds.)

safeguarding the surroundings. Nanotechnology is very crucial in achieving the goal of a green environment, and this technology has advanced significantly in the past 10 years, particularly in the field of nanoscale materials. The availability of uncontaminated water is diminishing because of commercialization, leading to a scarcity of pure water worldwide, particularly in emerging countries. Water can be contaminated by pollutants like organic substances, microbes, pathogens, coloring agents, and trace metals like Pb^{2+}, Cd^{2+}, Zn^{2+}, Ni^{2+}, Cr^{3+}, and Hg^{2+}. These contaminants do not decompose in the surroundings, and they even pose serious threats to living species and the environment. Trace metals can result in numerous harmful impacts, such as malignancy, renal impairment, liver inflammation, pregnancy loss, nephritic syndrome, blood deficiency, and brain disorder [2]. Pb^{2+} ions are typically discharged into the surroundings, often originating from the metallurgical sector, paper manufacturing industries, glass fabrication industries, and surface finishing industries [3]. Electrodeposition in battery design, solar cells, mining industries, and textile sectors are common sources of cadmium found in wastewater. Ni^{2+} ions can induce skin disorders upon contact with items such as jewelry, watches, zippers, and coins. Cr^{6+} ions can cause ailments such as liver impairment, renal inflammation, and abdominal discomfort. Additionally, Cr^{6+} ions are the predominant cause of mucosal lesions. Due to these severe negative consequences, the elimination of trace metals from aqueous media is critically important to safeguard humans from potential wellness problems. To eliminate heavy metals, various techniques can be employed, including charge transfer, bioaccumulation, flocculation, extraction, and reverse osmosis. Adsorption is regarded as the most commonly used approach due to its relatively cheap, high efficacy, and facile method of eliminating trace metals at low levels [4]. Fig. (**1**) illustrates various origins of trace metal ion pollution. Various NMs have been utilized for wastewater remediation, including organic and plant-based, particularly humic acid, which has found extensive use for wastewater treatment and the elimination of toxic metals. Material science plays a major role in various domains, including ecological studies and sustainability, the medical sector, supercapacitors/batteries, wastewater treatment facilities, catalyst development, power production, and storage of energy. Nano-sized materials offer unique characteristics for the treatment of polluted water, leveraging the expansive area of adsorbents and their capacity for chemical transformation and facile recyclability [5]. Nano-sized materials are increasingly utilized for the elimination of various contaminants, including inorganic contaminants, micro-organisms, and organic contaminants from polluted water. Different forms of carbon nanomaterial, *i.e.*, carbon nanotubes, graphene, fullerenes, graphene oxide, and activated carbons, have found extensive applications in detectors, storage of energy, electrical appliances, wastewater treatment, pharmaceuticals, and diagnostic evaluation due to their remarkable mechanical, physico-chemical, heat, and electrical properties. This chapter discusses the current achievements made in environmental remediation using carbon NMs as adsorptive agents, including graphene, carbon nanotubes, activated carbon, and fullerenes [6].

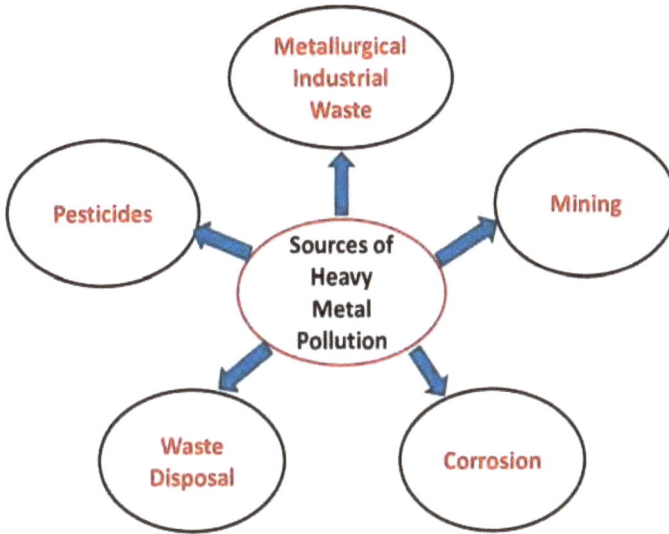

Fig. (1). Various origins of trace metal ion pollution [5].

TYPES OF CARBON NMS AND THEIR APPLICATIONS

Nanoparticles with all three dimensions below 100 nm are classified as zero-dimensional (0D) materials. Examples of such materials include quantum dots and fullerenes. Nanoparticles that have two dimensions smaller than 100 nm and one dimension larger than 100 nm are classified as one-dimensional (1D) materials [7]. Examples of such materials include carbon nanotubes. Nanoparticles with two dimensions larger than 100 nm are classified as two-dimensional (2D) materials. Examples of such materials include graphene. NMs characterized by all three dimensions, each exceeding 100 nm, are classified as three-dimensional (3D) NMs. Examples include certain NM composites and graphite.

Fullerenes and their application in wastewater treatment

Fullerenes were invented in the year 1985 from cosmic particles, and they exhibit an enclosed structure consisting of 5 and 6-membered rings. They are written by the chemical formula C20 + m, where m is an integral number. Fullerenes exhibit hydrophobicity, robust electron binding, large surface area, and surface irregularities. These distinctive chemical and physical characteristics render them suitable carbon NMs for different uses like electrical, medical biology semiconductors, electronics, biomedical sciences, photovoltaic cells, detectors, beauty products, photocatalysis, and finishings. Fullerenes have also been employed for remediating pollutants from groundwater or surface water through the utilization of light-induced catalysis. Fullerenes are also excellent environmentally friendly NMs for hydrogen fixation, as they can readily undergo conversion from carbon-carbon bonds to carbon-hydrogen bonds due to the reduced bond strengths of C-H bonds. They can exhibit an optimum hydrogen storage capacity of 6.1% owing to their chemical properties and three-dimensional structure. Moreover, they can be readily reverted due to the greater bond strengths of carbon-carbon bonds [8]. Carbon-based conductive coatings are employed on the capacitive electrode surface, and their capacitive performance depends on the external area, internal porosity, and conductance. Carbon materials offer greater conductance than conventional ones because of their increased outer surface region. The composite of fullerenes has been documented to exhibit a superior capacitance of 135.36 Fg^{-1} compared to the pristine graphene that was not amalgamated with fullerene. Additionally, the composite of fullerenes demonstrated an improved stability rate of 92.35%. Fullerenes have also been employed in batteries as anodic materials, offering enhanced efficacy while substituting enduring metal electrodes, hence demonstrating advantages concerning efficacy and sustainable material. The physical and chemical characteristics of these materials make them capable of removing various pollutants from water. Soluble fullerene complexes (SFCs) have been effectively utilized as catalysts to generate oxygen-free radicals in aqueous media upon exposure to UV light. These radicals can photochemically degrade water pollutants, and the SFCs also work as antiradicals. Interestingly, the SFCs can be readily extracted from aqueous media after the light-assisted degradation process. Fullerenes adsorb contaminants through the infiltration of adsorbed species into the gaps between the carbon nanostructures, and reduced accumulation property and an expansive area render these NMs valuable for effective extraction of contaminants from polluted water. Alekseeva and co-authors performed relative analyses of fullerene for the elimination of copper metal ions; their findings revealed that fullerenes exhibited superior efficacy [9]. They also observed that fullerenes conform to the Langmuir theory. While fullerenes can be used for extracting contaminants *via* the adsorption process, their elevated price serves as a constraint, limiting their practical usage. Nevertheless, the minute quantity of these NMs can be employed in the production of zeolites, lignin, and carbon to enhance their adsorption efficacy. The synthesis of fullerene enhances hydrophobicity and adsorption efficacy and also facilitates upcycling of materials. Fullerene bonded with polymer polyvinylpyrrolidone has been reported to show good efficacy in water purification. The membrane filtration process is becoming increasingly prominent for water disinfection. The efficacy of this process is contingent upon the material constituents, as it governs responsiveness, specificity, and durability. Fullerenes exhibit significant promise for utilization in the membrane process due to their facile usage, strong electron bonding, high tensile strength, and the capacity to customize dimensions. Fullerenes can prove advantageous in binding nanoscale adsorbent species to enhance their efficacy for adsorption.

Use of Carbon Nanotubes (CNTs) in photocatalytic systems

CNTs are NMs of cylindrical shape arranged in a tubular configuration. They are categorized into two classes, namely, single-walled carbon nanotubes (SWCNT) and multi-walled carbon nanotubes (MWCNT). SWCNTs consist of carbon monolayers, and MWCNTs consist of multiple layers of carbon [10]. CNTs have found extensive applications in the domain of water purification, biosensors, stimulators, and membranes due to their distinctive features, including extensive specific area, microporous structure, enclosed structure, layered composition, and robust bonding with contaminants. Various harmful contaminants have been eliminated from water through the utilization of CNTs, leveraging their unique features. Furthermore, CNTs can be synergistically combined with other carbon NMs to enhance their performance in eliminating a broad spectrum of contaminants from water [11]. Despite the extensive utilization of CNTs in the area of water purification, these materials may experience suboptimal activity because of adulteration, which may be carbon components used in the synthesis process. The adulteration leads to the modification of their surface characteristics by obstructing the reactive regions for the bonding of CNTs with the contaminants [12]. Nevertheless, the majority of the research studies indicated that surface modification can be averted by tailoring CNTs in acidic or basic solutions. This process also leads to the addition of unique functionality on the CNTs surface, which proved to be useful for contaminants removal. The addition of novel functionality to the CNTs surface, along with the characteristics of the CNTs, helps in assessing the removal rate of various pollutants. CNTs have effectively demonstrated as an exceptional NM in eliminating contaminants from water. This contributes to creating greener surroundings because of its remarkable adsorption efficiency and outstanding restoration capacities [13]. The photocatalytic process stands as a novel technique employed in water purification that employs semiconductors, specifically iron oxide, zinc oxide, and titanium oxide, but the photocatalytic efficacy of these nano-oxides is relatively low, and their response to UV light is sluggish. CNTs are novel materials for catalytic processes due to their enhanced photocatalytic efficacy, nanoscale dimensions, large durability, tubular hollow design, and expanded photo absorption capabilities resulting from their extensive specific area. Gupta and co-authors formulated SWCNTs and a titanium oxide photocatalytic framework for oil-water separation [14]. Park and co-authors formed titania-coated SWCNTs aerogel and effectively employed it to eliminate dye from water [15]. Xu and co-authors formulated a photocatalytic-based system of hydroxy MWCNTs and lead oxide, demonstrating application in the water treatment for pyridine removal. SWCNTs represent 1-D materials characterized by a tubular hollow structure with walls of just singular atoms in thickness [16]. These carbon NMs demonstrate extraordinary physical and chemical characteristics owing to their distinctive design. SWCNTs are extensively utilized for various applications, including semiconductor materials, electric devices, biomedicine, and detectors. They are also commonly employed in the remediation of ecological contamination due to their porosity, large area, facile usage, and nanoscale dimensions. These characteristics make SWCNTs highly promising for water purification. Alijani and co-authors engineered a composite based on SWCNTs by incorporating them with cobalt ferrite sulfide, and the resultant materials were employed for the elimination of mercury; findings demonstrated a remarkable adsorption rate exceeding 99.56% in a brief duration of 7 minutes [17]. In contrast to this, the adsorption rate for solely SWCNTs was observed to be 45.39% for mercury removal. Anitha and co-authors performed molecular modeling of pristine SWCNTs and their modified variants, such as hydroxy SWCNTs, amino SWCNTs, and carboxyl SWCNTs for the removal of trace metal ions, such as cadmium, lead, copper, and mercury from water [18]. The findings indicated that the carboxy SWCNTs exhibited a significantly higher adsorption rate, approximately 150–230% greater in contrast to pristine SWCNTs. The hydroxy SWCNTs and amino SWCNTs were identified to exhibit lower adsorption strength, displaying only 10 to 47% greater adsorption rate in contrast to pristine SWCNTs. The carboxy SWCNTs have also been explored for the removal of heavy metal ions like lead, copper, and cadmium. The findings indicated the adsorption efficacies for metal ions of 96.02, 55.89, and 77.00, respectively. In contrast to this, unmodified SWCNTs showed adsorption efficacies of 24.29, 33.55, and 24.07 mg/g for copper, lead, and cadmium, respectively. Zazouli and co-authors engineered composite materials of SWCNT with L-cysteine. They utilized these materials for extracting Hg^{2+} ions from aqueous media with adsorption efficacy of 95% [19]. Gupta and co-authors developed a carbon nanotube-reinforced polysulfone membrane, which was utilized for the elimination of trace metal ions. Integration of SWCNTs led to a decrease in the porosity of the membrane and a more even surface. The engineered polysulfone membrane exhibited significant adsorption capacity of 96.8%, 87.6%, and 94.2% for chromium, arsenic, and lead, respectively. The sole membrane exhibited adsorption capacities of 28.5%, 30.3%, and 28.3% for arsenic, chromium, and lead, respectively [20]. These findings demonstrate an enhancement in the membrane's efficacy attributed to the integration of SWCNTs. Dehghani and co-authors utilized SWCNTs in the elimination of chromium from aqueous media and assessed the impact of various factors such as time, acidity, and amount of chromium ions on the adsorption capability. It was noted that the removal rate varied with the acidity. The peak efficacy was identified at a pH value of 2.5, and the adsorption conforms to the Langmuir model [21]. These investigations indicated that SWCNTs are effective for water purification. The MWCNTs demonstrate distinctive features, including large areas, elevated heat, and electrical conductivity, as well as exceptional durability. Due to these physical and chemical characteristics, they find extensive applications in electric devices, photovoltaic cells, detectors, and biomedicine. MWCNTs have also been broadly utilized in water purification, particularly trace metals are removed through bonding with the surface moieties of MWCNTs. The oxygen-functionalized MWCNTs have a large adsorption capability and efficacy for the removal of metal ions like chromium, lead, and cadmium from aqueous media. The removal of metals relies on acidity, and this characteristic can be utilized for ion desorption by adjusting the initial pH value, allowing MWCNTs to be reused. Certain investigations have indicated that plasma-treated MWCNTs exhibit superior adsorption characteristics compared to acid-treated MWCNTs; this can be attributed to an abundance of oxygen-functionalized moieties on CNTs. Additionally, reports indicated that plasma-treated MWCNTs can be readily upcycled and reutilized. The MWCNTs composites with iron oxide, zirconium oxide, and aluminum oxide have been employed for the removal of trace metals like chromium, arsenic, nickel, lead, and copper from aqueous media. The adsorption properties of MWCNTs can be influenced by variables such as acidity and the amount of metallic ions. Their research findings align well with the Freundlich isotherm. The removal rate of modified MWCNTs exhibited a notable increase when compared to other metal oxides. Furthermore, the modified MWCNTs showed 20-fold more efficacy compared to unmodified MWCNTs in the removal of trace metal ions. It is widely accepted that the interaction between metal ions and the polarized surface of CNTs is generally considered the primary mechanism of adsorption. Functionalized MWCNTs have demonstrated elevated adsorption capability and efficacy for metal ions like lead, cadmium, and chromium from the aqueous media. The adsorption efficacy of acid-treated MWCNTs enhances the ability to eliminate Pb^{2+} and Cd^{2+} ions with oxygen-containing moieties, resulting in the precipitation of salts or ion complexes. The adsorption capacity of nitric acid-treated MWCNTs experiences a substantial increase primarily because of the presence of oxygen-containing moieties interacting with metallic ions, leading to the formation of ion complexes on the nanotube surface. The composite materials of MWCNTs with oxides, such as Fe_2O_3, ZnO, and Al_2O_3, were synthesized using a coprecipitation technique. The resulting composites have been effectively employed for the adsorption of Cu^{2+}, Ni^{2+}, and Pb^{2+} ions. The adsorption efficacy of these materials was observed to vary with the acidity and temperature. Based on the acidity and temperature, the adsorption capability of MWCNT composite materials ranged from 10 to 31 mg/g and the adsorption conforms to the Langmuir isotherm. The MWCNT composites with iron and manganese oxide were regarded as promising materials for their ability to eliminate chromium ions. The highest adsorption capability of 186.9 mg/g was observed with a peak elimination efficiency of 85% at the optimal pH value of 2.1. The impressive adsorption performance can be attributed to the polarized surface of the nanotubes. Additionally, research indicates that plasma-treated MWCNTs exhibit superior adsorption capabilities in comparison to MWCNTs having oxygen functionalities. The plasma-oxidation method has also been documented in the creation of CNTs incorporating manganese dioxide and titanium oxide for the extraction of Pb^{2+} ions from the aqueous media. On a heterogeneous adsorption site, certain regions combined twice align with the Langmuir-Freundlich model. This model was employed to distinguish between two types of adsorbent surfaces with enhanced and lower electron affinities for nickel ions. The adsorption of Ni^{2+} ions predominantly takes place at modified surfaces of MWCNTs composites, and the functionalization results in a 20% rise in the adsorption capability at low amounts of adsorbate. MWCNTs can also be modified with hydroxyquinoline, and this modification has been employed for the extraction of numerous toxic metal ions like Cu^{2+}, Pb^{2+}, and Cd^{2+}. CNTs, whether in their pristine state or with oxygen-bearing groups and in complex with others, exhibit remarkable capacity for adsorbing trace metals and extensive studies are currently underway for their utilization in water treatment. Elsehly and co-authors utilized available MWCNTs for eliminating Mn^{2+} and Fe^{2+} ions, achieving a removal efficiency of 71.5% and 52%, respectively, at a 50 ppm concentration of metal ions. Another group utilized CNT nanocomposite materials for extracting Fe^{2+} and Mn^{2+} from aqueous media [22]. Table **1** shows the adsorption capacities of modified and un-modified CNTs for various toxic metal ions.

Table 1. Application of modified and raw carbon nanotubes for toxic metals removal [23].

Sorbent Material	Pollutant	Highest adsorptive capability (mg.g^{-1})	Type of bonding	Adsorption isotherm
Modified MWCNTs	Zn^{2+}	1.05	Coulombic	Freundlich, langmuir
SWCNTs	Hg^{2+}	40.1	Chemisorption	Freundlich, langmuir
SWNTs with thiol groups	Hg^{2+}	131.5	Chemisorption	Langmuir

(Table 1) cont.....

Sorbent Material	Pollutant	Highest adsorptive capability (mg.g⁻¹)	Type of bonding	Adsorption isotherm
MWCNTs	As^{5+}	2.9	Chemisorption, coulombic	Combined langmuir-freundlich
Oxidized MWCNTs	As^{5+}	3.61	Specific adsorption, chemisorption, coulombic	Combined langmuir-freundlich
Ethylenediamine MWCNTs	As^{5+}	12.17	Specific adsorption, chemisorption, coulombic	Freundlich
MWCNTs	Cu^{2+}	3.19×10^{-5} mol/g	Coulombic, deposition, surface bonding	Langmuir
SWCNT	Cr^{6+}	20.3	Chemisorption, coulombic, physiosorption	Langmuir
MWCNTs	Cr^{6+}	2.4	Chemisorption, coulombic, physiosorption	Langmuir
MWCNTs-hydroxyapatite	Co^{2+}	16.2	Surface binding	Langmuir

Application of Graphene in Environmental Remediation

Graphene is regarded as a cutting-edge material and has garnered immense interest at the time of its invention in the year 2004. It is a 2D single-layer carbon nanostructure composed of carbon atoms with sp^2 hybridization structured in a hexagonal crystal lattice. It has attracted substantial interest as an innovative carbon NM for ecological uses owing to its large area of around 2630 m²/g, large heat transfer capability, and fast electron mobility [24]. Moreover, their exceptional tensile strength and certain remarkable attributes like numerous functional moieties, large electron charge density, and high water-attracting features render them attractive materials in polluted water treatment. Graphene and reduced graphene oxide (RGO) can be easily synthesized by graphene oxide reduction through an exfoliation process. Furthermore, graphene can undergo functionalization to produce graphene oxide using the Hummers technique, resulting in a significant amount of oxygen functionalities (such as carboxyl acid group, hydroxy group, carbonyl functional group, and epoxide group) within the carbon framework [25]. This characteristic is especially noteworthy for the removal of diverse contaminants from water. Graphene oxide can be produced through facile and cost-effective procedures, like the oxidative transformation from graphite to graphene oxide, followed by ultrasonic separation [26]. Because of its water-attracting characteristics, large surface area, and abundant density of functional moieties, it can serve as an adsorbent and function as a catalyst for eliminating various contaminants from polluted water. Utilization of graphene in the area of wastewater treatment is presently impeded by limited recyclability and isolation, challenges that can be surmounted by hybridization and functionalization. Moreover, graphene is generally combined with other materials for appropriate uses because of its no band energy difference and vulnerability to oxidation. When paired with other materials, it can boost the photocatalysis of coupling materials by serving as an electron receptor and carrier. Carbon emissions have become a significant ecological issue because it causes global warming. Carbon materials have been regarded as potential NMs when compared to traditional materials, both in terms of efficacy and expenses. Graphene NMs have been employed for the removal of environmental pollutants. Graphene materials can be used for the elimination of carbon dioxide and other toxic pollutants. Graphene monolayer can eliminate 37.93% of carbon dioxide. Graphene-based materials have been noted to preferentially eliminate carbon dioxide in contrast to nitrogen and methane. The adsorption of only carbon dioxide by graphene oxide can be ascribed to the increased polarity of CO_2, facilitating its facile interaction with the oxygen-containing functional moieties of carbon dioxide. Additional investigations have also been documented for modifying the graphene interaction to enhance the specificity of the targeted air pollutant. Graphene NMs are outstanding NMs for eliminating various pollutants from aqueous media due to their small size, large area, and capability for pi bonding, H-bonding, and coulombic forces. In comparison to graphite, graphene NMs showed better adsorptive performance in dye removal. Graphene oxide has also been employed for the elimination of basic dyes, specifically crystal violet, methylene blue, and rhodamine B from aqueous media. It was observed that an increase in the dye amount resulted in increased adsorption, with adsorption efficiency of 195.4, 199.2, and 154.8 mg g^{-1} for crystal violet, methylene blue, and rhodamine B, respectively. Graphene oxide has also been effectively utilized for the elimination of acidic dyes such as Direct Red 23 and Acid Orange 8 from water. While the adsorption method can eliminate pollutants from the aqueous media, it cannot decompose the pollutants, necessitating an elimination process. The photocatalytic process is a valuable method for water purification, facilitating the decomposition and breakdown of organic pollutants. Graphene NMs have been regarded as promising materials for their enhanced photocatalytic performance due to their large area and small size in contrast to conventionally employed NMs. Rommozzi and co-authors formulated an RGO photocatalyst using a more environmentally friendly method involving ammonium hydroxide and glucose, effectively designing a catalyst visible through the incorporation of titanium oxide [27]. The engineered RGO-titanium oxide composite was effectively utilized for the degradation of Alizarin Red S dye. The grafting of graphene oxide over titanium oxide and zinc oxide demonstrated significantly enhanced removal of dye compared to the individual use of titanium oxide and zinc oxide alone. Graphene is composed of hexagonal layers, forming a material with walls of just one atom thick, and is recognized as the finest NM, possessing a tensile strength 200 times greater than that of steel. Graphene was identified in 2004 by Konstantin Novoselov and Andre Geim, who were given a Nobel Prize in 2010 for their innovation. Graphene, a two-dimensional NM, is extensively employed in nearly every sector, including touch panels, mobile devices, liquid crystal displays, semiconductor devices, microprocessors, energy storage devices, energy production, water purification, supercapacitor devices, solar panels, biomedicine, and ecological research. The increasing focus on graphene NMs in water purification is driven by their distinctive physical and chemical properties, particularly conductivity, large area, heat conductivity, tensile strength, and adjustable surface properties. Tabish and co-authors developed graphene NMs and employed them as an adsorbent for effectively eliminating pollutants and trace metals from aqueous media [28]. Moreover, this graphene NM was utilized for the elimination of arsenic from aqueous media, demonstrating a removal efficacy of 80%. The material maintained its water purification efficacy even after repeated cycles and reutilization. Guo and co-authors engineered a graphene oxide nanocomposite using iron oxide through an on-site co-precipitation technique. Subsequently, they employed this composite for the elimination of lead ions from aqueous media [29]. The formulated material demonstrated outstanding adsorptive performance in extracting lead ions from water, exhibiting an outstanding adsorption capability of 373.14 mg/g. Zhang and co-authors modified graphene oxide with 4-sulfophenylazo and utilized it to eliminate various trace metals from water [30]. The engineered NM exhibited the highest adsorption capability of 59, 689, 267, 66, and 191 mg/g for copper, lead, cadmium, nickel, and chromium ions, respectively. Diana and co-authors engineered an autonomous microscopic robot based on graphene, comprising a layered composition made of nickel, graphene oxide, and platinum [31]. Each layer served a distinct usage; for instance, graphene oxide extracted trace metals, while the intermediate layer of nickel facilitated microscopic robot control through exterior magnetic force. Additionally, the inside platinum layer contributed to the motor in an autonomous system. The devised system successfully eliminated 80% of lead ions from the aqueous media. Yang and co-authors formulated hydrogen beads employing sodium alginate and graphene oxide and effectively employed hydrogen beads for extracting manganese ions from water, demonstrating an outstanding adsorption capacity of 56.49 mg/g [32]. Zheng and co-authors engineered a material composed of tea polyphenol, zinc oxide, and RGO. The developed NM was utilized for the elimination of trace metals, offering the additional benefit of antimicrobial characteristics. They employed this NM for extracting lead ions from water, achieving an adsorption capacity of 98.9%, and the nanocomposite was discovered to exhibit antimicrobial characteristics against Streptococcus mutans, resulting in a 99% removal rate [33]. Mousavi and co-authors engineered graphene oxide nanocomposites with Fe_3O_4 and utilized them for extracting lead ions from an aqueous solution. The NM demonstrated a 98% adsorption efficacy with an adsorption capability of 126.6 mg/g. Modified graphene NM can be used as an adsorbent for eliminating lead ions from water, the maximum adsorption capacity recorded for the removal of lead ions by graphene is 406.6 mg/g at a pH value of 5.0 in a duration of 40 minutes [34]. Graphene-hydrogel nanocomposite modified with lingo sulfonate with oxygen-bearing functional moieties that make the highly polarizable surface was identified to enhance the speed of lead ion adsorption. The highest adsorption efficiency obtained was 1308 mg/g in a duration of 40 minutes.

Activated Carbon (AC) and its application in water purification

AC is a promising NM for adsorption due to its large area, large porosity, and facile synthesis method using a range of precursors. Due to its excellent physical and chemical characteristics, it finds broad use in the elimination of pollutants generated by various factories such as medicinal, agricultural input manufacturing sites, oil and petroleum, beauty products, transport, and clothing. It is also extensively utilized for the adsorption of air pollutants, retrieval of solvents, and water purification, particularly for eliminating dyes and other organic contaminants. Moreover, it also serves as a catalyst in biodiesel generation. It is also utilized as an economical carbon NM for eliminating pollutants such as chemical oxygen demand, biochemical oxygen demand, and total suspended solids from water while sustaining the ideal pH value for biochemical applications. Maguana and co-authors conducted a study wherein they prepared pear seed cake-derived AC and employed it for the elimination of methylene blue. The achieved adsorption capability was 260 mg/g [35]. Antonio and co-authors synthesized AC derived from the kenaf plant [36]. They effectively used this material for the management and removal of pharmaceutical pollutants from water. AC is commonly referred to as activated charcoal. This NM is produced through specific synthesis methods, leading to the formation of high porosity and a maximum area greater than 3000 m^2. This carbon NM is manufactured extensively using timber, agro-waste, and coal. Aside from its porous characteristics, this material also possesses notable resilience. This property facilitates its utilization in various applications, including catalyst substrates, supercapacitors, batteries, hydrogen storage, and adsorbents for the elimination of various pollutants from water. Due to the high strength of AC, it can be recycled and reused. Abeer and co-authors synthesized AC derived from apricot stones and employed it for the elimination of zinc and aluminum ions, achieving an adsorption efficacy of 92% [37]. Ibrahim and co-authors synthesized AC using sewage sludge, employed it for extracting copper ions from an aqueous solution, and observed that the developed NM exhibited the highest removal efficiency exceeding 50% [38]. Li and co-authors synthesized sludge-derived AC and modified it with sulfur groups. They utilized sulfur-modified AC for the extraction of lead, cadmium, nickel, and copper ions from micropollutant water [39]. The adsorption capabilities were determined to be 96.2 mg/g, 238.1 mg/g, 52.4 mg/g, and 87.7 mg/g for cadmium, lead, nickel, and copper ions, respectively. Cao and co-authors developed highly porous AC from long-root Eichhornia crassipes and employed them for the extraction of trace metals, including lead, cadmium, copper, nickel, and zinc [40]. Dong and co-authors explored the utilization of AC for the elimination of trace metals from aqueous media. They discovered maximum adsorption capabilities of 95% and 86% for lead and cadmium ions, respectively [41]. M. Bali and co-authors utilized industrial AC for the extraction of trace metals and observed that the complete removal of cadmium ions was achieved in 15 minutes, and for metal ions like lead, zinc, and copper, the adsorption equilibrium was attained in 45 minutes with an adsorption efficiency of 64%, with cadmium ions exhibiting the maximum removal percentage [42]. Kongsuwan and co-authors synthesized eucalyptus bark-derived AC [43]. They employed it for extracting lead and copper ions from polluted water, with the highest adsorption capabilities of 0.53 mmol/g and 0.45 mmol/g, respectively. AC produced from latex tree sawdust has been used for its effectiveness in removing trace metals like chromium ions from an aqueous media. The adsorption capability of 44 mg/g was observed. AC derived from Moso and Ma bamboo demonstrated high efficiency in eliminating trace metals, including lead, copper, chromium, and cadmium ions, with the highest adsorption capability greater than 90%. In another study, researchers reported the utilization of microwave-prepared AC derived from olive stones for the elimination of various metals from water, including iron, lead, copper, zinc, nickel, and cadmium ions, achieving an adsorption efficacy of greater than 98% [44]. AC is facile to produce, is cost-effective, and stands out as the most prospective NM for the extraction of organic pollutants. It can be synthesized extensively using diverse precursors of carbon, particularly from agro-waste. Besides its simplified synthesis process, AC can be easily subjected to functionalization.

MECHANISM OF ADSORPTION

The adsorption method is a highly effective technique for the elimination of various contaminants from wastewater. Adsorbents with high adsorption capability are the most favored. To attain optimal adsorption efficacy for the elimination of contaminants, a more profound knowledge of the adsorption pathways between the adsorptive material and adsorbed species is the crucial aspect [45]. Various mechanical interactions typically determine adsorption pathways. The primary interactions influencing the adsorption mechanism include water-repelling interactions, pi-stacking, coulombic forces, and H-bonds. Various adsorptive materials interact with adsorbed species in diverse ways, contingent upon their characteristics like the composition and characteristics of contaminants and the corresponding functionalities that exist on the adsorptive material surface [46]. When addressing the aspect of adsorption involving functionalities like the carboxyl acid group, hydroxy group, amino group, and carbonyl functional group, H-bonding comes into play, while for adsorptive materials having electrically charged moieties, coulombic forces are primary, and non-polar or water-repelling forces typically prevails for adsorbents containing aliphatic compounds. Carbon materials have demonstrated notable efficacy in the elimination of various contaminants from polluted water. CNTs exhibit a large removal efficiency for organic contaminants because of their extensive specific area and strong bonds with organic contaminants. The adsorptive pathways between CNTs and organic contaminants typically involve interactions like water-repelling forces, pi bonding, London forces, coulombic forces, and H-bonding, which can take place independently or concurrently [47]. Moreover, the oxidative functionalization of CNTs can greatly improve the adsorption performance of inorganic contaminants because of the presence of carboxyl acid group, hydroxy group, and phenolic groups on the CNTs surface. The adsorptive pathway between CNTs and oxidized CNTs is illustrated in Fig. (**2**). For unfunctionalized CNTs, adsorption takes place at the adsorbent's surface, whereas for oxidized CNTs, adsorption takes place both at the material's surface and *via* the functionalities present on the surface. These functionalities serve as the primary sorption region of many inorganic contaminants *via* ionic interaction and molecular bonding. In the context of the removal of organic contaminants by graphene materials, the adsorption is governed by the interaction between the graphene-based NMs and contaminants; hence, the greater the specific surface area, the higher the removal rate of pollutants [48]. Release processes typically take place in the introduction of acidic or basic solutions, contingent on the characteristics and suitability of NMs. The prevailing bonding between graphene materials and organic contaminants is pi stacking. For the elimination of inorganic contaminants, the presence of functionalities on the material's surface is a crucial element that determines the removal rate [49]. Hence, functionalized graphene and graphene oxide exhibit large adsorption capabilities for inorganic contaminants compared to untreated graphene. The adsorptive pathway between graphene-based NMs and inorganic pollutants typically involves coulombic forces.

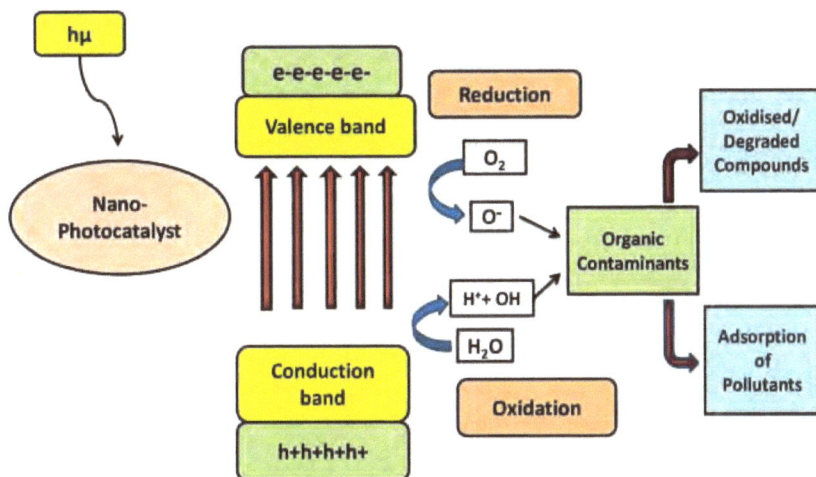

Fig. (2). Adsorption mechanism between carbon nanotubes and oxidized carbon nanotubes for pollutant removal [2].

CONCLUSION AND FUTURE RECOMMENDATIONS

The existence of metal ions, pathogens, and organic contaminants in potable water is a matter of significant concern worldwide. These contaminants can cause severe danger to both the environment and human health; therefore, their elimination from the aqueous medium is a key challenge. Numerous methods have been utilized for the removal of contaminants from water, and the application of nanoscience *i.e.*, photocatalysis and adsorption, has been gaining the interest of researchers in this domain. The carbon materials have been effectively employed in the decontamination of polluted water with trace metal ions. This can be attributed to their remarkable features, such as extensive specific area, regenerative properties, microporous structure, enclosed structure, and robust bonding with contaminants. In addition to these characteristics, carbon materials can make composites with other materials and can be easily modified, leading to the formation of a versatile sorbent material. Carbon NMs exhibit high biocompatibility with surroundings and living species. Various factors, including acidity, time, and the sorbent type, significantly influence the adsorption mechanism.

REFERENCES

[1]   Wu Y, Pang H, Liu Y, *et al.* Environmental remediation of heavy metal ions by novel-nanomaterials: A review. Environ Pollut 2019; 246: 608-20.
[http://dx.doi.org/10.1016/j.envpol.2018.12.076] [PMID: 30605816]

[2]   Madima N, Mishra SB, Inamuddin I, Mishra AK. Carbon-based nanomaterials for remediation of organic and inorganic pollutants from wastewater. A review. Environ Chem Lett 2020; 18(4): 1169-91.
[http://dx.doi.org/10.1007/s10311-020-01001-0]

[3]   Nimibofa A, Newton EA, Cyprain AY, Donbebe W. Fullerenes: synthesis and applications. J Mater Sci 2018; 7: 22-33.

[4]   Zhang X, Huang Q, Deng F, *et al.* Mussel-inspired fabrication of functional materials and their environmental applications: Progress and prospects. Appl Mater Today 2017; 7: 222-38.
[http://dx.doi.org/10.1016/j.apmt.2017.04.001]

[5]   Baby R, Saifullah B, Hussein MZ. Carbon NMs for the treatment of heavy metal-contaminated water and environmental remediation. Nanoscale Res Lett 2019; 14(1): 341.
[http://dx.doi.org/10.1186/s11671-019-3167-8]

[6]   Baby Shaikh R, Saifullah B, Rehman F. Greener method for the removal of toxic metal ions from the wastewater by application of agricultural waste as an adsorbent. Water 2018; 10(10): 1316.
[http://dx.doi.org/10.3390/w10101316]

[7]   Han X, Li S, Peng Z, *et al.* Interactions between carbon NMs and biomolecules. J Oleo Sci 2016; 65(1): 1-7.
[http://dx.doi.org/10.5650/jos.ess15248] [PMID: 26666276]

[8]   Mohajeri A, Omidvar A. Fullerene-based materials for solar cell applications: design of novel acceptors for efficient polymer solar cells – a DFT study. Phys Chem Chem Phys 2015; 17(34): 22367-76.
[http://dx.doi.org/10.1039/C5CP02453F] [PMID: 26248255]

[9]   Alekseeva OV, Bagrovskaya NA, Noskov AV. Sorption of heavy metal ions by fullerene and polystyrene/fullerene film compositions. Prot Met Phys Chem Surf 2016; 52(3): 443-7.
[http://dx.doi.org/10.1134/S2070205116030035]

[10]   Shan S, Zhao Y, Tang H, Cui F. A mini-review of carbonaceous NMs for removal of contaminants from wastewater. IOP Conference Series: earth and environmental science: 2017. IOP Publishing 2017; p. 012003.

[11]   Das R, Leo BF, Murphy F. The toxic truth about carbon nanotubes in water purification: a perspective view. Nanoscale Res Lett 2018; 13(1): 183.
[http://dx.doi.org/10.1186/s11671-018-2589-z] [PMID: 29915874]

[12]   Sarkar B, Mandal S, Tsang YF, Kumar P, Kim KH, Ok YS. Designer carbon nanotubes for contaminant removal in water and wastewater: A critical review. Sci Total Environ 2018; 612: 561-81.
[http://dx.doi.org/10.1016/j.scitotenv.2017.08.132] [PMID: 28865273]

[13]   Thines RK, Mubarak NM, Nizamuddin S, Sahu JN, Abdullah EC, Ganesan P. Application potential of carbon nanomaterials in water and wastewater treatment: A review. J Taiwan Inst Chem Eng 2017; 72: 116-33.
[http://dx.doi.org/10.1016/j.jtice.2017.01.018]

[14]   Gupta RK, Dunderdale GJ, England MW, Hozumi A. Oil/water separation techniques: a review of recent progresses and future directions. J Mater Chem A Mater Energy Sustain 2017; 5(31): 16025-58.
[http://dx.doi.org/10.1039/C7TA02070H]

[15]   Park HA, Liu S, Salvador PA, Rohrer GS, Islam MF. High visible-light photochemical activity of titania decorated on single-wall carbon nanotube aerogels. RSC Advances 2016; 6(27): 22285-94.
[http://dx.doi.org/10.1039/C6RA03801H]

[16]   Xu Z, Liu H, Niu J, Zhou Y, Wang C, Wang Y. Hydroxyl multi-walled carbon nanotube-modified nanocrystalline PbO_2 anode for removal of pyridine from wastewater. J Hazard Mater 2017; 327: 144-52.
[http://dx.doi.org/10.1016/j.jhazmat.2016.12.056] [PMID: 28064142]

[17]   Alijani H, Shariatinia Z. Synthesis of high growth rate SWCNTs and their magnetite cobalt sulfide nanohybrid as super-adsorbent for mercury removal. Chem Eng Res Des 2018; 129: 132-49.
[http://dx.doi.org/10.1016/j.cherd.2017.11.014]

[18] Anitha K, Namsani S, Singh JK. Removal of heavy metal ions using a functionalized single-walled carbon nanotube: a molecular dynamics study. J Phys Chem A 2015; 119(30): 8349-58. [http://dx.doi.org/10.1021/acs.jpca.5b03352] [PMID: 26158866]

[19] Zazouli MA, Yousefi Z, Yazdani Cherati J, Tabarinia H, Tabarinia F, Akbari Adergani B. Evaluation of L-Cysteine functionalized single-walled carbon nanotubes on mercury removal from aqueous solutions. J Mazandaran Univ Med Sci 2014; 24(111): 10-21.

[20] Gupta S, Bhatiya D, Murthy CN. Metal removal studies by composite membrane of polysulfone and functionalized single-walled carbon nanotubes. Sep Sci Technol 2015; 50(3): 421-9. [http://dx.doi.org/10.1080/01496395.2014.973516]

[21] Dehghani MH, Taher MM, Bajpai AK, *et al.* Removal of noxious Cr (VI) ions using single-walled carbon nanotubes and multi-walled carbon nanotubes. Chem Eng J 2015; 279: 344-52. [http://dx.doi.org/10.1016/j.cej.2015.04.151]

[22] Elsehly EMI, Chechenin NG, Bukunov KA, *et al.* Removal of iron and manganese from aqueous solutions using carbon nanotube filters. Water Sci Technol Water Supply 2016; 16(2): 347-53. [http://dx.doi.org/10.2166/ws.2015.143]

[23] Ouni L, Ramazani A, Taghavi Fardood S. An overview of carbon nanotubes role in heavy metals removal from wastewater. Front Chem Sci Eng 2019; 13(2): 274-95. [http://dx.doi.org/10.1007/s11705-018-1765-0]

[24] Xu L, Wang J. The application of graphene-based materials for the removal of heavy metals and radionuclides from water and wastewater. Crit Rev Environ Sci Technol 2017; 47(12): 1042-105. [http://dx.doi.org/10.1080/10643389.2017.1342514]

[25] Jilani A, Othman MHD, Ansari MO, *et al.* Graphene and its derivatives: synthesis, modifications, and applications in wastewater treatment. Environ Chem Lett 2018; 16(4): 1301-23. [http://dx.doi.org/10.1007/s10311-018-0755-2]

[26] Wang S, Sun H, Ang HM, Tadé MO. Adsorptive remediation of environmental pollutants using novel graphene-based nanomaterials. Chem Eng J 2013; 226: 336-47. [http://dx.doi.org/10.1016/j.cej.2013.04.070]

[27] Rommozzi E, Zannotti M, Giovannetti R, *et al.* Reduced graphene oxide/TiO2 nanocomposite: from synthesis to characterization for efficient visible light photocatalytic applications. Catalysts 2018; 8(12): 598. [http://dx.doi.org/10.3390/catal8120598]

[28] Tabish TA, Memon FA, Gomez DE, Horsell DW, Zhang S. A facile synthesis of porous graphene for efficient water and wastewater treatment. Sci Rep 2018; 8(1): 1817. [http://dx.doi.org/10.1038/s41598-018-19978-8] [PMID: 29379045]

[29] Guo T, Bulin C, Li B, *et al.* Efficient removal of aqueous Pb(II) using partially reduced graphene oxide-Fe$_3$O$_4$. Adsorpt Sci Technol 2018; 36(3-4): 1031-48. [http://dx.doi.org/10.1177/0263617417744402]

[30] Zhang CZ, Chen B, Bai Y, Xie J. A new functionalized reduced graphene oxide adsorbent for removing heavy metal ions in water *via* coordination and ion exchange. Sep Sci Technol 2018; 53(18): 2896-905. [http://dx.doi.org/10.1080/01496395.2018.1497655]

[31] Vilela D, Parmar J, Zeng Y, Zhao Y, Sánchez S. Graphene-based microbots for toxic heavy metal removal and recovery from water. Nano Lett 2016; 16(4): 2860-6. [http://dx.doi.org/10.1021/acs.nanolett.6b00768] [PMID: 26998896]

[32] Yang X, Zhou T, Ren B, Hursthouse A, Zhang Y. Removal of Mn (II) by sodium alginate/graphene oxide composite double-network hydrogel beads from aqueous solutions. Sci Rep 2018; 8(1): 10717. [http://dx.doi.org/10.1038/s41598-018-29133-y] [PMID: 30013177]

[33]   Zheng S, Hao L, Zhang L, *et al.* Tea polyphenols functionalized and reduced graphene oxide-ZnO composites for selective Pb2+ removal and enhanced antibacterial activity. J Biomed Nanotechnol 2018; 14(7): 1263-76.
[http://dx.doi.org/10.1166/jbn.2018.2584] [PMID: 29944100]

[34]   Mousavi SM, Hashemi SA, Amani AM, *et al.* Pb (II) removal from synthetic wastewater using Kombucha Scoby and graphene oxide/Fe3O4. Physical Chemistry Research 2018; 6(4): 759-71.

[35]   El Maguana Y, Elhadiri N, Bouchdoug M, Benchanaa M, Jaouad A. Activated carbon from prickly pear seed cake: optimization of preparation conditions using experimental design and its application in dye removal. International Journal of Chemical Engineering 2019.
[http://dx.doi.org/10.1155/2019/8621951]

[36]   Macías-García A, García-Sanz-Calcedo J, Carrasco-Amador JP, Segura-Cruz R. Adsorption of paracetamol in hospital wastewater through activated carbon filters. Sustainability (Basel) 2019; 11(9): 2672.
[http://dx.doi.org/10.3390/su11092672]

[37]   El-Saharty A, Mahmoud SN, Manjood AH, Nassar AAH, Ahmed AM. Effect of apricot stone activated carbon adsorbent on the removal of toxic heavy metals ions from aqueous solutions. Int J Ecotoxicol Ecobiol 2018; 3(2): 51.
[http://dx.doi.org/10.11648/j.ijee.20180302.13]

[38]   Salihi IU, Kutty SRM, Ismail HHM. Copper Metal Removal using Sludge Activated Carbon Derived from Wastewater Treatment Sludge.
[http://dx.doi.org/10.1051/matecconf/201820303009]

[39]   Li J, Xing X, Li J, *et al.* Preparation of thiol-functionalized activated carbon from sewage sludge with coal blending for heavy metal removal from contaminated water. Environ Pollut 2018; 234: 677-83.
[http://dx.doi.org/10.1016/j.envpol.2017.11.102] [PMID: 29227953]

[40]   Cao F, Lian C, Yu J, Yang H, Lin S. Study on the adsorption performance and competitive mechanism for heavy metal contaminants removal using novel multi-pore activated carbons derived from recyclable long-root Eichhornia crassipes. Bioresour Technol 2019; 276: 211-8.
[http://dx.doi.org/10.1016/j.biortech.2019.01.007] [PMID: 30640014]

[41]   Dong L, Hou L, Wang Z, Gu P, Chen G, Jiang R. A new function of spent activated carbon in BAC process: Removing heavy metals by ion exchange mechanism. J Hazard Mater 2018; 359: 76-84.
[http://dx.doi.org/10.1016/j.jhazmat.2018.07.030] [PMID: 30014917]

[42]   Bali M, Tlili H. Removal of heavy metals from wastewater using infiltration-percolation process and adsorption on activated carbon. Int J Environ Sci Technol 2019; 16(1): 249-58.
[http://dx.doi.org/10.1007/s13762-018-1663-5]

[43]   Kongsuwan A, Patnukao P, Pavasant P. Binary component sorption of Cu(II) and Pb(II) with activated carbon from Eucalyptus camaldulensis Dehn bark. J Ind Eng Chem 2009; 15(4): 465-70.
[http://dx.doi.org/10.1016/j.jiec.2009.02.002]

[44]   Alslaibi TM, Abustan I, Ahmad MA, Foul AA. Application of response surface methodology (RSM) for optimization of Cu $^{2+}$, Cd $^{2+}$, Ni $^{2+}$, Pb $^{2+}$, Fe $^{2+}$, and Zn $^{2+}$ removal from aqueous solution using microwaved olive stone activated carbon. J Chem Technol Biotechnol 2013; 88(12): 2141-51.
[http://dx.doi.org/10.1002/jctb.4073]

[45]   Ihsanullah , Abbas A, Al-Amer AM, *et al.* Heavy metal removal from aqueous solution by advanced carbon nanotubes: Critical review of adsorption applications. Separ Purif Tech 2016; 157: 141-61.
[http://dx.doi.org/10.1016/j.seppur.2015.11.039]

[46]   Tong Y, McNamara PJ, Mayer BK. Adsorption of organic micropollutants onto biochar: a review of relevant kinetics, mechanisms and equilibrium. Environ Sci Water Res Technol 2019; 5(5): 821-38.
[http://dx.doi.org/10.1039/C8EW00938D]

[47]   Madhura L, Singh S, Kanchi S, Sabela M, Bisetty K, Inamuddin . Nanotechnology-based water quality

management for wastewater treatment. Environ Chem Lett 2019; 17(1): 65-121.
[http://dx.doi.org/10.1007/s10311-018-0778-8]

[48]   Mashkoor F, Nasar A, Inamuddin . Carbon nanotube-based adsorbents for the removal of dyes from waters: A review. Environ Chem Lett 2020; 18(3): 605-29.
[http://dx.doi.org/10.1007/s10311-020-00970-6]

[49]   Crini G, Lichtfouse E, Wilson LD, Morin-Crini N. Conventional and non-conventional adsorbents for wastewater treatment. Environ Chem Lett 2019; 17(1): 195-213.
[http://dx.doi.org/10.1007/s10311-018-0786-8]

CHAPTER 5

Environmental Remediation by Copolymer Nanocomposites

W. B. Gurnule[1,*], Rashmi R. Dubey[1], Yashpal U. Rathod[2] and Anup K. Parmar[2]

[1] *Department of Chemistry, Kamla Nehru Mahavidyalaya, Nagpur 440024, India*

[2] *Department of Chemistry, C. J. Patel College, Tirora, Dist. Gondia, Maharashta, India*

Abstract: Environmental pollution due to human activities has become a serious problem around us, which has affected various living organisms worldwide. Therefore, there is an urgent need for new materials to remediate the polluted environment. Activated charcoal and copolymer were used to create a composite material. The material was spectrally characterized, and scanning electron microscopy (SEM) was used to examine the material's morphology. The composite material has effectively eliminated the chosen metal ions from the aqueous solution, according to the metal ion sorption data. This might be because of the composite's higher specific surface area and very porous nature. The cation exchange and synthesis of the 2-Amino-6-nitrobenzothiazole-adipamide-formaldehyde copolymer are described in this study. The condensation of 2-Amino-6-nitrobenzothiazole, adipamide, and formaldehyde with an acid catalyst in the presence of 1:1:2 molar proportions of the reacting monomers at 124°C produced the copolymer. The average molecular weight of this copolymer was determined by gel permeation chromatography, and the elemental analysis of the copolymer was used to determine its composition. The UV-visible, FTIR, and 1H NMR methods were used to characterize the newly synthesized copolymer and its nanocomposites. This copolymer nanocomposites ion-exchange properties for Cu^{2+}, Ni^{2+}, Zn^{2+}, Co^{2+}, and Pb^{2+} ions were examined using the batch equilibrium method in fluids with varying ionic strengths and a pH range of 2.0 to 6.0. The removal of these ions by copolymer nanocomposites followed the order of $Cu^{2+} > Ni^{2+} > Pb^{2+}$. According to the analysis ratio of distribution as a function of pH, this research could be used to treat industrial wastewater because resin uses more metal ions as the medium's pH rises. The emerging idea of environmental remediation using polymeric nanocomposites will be the focus of this chapter.

Keywords: Copolymer, Nanocomposites, Environment remediation, Batch equilibration method, Synthesis, Toxic metal ions.

---

* **Corresponding author W. B. Gurnule:** Department of Chemistry, Kamla Nehru Mahavidyalaya, Nagpur 440024, India; E-mail: wbgurnule@gmail.com

INTRODUCTION

Heavy hazardous metal ions tend to accumulate in living things. There has been a lot of focus in recent decades on removing residues of these ions from household, industrial, and nuclear wastes, which has detrimental consequences on the environment. As a result, an effort has been undertaken to create a new chelating copolymer and evaluate its ion-exchange properties. Electrical appliances, materials with great heat resistance, and ion exchangers are only a few of the research applications for copolymers. Heavy metal ions, including Pb^{2+}, Cd^{2+}, Zn^{2+}, Ni^{2+}, and $Hg^{2+,}$ have the most hazardous qualities of all water contaminants and can seriously harm both human and animal health.

Around the world, certain industries have experienced significant growth, including those related to mining, refining, battery production, paint manufacturing, chemicals, dyes, and pharmaceuticals. Among them, industries are widely contaminating the environment, particularly with regard to the elevated concentrations of heavy metals like cobalt, lead, cadmium, and mercury in the aquatic system and soil. These heavy metals are extremely dangerous and poisonous, and they may have an impact on biological processes and human health. Living systems in close proximity to industry typically absorb heavy metal toxicity in many ways, which can result in health risks and hazards such as cancers of the skin, lungs, and liver [1]. In addition to having renal effects, cadmium increases diarrhea. Lead and mercury have an adverse effect on brain function, memory, blood pressure, skin rashes, miscarriages in pregnant women, and other health issues [2]. Controlling or eliminating these heavy metals from wastewater and industrial effluents is the current environmental concern. Taking into account every factor, many traditional techniques, including chemical precipitation, electrochemical procedure, membrane filtration, biological treatment, coagulation, flocculation, and ion exchange processes, have been used to remove heavy metals. Batch separation ion-exchange analysis, with its better adsorption efficiency, selectivity, and reusability, is a powerful tool for sorting out such issues [3].

There have been several reports on the use of polymeric resins and composites in the ion-exchange process to remove heavy and hazardous metal ions from wastewater, industrial effluents, *etc.*, utilising the ion-exchange technique. The chelating characteristics of a synthetic resin were successfully analysed utilising p-aminophenol, dithiooxamide, and formaldehyde terpolymer. The stated findings were dependent on physical parameters, including diffusion counter ion, particle size, and pore size [4].

The radiotracer approach has been used to study the polyaniline-polystyrene composite with the purpose of recovering Hg^{2+} metal ions. The findings demonstrated that a rise in temperature causes a rise in metal sorption. PGME copolymer was used in a batch equilibrium approach at 25–70°C to effectively adsorb the hexavalent chromium (VI) metal ion [5]. For the purpose of removing heavy metals, a unique comparison study was conducted between a synthetic terpolymer and its composite based on metals and heat deterioration. Both the terpolymer and the composite exhibit selectivity in the following order: $Zn^{2+}> Cu^{2+}> Co^{2+}> Pb^{2+}> Cd^{2+}$ and $Pb^{2+}> Cd^{2+}> Cu^{2+}> Co^{2+}> Zn^{2+}$ respectively [6].

To remove cadmium ions from the aqueous solution, a comparison of chitosan, commercial activated carbon, and chitosan/activated carbon composite was conducted. The most crucial element in the removal of heavy metal ions is the pH solution of the adsorbent [7].

Using a straightforward solution-evaporation technique, chitosan and charcoal were combined to create a chitosan/charcoal composite. Chromium from wastewater may be efficiently treated using the produced composite. The composite was used to adsorb chromium at different adsorbent doses, rates, and pH values. The maximum pH has an impact on the adsorption capacity [8]. Selective removal of heavy metal ions from an aqueous solution was achieved by combining magnesium and activated carbon from coconut shells. The generated magnesium and coconut shell-activated carbon composite successfully extracted Zn(II) and Cd(II) ions from the waste aqueous solution by utilizing the ion-exchange method [9].

There have been a lot of reports [10 - 12] on the sorption of various heavy metal ions by urea-based chelate polymers using the batch equilibrium method. These studies involved thiourea/urea with formaldehyde and o-aminophenol/o-cresol with formaldehyde. These polymers contain a variability of vigorous chelating functional groups. The >C=O group in urea is particularly significant due to its frequent presence in the separation of certain metal ions, including Fe^{3+}, Cu^{2+}, Ni^{2+}, Co^{2+}, Zn^{2+}, Cd^{2+} and Pb^{2+} ions [13]. Gurnule and colleagues [14 - 17] investigated the chelation ion exchange characteristics of a number of copolymer/terpolymer resins. Additionally, they have investigated and documented the ion exchange characteristics of formaldehyde, biuret, and 4-hydroxyacetophenone terpolymer resins [18]. By employing the solution condensation process to synthesize semicarbazide with formaldehyde and 8-hydroxyquinoline5-sulfonic acid, a new terpolymer was created that functions as an efficient chelating ion exchanger. Its ion-exchange characteristics were assessed utilizing the batch equilibrium approach with various electrolyte concentrations, pH ranges, and time intervals against certain metal ions, including Fe^{3+}, Cu^{2+}, Ni^{2+}, Co^{2+}, Zn^{2+}, and Pb^{2+} [19].

A batch sorption system with environmentally friendly conducting polymer/biopolymer composites like polyaniline/chitosan (PANi/Ch) and polypyrrole/chitosan (PPy/Ch) as adsorbents was studied in order to capture fluoride ions from aqueous solutions [20]. The batch equilibrium method was used to investigate the chelation ion-exchange characteristics of 8-hydroxyquinoline-5-sulfonic acid–thiourea–formaldehyde copolymer resins following their synthesis. It was demonstrated that the resins were selective chelating ion-exchange copolymers for particular metals and that their selectivity for Fe3+ ions was superior to that of Cu^{2+}, Ni^{2+}, $Co^{2+,}$ and Pb^{2+} ions [21].

Thus, it is concluded that there were no prior reports for the copolymer nanocomposites based on the literature. As a result, 2-amino 6-nitrobenzothiazole and biuret were combined with formaldehyde to create a novel copolymer, which was then combined with activated charcoal to create a novel copolymer nanocomposite. Using copolymer and its nanocomposite, the batch separation approach was used and reported for the removal of heavy metal ions. After that, a comparison of the copolymer's ion-exchange tests with those of its composite was made. The elimination of metal ions and thermal stability were compared between the copolymer and its nanocomposite.

EXPERIMENTAL

Materials

Formaldehyde, 2-amino6-nitrobenzathiazole, and adipamide acid were acquired from Merck, India. Cu^{2+}, Zn^{2+}, Co^{2+}, Pb^{2+}, and Cd^{2+} metal ion solutions were made by dissolving their nitrate salts in deionized water. Without additional purification, the other chemicals and solvents that were purchased from Merck were utilized just as they were.

Synthesis of Copolymer and Composite

Copolymer was prepared by the condensation polymerization of 2-amino-6-niro-benzothiazole (0.1mol, 1.9520 gm) and adipamide (0.1mol, 0.088065 g) with formaldehyde (0.2 mol) in the molar proportion of 1:1:2 in hydrochloric acid medium. This mixture was heated for 5 hrs at 124 ^{0}C in an oil bath with periodic shaking to ensure exhaustive blending followed by the methodology in view of prior writing [23]. In a 1:2 ratio, the new copolymer/activated charcoal nanocomposite was made. After dissolving the copolymer in 25 milliliters of DMF, activated charcoal was added, and the mixture was ultrasonically sonicated for three hours by continuous stirring for a full day. Ultimately, the resulting black composite was dried for 24 hours at 70°C in an air oven. The synthesis of the copolymer is shown in Fig. (**1**).

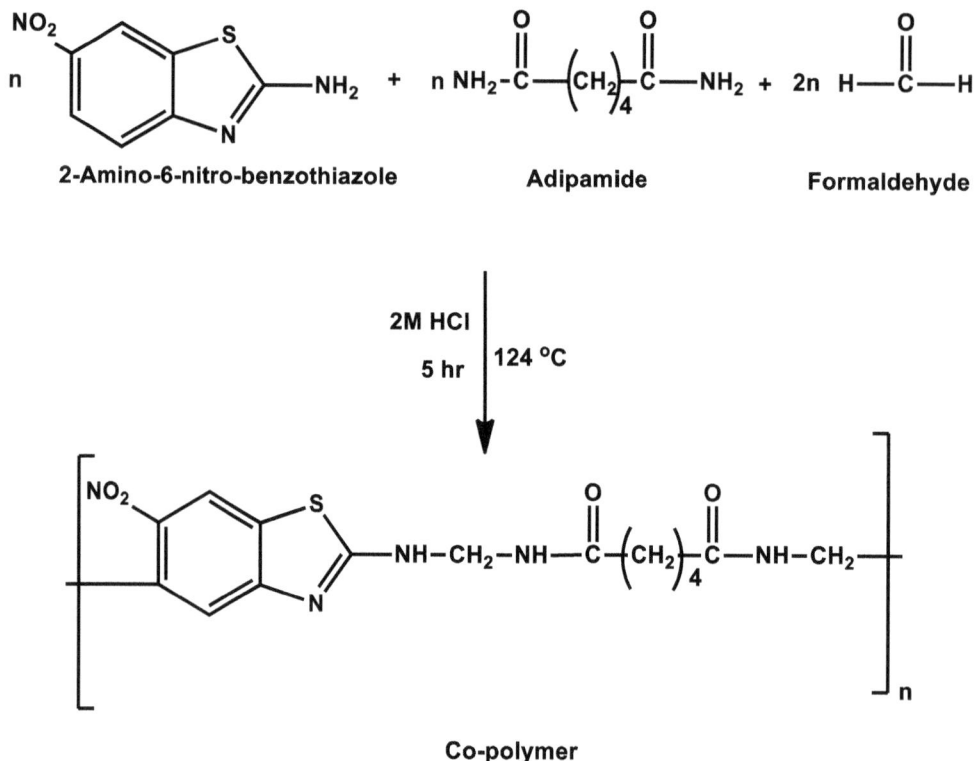

Fig. (**1**). Synthesis of Copolymer ANBAF.

Ion Exchange Studies

Metal Ion Uptake in the Presence of Various Electrolytes of Different Concentrations

The following method was utilized with a 0.1M metal nitrate solution and four distinct electrolytes, each of which had five distinct concentrations on the quantity of metal ion uptake by the copolymer samples. In a known-concentration electrolyte solution of 25 milliliters, the 25-milligram copolymer was suspended. 0.1 M NaOH or 0.1 M HNO_3 was added to the solution to bring it to the desired pH level. At room temperature, the suspension was constantly stirred for 24 hours. The pH was adjusted to the required level, and metallic nitrate solution (2 ml, 0.1 M) was added to this suspension. After being automatically stirred for 24 hours, the contents were filtered and rinsed with distilled water. The filtrate was then collected in a conical flask and titrated with standard EDTA solution with the appropriate buffer and indicators to determine the steel ion content. The same procedure was followed in a blank experiment without the polymer pattern. A straightforward solution for the metal ion content material was envisioned. The difference between the analysis of the actual test and the analysis of the blank test was used to calculate the amount of metallic ions adsorbed by the copolymer resins in the presence of the known concentration of the electrolyte. This value was expressed in millimoles per gram of the copolymer sample. The test was repeated with different electrolytes of known concentration and seven unique metal ions, namely Zn(II), Pb(II), Cu(II), Ni(II), Co(II), etc., and all of the copolymer samples' molar ratios were treated in the same manner.

Determination of Rate of Metal Ion Uptake at Different Time Intervals

The copolymer (25mg) was suspended in an electrolyte solution (25 ml) of recognized concentration. The pH of the solution was adjusted to the required value by the addition of both 0.1M HNO_3 and 0.1 M NaOH. The suspension was routinely stirred for 24 hrs at room temperature. To this suspension, metallic nitrate solution 2ml of 0.1M was introduced and the pH was adjusted to the required value. The contents were automatically stirred for 24 hrs and then filtered and washed with the distilled water. The filtrate was then amassed in a conical flask and anticipated for steel ion contents by using titration with standard EDTA solution, appropriate buffer, and appropriate indicators. A blank experiment was also carried out in the same way without including a polymer pattern. The quantity of metallic ion adsorbed through the copolymer resins in the presence of the given electrolyte of known concentration was calculated from the difference between analyzing the blank test and analyzing the actual test and was expressed in terms of milli mole per gram of the copolymer samples. The test is repeated in the presence of numerous electrolytes of recognized concentration with seven one-of-a-kind metal ions *viz.* Cu(II), Ni(II), Co(II), Zn(II) and Pb(II). The identical manner was implemented to all the molar ratios of the copolymer samples. The absorption of various metal ions by resin can be determined using this equation.

The rate of metal ion uptake is measured as a percentage of the amount of metal ions taken up after a certain period compared to the time spent in equilibrium. The amount of metal ions taken up in percent at different times is defined as.

$$\text{Percentage of metal ion uptake at different time} = \frac{\text{Amount of metal ion adsorbed}}{\text{Amount of metal ion adsorbed at equilibrium}} \times 100$$

The amount of metal adsorbed by polymer after various time intervals was calculated using this expression. This experiment was performed using a 0.1M metal nitrate solution of Cu^{2+}, Ni^{2+}, Co^{2+}, Zn^{2+} and Pb^{2+}.

Determination of the Distribution of Metal Ions at Different pH

At 25 ^{0}C and in the presence of a 1M $NaNO_3$ solution, the distribution of each of the five metal ions, Cu^{2+}, Ni^{2+}, Co^{2+}, Zn^{2+} and Pb^{2+}, between the copolymer phase and the aqueous phase was determined. The trials were carried out at various pH levels, as indicated above. The following relationship determined the distribution ratio (D):

$$D= \frac{\text{Weight (mg) of metal ion taken up by 1g of polymer}}{\text{Weight (mg) of metal ion present in 1 ml of solution}}$$

If the difference between the actual experiment reading and the blank reading is defined as 'Z,' the amount of metal ion in 2ml of 0.1M metal nitrate solution is 'C' gram. After absorption, 'Y' gram of metal ion was added to 2ml of metal nitrate solution. Metal ion adsorbed (uptake) by the resin

$$=\frac{ZX}{Y} \frac{2}{0.025}=\left(\frac{ZX}{Y}\right)80$$

RESULTS AND DISCUSSION

Elemental Analysis

The amounts of carbon, hydrogen, and nitrogen in the copolymer molecular structures that were made have been looked at. The empirical formula and its weight were used to assign the elemental analysis result that can be seen in Table **1**, and they were found to be in good agreement with both the calculated and observed elemental analysis [24].

Table 1. Elemental Analysis and Empirical Formula of ANBAF copolymer.

Copolymer Resins	% of C observed (Cal.)	% of H observed (Cal.)	% of N observed (Cal.)	% of O observed (Cal.)	%of S Observed (Cal)	Empirical Formula of Repeated Unit	Empirical Formula Weight
ANBAF-I	49.01 (49.59)	4.22 (4.68)	19.20 (19.28)	17.21 (17.63)	8.10 (8.82)	$C_{15}H_{17}O_4N_5S_1$	363

Determination of Molecular Weight

Using gel permeation chromatography, the ANBAF copolymer's average molecular weight was ascertained. The copolymer's molecular weights, measured as the weighted average $(\overline{M_w})$ and number average $(\overline{M_n})$, were discovered to be 3401 and 3990, respectively. $(\overline{M_w})/(\overline{M_n})$ was determined to have a polydispersity index of 1.0032. The copolymer's average molecular weight was found to be 3401. It was discovered that the $(\overline{M_z})/(\overline{M_w})$ polydispersity index was 1.0037. The polydispersity index values for $(\overline{M_w})/(\overline{M_n})$ and $(\overline{M_z})/(\overline{M_w})$ show that (Table 2) the copolymer has a limited molecular weight dispersion. According to the statistics, the ANBAF copolymer has a low molecular weight.

Table 2. Molecular weight of copolymer

Copolymer	Weight Average Molecular Weight $(\overline{M_w})$	Number Average Mn-Molecular Weight	Size Average $(\overline{M_z})$ molecular weight	The Polydispersity Index $(\overline{M_w})/(\overline{M_n})$	The Polydispersity Index $(\overline{M_z})/(\overline{M_w})$
ANBOF	3401	3990	3411	1.0032	1.0029

FTIR Spectra

The ANBAF copolymer's FTIR spectra are displayed in Fig. (2), and the information from the literature is used to compute the vibration frequency. One possible explanation for the stretching vibration of the -NH group is a broad, strong band seen at about 3290 cm^{-1} [25]. The presence of nitro produces a strong and noticeable peak at 2970 cm^{-1}. The sharp and weak band at 2939 cm^{-1} may be explained by the -NH- in the adipamide molecule. The strong band at 1628 cm^{-1} might be caused by the stretching vibration of the Ar-CO- group [32]. The presence of the aromatic group >C=C is shown by a distinct band at 1482 cm^{-1}. The appearance of a strong and distinct band at 1373 cm^{-1} indicates that the copolymer chain has a -CH$_2$-methylene bridge [26]. A faint band that appears in the 905-860 cm^{-1} area indicates the presence of 1, 2, 3, 4, and 5-pentasubstituted aromatic rings [33 - 35].

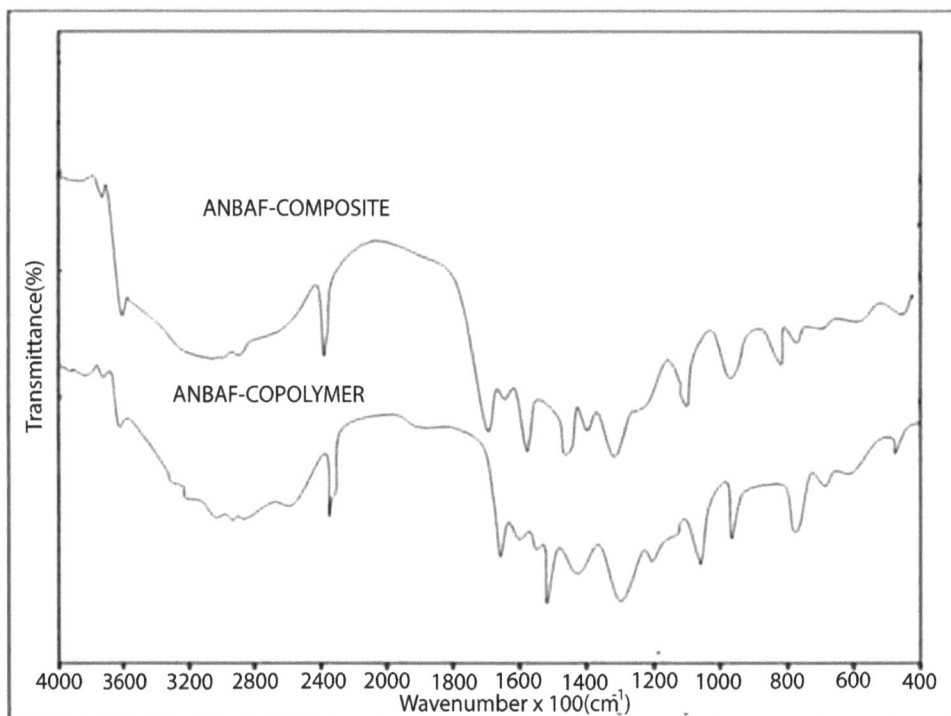

Fig. (2). FTIR Spectra of Copolymer ANBAF and its Nanocomposites.

In Fig. (**2**), the composite's FTIR spectrum is also displayed. According to the figure, the copolymer's -NH bridge was the cause of the particular adsorption band at 3320 cm^{-1}. It is possible to explain the band's adsorption at around 2851.95 to the copolymer's aromatic ring stretching (C–H). The copolymer's -CH$_2$ stretching is shown by a band at 2800 cm^{-1}.

The aforementioned information confirms that the copolymer adheres to the charcoal to form a composite.

^{1}H NMR Spectra

The ^{1}H NMR spectra of the ANBAF copolymer are displayed in Fig. (**3**), containing the spectral data. As a result of each of them having a distinct proton environment for its set of protons, the spectrum displays a distinctive pattern of peaks. Weak multiplicity signals (unsymmetrical pattern) in the range of 7.1 ppm (Ar-H) are caused by aromatic proton [27]. At 6.4 ppm, there is a significant singlet signal caused by the triazole proton. The copolymer chain's amido proton Ar-CH$_2$-NH- could be the cause of the medium triplet signal at 7.6 ppm. To determine the proton of the methylene bond Ar-CH$_2$-NH-, doublet signals in the 2.6 ppm range can be utilized [28]. Four peaks in the 3.0 ppm range suggest that the methylene proton is present.

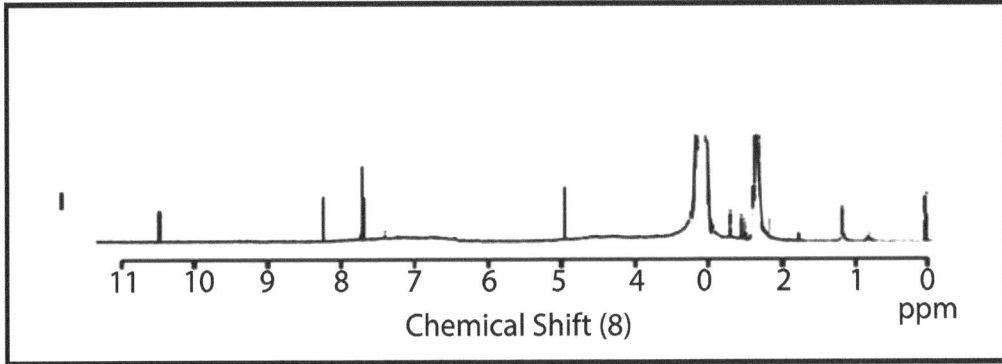

Fig. (3). ^{1}H NMR Spectra of Copolymer ANBAF.

Scanning Electron Microscopy (SEM)

A scanning electron microscope operating at a maximum magnification was used to examine the surface characteristics of the ANBAF copolymer; the results are shown in Fig. (**4**). It provides information about surface topography and internal structural flaws. The copolymer under investigation had a hue that seemed to be dark brown. When magnification is reduced, it indicates spherules, in which the crystals are arranged extremely densely in a smaller surface area. It implies that the resin is crystalline and that this item has a poor ability to exchange ions for longer, better-hydrated metallic ions. Higher magnification shows more amorphous characteristics with a surface that is significantly less densely packed and has deep crevices. The resin's amorphous structure demonstrates improved metallic ion exchange capacity. Resin is hence both crystalline and amorphous, or transitional between showing much less or more exact ion alternate ability [29, 30].

Fig. (**4**) displays the copolymer nanocomposite's SEM picture. The composite contains a wide variety of pores with a more amorphous structure. The photo demonstrates fresh active site adhesive on the sputter cluster surface, indicating that the copolymer and activated charcoal have been composited. It is discovered that the composite's surface area increases significantly more than the copolymers or the metal ion uptake.

Fig. (4). SEM Images of Copolymer ANBAF and its Nanocomposites.

ION EXCHANGE STUDIES

Effect of Metal Ion Uptake in Different Electrolytes with Variation in Concentrations

Ion exchange was used to compare the effects of the quantity of metal ion uptake of the ANBAF copolymer and the copolymer/activated charcoal composite. It investigated how various electrolyte solutions, such as Cl^-, NO_3^-, and ClO_4^{2-}, affected the interactions between various metal ions and the adsorbent at concentrations of 0.1, 0.5, and 1.0 M. Table **3** presents the findings of the metal ion uptake experiments. When the concentration of the electrolytes was increased for both adsorbents, the sorption capacity was improved. The quantity of metal ions in relation to the accessible surface cavities is lower at lower concentrations.

The amount of heavy metal adsorbed by the ANBAF copolymer and its composite is determined by the electrolyte concentration and type, the pore size's physical structure, a number of physical characteristics, and the diffusion of counter ions. Both the copolymer/charcoal composite and the ANBAF copolymer are impacted by the aforementioned parameters. The quantity of metal ions (Cu^{2+}, Zn^{2+}, Co^{2+}, Pb^{2+}, and Cd^{2+}) absorbed rises as the electrolyte concentration does. Zn^{2+} and Cu^{2+} ions were shown to be more readily absorbed than Co^{2+}, Pb^{2+}, and Cd^{2+} ions, according to the copolymer's ion exchange data [31]. They discovered that the diffusion of big-size metal ions to the copolymer surface cavities takes place at a very sluggish pace, leading to the cation exchange. The Zn^{2+} readily diffuses due to its tiny size. The heavy metal adsorbed by the ANBAF copolymer through the accessible cavity is dependent on the diffusion of counterions and the distribution of pore size. The interionic force of attraction determines the value of the cation exchange capacity. Pb2+ and Cd^{2+} have a better selectivity of metal ion uptake for copolymer composite. The adsorption capacity was mostly determined by the metal ion's outermost electronic configuration, hydrated ionic radius, and ionic charges.

Table 3. Evaluation of Influence of Different Electrolytes on the Uptake of Several Metal Ions of ANBAF Copolymer and its Nanocomposites

Metal ions	Conc.	Weight of the metal uptake in the presence of electrolyte (m.mol/g)							
		$NaClO_4$		NaCl		$NaNO_3$		Na_2SO_4	
		Copolymer	Composite	Copolymer	Composite	Copolymer	Composite	Copolymer	Composite
Cu^{2+}	0.01	1.35	1.86	1.1	1.47	1.46	1.69	3.28	3.63
	0.05	1.88	2.14	1.66	2.12	1.7	1.95	2.54	2.95
	0.1	2.24	2.65	2.63	2.74	2.14	2.25	1.73	2.12
	0.5	2.79	3.3	3.03	3.12	2.58	2.83	1.22	1.43
	1	3.9	4.16	3.3	3.33	3.14	3.66	0.6	0.83
Ni^{2+}	0.01	1.43	1.79	1.13	1.24	1.04	1.14	2.86	2.99
	0.05	1.84	2.26	1.25	1.38	1.47	1.82	2.34	2.53
	0.1	2.42	3.06	1.34	1.46	1.84	2.12	1.59	1.75
	0.5	2.85	3.18	2.55	2.76	2.57	2.8	1.35	1.44
	1	3.32	3.55	2.92	3.12	3.19	3.39	0.62	0.84
Co^{2+}	0.01	1.62	1.75	1.62	1.79	1.8	1.94	1.65	1.76
	0.05	1.5	1.59	1.29	1.37	1.55	1.75	1.38	1.53
	0.1	1.23	1.32	1.14	1.25	1.28	1.39	1.2	1.28
	0.5	0.82	0.98	0.86	0.93	1.02	1.15	0.93	0.99
	1	0.58	0.73	0.63	0.8	0.73	0.86	0.85	0.92
Zn^{2+}	0.01	2.03	2.15	1.74	1.82	2.24	2.42	1.88	2.04
	0.05	1.64	1.88	1.43	1.57	1.84	2.05	1.56	1.76
	0.1	1.34	1.44	1.15	1.24	1.48	1.74	1.26	1.45
	0.5	1.08	1.16	0.79	0.87	0.93	1.56	0.99	1.13
	1	0.57	0.68	0.3	0.59	0.72	0.85	0.77	0.9

(Table 3) cont.....

Metal ions	Conc.	Weight of the metal uptake in the presence of electrolyte (m.mol/g)							
		$NaClO_4$		NaCl		$NaNO_3$		Na_2SO_4	
		Copolymer	Composite	Copolymer	Composite	Copolymer	Composite	Copolymer	Composite
Pb^{2+}	0.01	1.5	1.69	1.48	1.79	1.74	1.83	1.86	1.95
	0.05	1.18	1.4	1.24	1.44	1.63	1.74	1.67	1.72
	0.1	0.97	1.17	1.09	1.18	1.24	1.37	1.33	1.39
	0.5	0.73	0.9	0.83	0.95	0.98	1.06	1.13	1.09
	1	0.47	0.65	0.4	0.53	0.86	0.95	0.73	0.76

[a]$[M(NO_3)_2] = 0.1$ mol/l; Volume = 2 ml; Volume of electrolyte solution: 25 ml
Weight of resin = 25 mg; time: 24 h: Room temperature.

The outermost electronic configuration and the hydrated ionic radius of Pb^{2+} were dissimilar from those of Cd^{2+}, Cu^{2+}, $Co^{2+,}$ and Zn^{2+}. Furthermore, in comparison to other metal ions, Pb^{2+} had a greater hydrated ionic radius. These lead to the conclusion that, in comparison to the ANBAF copolymer, there is a greater removal of Pb^{2+} metal ions onto the composites [32]. However, because of the ionic radius of Cu^{2+} (ionic radius 0.79 Å), which is larger than that of Zn^{2+} and Co^{2+} because of its intermediate behavior of high electro-negativity, Cu^{2+} metal ion absorption is higher than Zn^{2+} and Co^{2+}.

According to the findings, the surface adsorbent's loading capacity caused less Cu^{2+} and Co^{2+} to be adsorbed at the beginning concentration. The surface of the copolymer and activated charcoal is more porous and amorphous. As a result, they may readily accept the metal ions into the cavities depending on the particular particle size; that is, as the particle size grows, so does the surface area. As a result, the composite has a larger metal ion uptake capacity than the copolymer when compared to it. $Cu^{2+} > Ni^{2+} > Co^{2+} > Zn^{2+} > Pb^{2+}$ is the order of selectivity for metal ion uptake for the copolymer, while $Pb^{2+} > Cd^{2+} > Cu^{2+} > Co^{2+} > Zn^{2+}$ is the order for the copolymer composite. Additionally, it has been discovered that the copolymer/activated charcoal composite has a greater metal-binding property than the other previously reported [33 - 35].

Analysis of Metal Ions at Different pH

The results shown in Figs. (**5** and **6**) demonstrate how pH affects the quantity of metal ions that are absorbed between the copolymer and its composite phases as well as liquid phases. The resultant produced shows that relative to a rise in medium pH, the copolymer and its composite absorb more metal ions. The graph suggests that the greater concentration and mobility of H+ ions may be connected to the lowest adsorption seen at low pH. The adsorbent's surface is intimately linked to hydronium ions (H_3O^+) at low pH levels, which would decrease the active site.

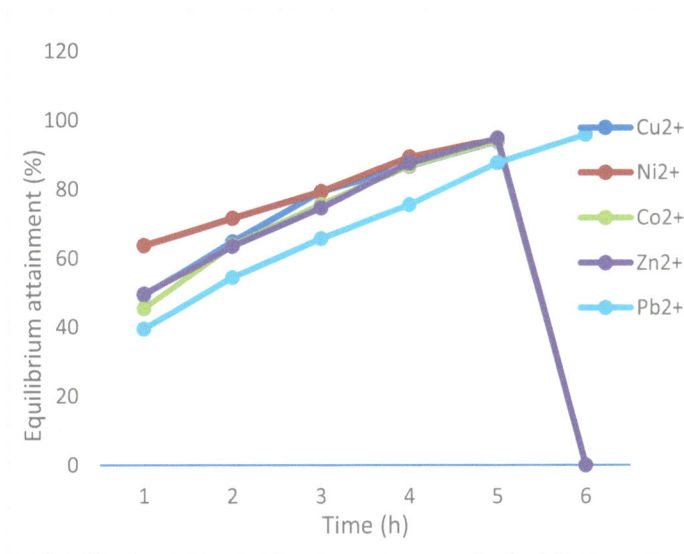

Fig. (5). Rate of metal ion uptake by ANBAF copolymer.

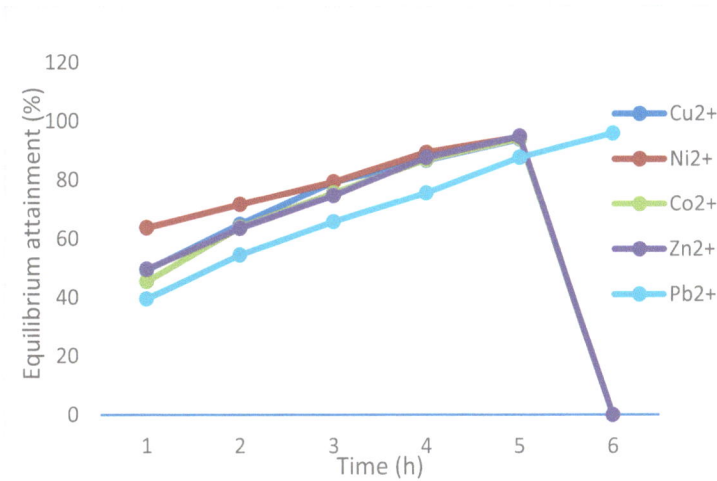

Fig. (6). Rate of metal ion uptake by ANBAF copolymer composites.

On the other hand, the adsorption of metal ions increased with increasing pH. Due to the stability of the complex produced during the adsorption process, Zn^{2+} and Cu^{2+} metal ion uptake in copolymers is highly high when compared to other metal ions. Greater metal ion absorption is seen for Zn^{2+} and Cu^{2+}, which may be related to greater copolymer and metal complex stability constants (Tables **4** and **5**). Pb^{2+} and Cd^{2+} have the maximum metal ion uptake in copolymer composite. It results from the composite's electrostatic interaction with cations and negative surface charges throughout a wide pH range [36 - 38].

Table 4. Comparison of the Rate of Metal Ion Uptake[b] of ANBAF copolymer.

Metal ions	pH	Percentage of the amount of metal ion[a] taken up[b] at different times (hrs)					
		1	2	3	4	5	6
Cu^{2+}	4.5	46.4	59.6	73.9	83.8	92.6	-
Ni^{2+}	4.5	52.4	68.4	75.8	83.8	92.2	-
Co^{2+}	5	42.6	59.6	73.3	84.5	92.6	-
Zn^{2+}	5	46.3	59.4	72.9	83.4	92.3	-
Pb^{2+}	6	37.3	52.5	62.5	73.4	85.8	93.5

Table 5. Comparison of the rate of metal ion uptake[b] of ANBAF copolymer composite.

Metal ions	pH	Percentage of the amount of metal ion[a] taken up[b] at different times (hrs)					
		1	2	3	4	5	6
Cu^{2+}	4.5	49.3	64.8	79.3	86.6	93.8	-
Ni^{2+}	4.5	63.7	71.6	79.3	89.3	94.6	-
Co^{2+}	5	45.4	63.8	75.6	86.8	94.0	-
Zn^{2+}	5	49.6	63.5	74.6	87.6	94.8	-
Pb^{2+}	6	39.5	54.4	65.7	75.5	87.6	95.8

[a] $[M(NO_3)_2]= 0.1$ mol/l; volume: 2ml; $NaNO_3 = 1.0$ mol/l; volume: 25ml, Room temperature.
[b] Metal ion uptake = (Amount of metal ion absorbed x 100) / amount of metal ion absorbed at equilibrium.

Analysis of Metal Ion Uptake in Different Electrolytes with the Variation in Rate

The copolymer and copolymer/charcoal composite are compared to an additional crucial parameter: the influence of rate for the various metal ions, such as Cu^{2+}, Zn^{2+}, Co^{2+}, $Pb^{2+,}$ and Cd^{2+}. Tables 6 and 7, as well as Fig. (7), display the results. The metal ion uptake rate that takes the lowest amount of time to reach the near equilibrium condition during the ion exchange process was found to be the rate of metal ion adsorption. Based on the obtained data, 4 hours is the least amount of time needed to reach equilibrium for Zn^{2+} ion uptake. The difference in the metal ions' ionic radii, their affinity for different active groups on the ANBAF adsorbent, and the makeup of the metal ions' anions in their salt have all been linked to variations in the rate of adsorption of heavy metal ions. Kitchener, Greger, and colleagues discovered that a large-sized metal ion diffuses slowly within polymer lattices. Due to the effects of various electrolytes and adsorbents, Pb^{2+} and Cd^{2+} metal ions in the copolymer composite took 4 hours to approach equilibrium. Moreover, the smaller size metal ions are readily adsorbed. Larger and bulkier metal ions behave differently from Pb^{2+} ions in that they migrate more slowly into aqueous solutions due to their reduced hydration. The low of mass action provides an explanation for the rapid pace of exchange [38].

Table 6. Distribution Ratio 'D'a of Different Metal Ions[b] as a Function of Different pH of ANBAF copolymer

Metal Ions	Distribution Ratios of Different Metal Ions at Different pH							
	1.5	2	2.5	3	3.5	4	5	6
Cu^{2+}	-	-	71.3	77.4	93.6	126.3	367.4	936.7
Ni^{2+}	-	-	64.4	69.3	82.2	141.5	287.8	649.5
Co^{2+}	-	-	37.4	59.6	84.7	121.7	214.4	323.6
Zn^{2+}	-	-	35.3	53.6	74.8	86.7	129.9	181.4
Pb^{2+}	-	-	29.5	51.4	71.6	93.8	121.5	186.3

Table 7. Distribution Ratio 'D' of Different Metal Ions[b] as a Function of Different pH of ANBAF Copolymer Composites

Metal ions	Distribution ratios of different metal ions at different pH							
	1.5	2	2.5	3	3.5	4	5	6
Cu^{2+}	-	-	77.0	86.6	98.8	163.8	452.3	1213.3
Ni^{2+}	-	-	68.5	74.5	93.5	167.4	334.6	756.4
Co^{2+}	-	-	41.2	67.4	92.5	134.5	231.6	368.5
Zn^{2+}	-	-	37.7	56.7	77.4	89.5	143.6	246.3
Pb^{2+}	-	-	34.4	56.5	76.6	97.2	128.3	223.6

[a] D = weight (in mg) of metal ions taken up by 1g of copolymer/weight (in mg) of metal ions present in 1 ml of solution.
[b] $[M(NO_3)_2]$= 0.1 mol/l; volume: 2ml; $NaNO_3$ = 1.0 mol/l; volume: 25 ml, time 24h (equilibrium state) at room temperature.

According to the data above, $Zn^{2+} > Cu^{2+} > Co^{2+} > Pb^{2+} > Cd^{2+}$ is the order of metal ion absorption rate by the ANBAF copolymer, and $Pb^{2+} > Cd^{2+} > Cu^{2+} > Co^{2+} > Zn^{2+}$ is the rate of metal ion uptake by its composite.

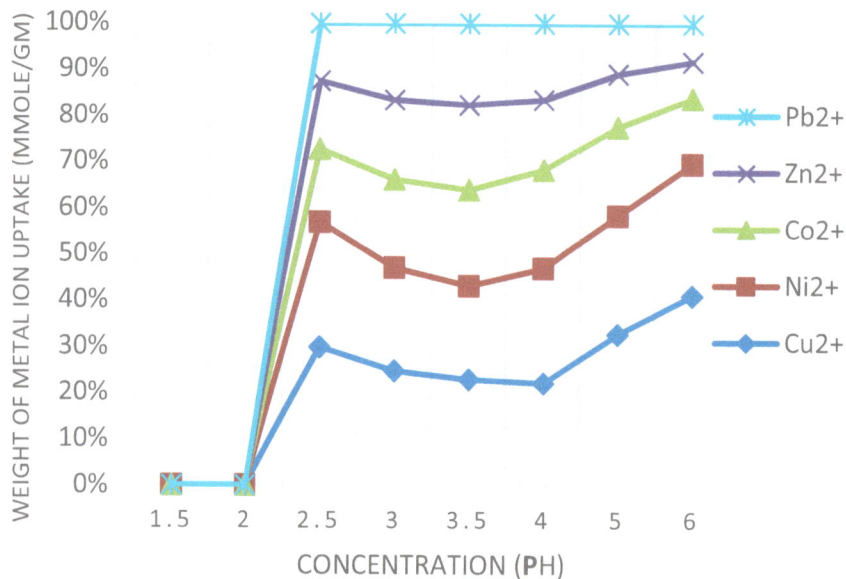

Fig. (7). Metal ion uptake at different pH by copolymer.

CONCLUSION

This study outlines the latest developments in synthetic techniques for the production of copolymer nanocomposites and highlights their potential uses in environmental remediation. In general, there will be a significant increase in the potential applications of copolymer nanocomposites, such as conductive polymer-based nanocomposites, as innovative adsorbents for environmental remediation in the future. For the sake of ion-exchange research, a unique comparison between the copolymer and its composite was made. The spectroscopic, morphological, and elemental analyses verified the copolymer's and its composite's structure and characteristics. The copolymer and its composite, which are more porous and amorphous in nature, can function as an efficient metal ion adsorbent, according to the ion-exchange data. Because of its larger surface area for the formation of new active sites, the copolymer composite is proven to be a better metal ion exchanger than the copolymer. SEM pictures corroborate this again. The outcomes demonstrated that the composite's metal ion is Pb^{2+} and Cd^{2+} due to the distinct metal ion selectivity of the copolymer for Zn^{2+} and Cu^{2+}.

REFERENCES

[1]   Ali SW, Waqar F, Malik MA, Yasin T, Muhammad B. Study on the synthesis of a macroporous ethylacrylate-divinylbenzene copolymer, its conversion into a bi-functional cation exchange resin and applications for extraction of toxic heavy metals from wastewater. J Appl Polym Sci 2013; 129(4): 2234-43.
[http://dx.doi.org/10.1002/app.38940]

[2]   Wang X, Wang A. Adsorption characteristics of chitosan-g-poly (acrylic acid)/attapulgite hydrogel composite for Hg(II) ions from aqueous solution. Sep Sci Technol 2010; 45(14): 2086-94.
[http://dx.doi.org/10.1080/01496395.2010.504436]

[3]   Valle H, Sanchez J, Rivas BL. Poly (N-vinylpyrrolidine-co-2-methylpropanesulfonate sodium): synthesis, characterization and its potential application for the removal of metal ions from aqueous solution. J Appl Polym Sci 2014.

[4]   Gurnule WB, Katkamwar SS. Analytical applications of newly synthesized copolymer resin derived from *p* -aminophenol, dithiooxamide, and formaldehyde. J Appl Polym Sci 2012; 123(3): 1421-7.
[http://dx.doi.org/10.1002/app.33726]

[5]   Gupta R, Singh R, Dubey S. Removal of mercury ions from aqueous solutions by composite of polyaniline with polystyrene. Separ Purif Tech 2004; 38(3): 225-32.
[http://dx.doi.org/10.1016/j.seppur.2003.11.009]

[6]   Velmurugan G, Ahamed KR, Azarudeen RS. A novel comparative study: synthesis, characterization and thermal degradation kinetics of a terpolymer and its composite for the removal of heavy metals. Iran Polym J 2015; 24(3): 229-42.
[http://dx.doi.org/10.1007/s13726-015-0315-6]

[7]   Hydari S, Sharififard H, Nabavinia M, Parvizi M. A comparative investigation on removal performances of commercial activated carbon, chitosan biosorbent and chitosan/activated carbon composite for cadmium. Chem Eng J 2012; 193-194: 276-82.
[http://dx.doi.org/10.1016/j.cej.2012.04.057]

[8]   Siraj S, Islam MM, Das PC, *et al.* Removal of chromium from tannery effluent using chitosan–charcoal composite. Journal of the Bangladesh Chemical Society 2012; 25(1): 53-61.
[http://dx.doi.org/10.3329/jbcs.v25i1.11774]

[9]   Yanagisawa H, Matsumoto Y, Machida M. Adsorption of Zn(II) and Cd(II) ions onto magnesium and activated carbon composite in aqueous solution. Appl Surf Sci 2010; 256(6): 1619-23.
[http://dx.doi.org/10.1016/j.apsusc.2009.10.010]

[10]   Rahangdale SS. Synthesis and biological activity of o-Cresol-Adipamide-Formaldehyde copolymer resin. Arch Appl Sci Res 2012; 4: 2280-8.

[11]   Gurnule WB, Patle DB. Study of chelation ion-exchange properties of new copolymer resin derived from o-aminophenol, urea and formaldehyde. Arch Appl Sci Res 2010; 2: 261-76.

[12]   Singru RN, Gurnule WB, Khati VA, Zade AB, Dontulwar JR. Eco-friendly application of p-cresol–melamine–formaldehyde polymer resin as an ion-exchanger and its electrical and thermal study. Desalination 2010; 263(1-3): 200-10.
[http://dx.doi.org/10.1016/j.desal.2010.06.060]

[13]   Kırcı S, Gülfen M, Aydın AO. Separation and Recovery of Silver(I) Ions from Base Metal Ions by Thiourea- or Urea-Formaldehyde Chelating Resin. Sep Sci Technol 2009; 44(8): 1869-83.
[http://dx.doi.org/10.1080/01496390902885163]

[14]   Das NC, Rathod YU, Pandit VU, Gurnule WB. Studies of chelation ion-exchange properties of copolymer resin derived from 1,5-diaminonaphthalene, 2,4-dihydroxy- propiophenone and formaldehyde. Mater Today Proc 2022; 53: 80-5.
[http://dx.doi.org/10.1016/j.matpr.2021.12.370]

[15] Mandavgade DSK, Gurnule DWB. Synthesis and chelate ion exchange properties of copolymer resin: 8-hydroxyquinoline-5-sulphonic acid-catechol-formaldehyde. Mater Today Proc 2022; 60: 1814-8. [http://dx.doi.org/10.1016/j.matpr.2021.12.494]

[16] Gurnule WB, Vajpai K, Belsare AD. Selective removal of toxic metal ions from waste water using polymeric resin and its composite. Mater Today Proc 2021; 36: 642-8. [http://dx.doi.org/10.1016/j.matpr.2020.04.372]

[17] Rahangdale SS, Gurnule WB. Chelation ion-exchange properties of copolymer resin derived from 2, 2'-dihydroxybiphenyl, biuret, and formaldehyde. Desalination Water Treat 2015; 55(7): 1806-15. [http://dx.doi.org/10.1080/19443994.2014.931533]

[18] Singru RN, Gurnule WB. Sorption behavior of ion-exchange terpolymer resin with environmental impact: synthesis, characterization and isotherm models. Iran Polym J 2010; 19: 169-89.

[19] Khobragade JV, Gurnule WB. Removal of Toxic Metal Ions Using Ion-Binding Copolymer Resin by Batch Equilibrium Techniques. Int J Res Biosci Agric Technol 2017; 7: 519-26.

[20] Riswan Ahamed MA, Azarudeen RS, Subha R, Burkanudeen AR. Sorption behavior of ion-exchange terpolymer resin with environmental impact: synthesis, characterization and isotherm models. Polym Bull 2014; 71(12): 3209-35. [http://dx.doi.org/10.1007/s00289-014-1246-7]

[21] Karthikeyan M, Kumar KKS, Elango KP. Batch sorption studies on the removal of fluoride ions from water using eco-friendly conducting polymer/bio-polymer composites. Desalination 2011; 267(1): 49-56. [http://dx.doi.org/10.1016/j.desal.2010.09.005]

[22] Mane VD, Wahane NJ, Gurnule WB. Copolymer resin. VII. 8-hydroxyquinoline-5-sulfonic acid–thiourea–formaldehyde copolymer resins and their ion-exchange properties. J Appl Polym Sci 2009; 111(6): 3039-49. [http://dx.doi.org/10.1002/app.29369]

[23] Burkanudeen AR, Azarudeen RS, Ahamed MAR, Gurnule WB. Kinetics of thermal decomposition and antimicrobial screening of terpolymer resins. Polym Bull 2011; 67(8): 1553-68. [http://dx.doi.org/10.1007/s00289-011-0497-9]

[24] Gupta PM, Rathod YU, Pandit VU, Gupta RH, Gurnule WB. Non-Isothermal Decomposition Study of Copolymer Derived from 2-Amino 6-nitrobenzothiazole. Melamine, and Formaldehyde Mat Tod Proc 2022; pp. 53101-6.

[25] Rathod YU, Pandit VU, Bhagat DS, Gurnule WB. Synthesis of copolymer and its composites with carbon and their photoluminescence studies. Mater Today Proc 2022; 53: 123-9. [http://dx.doi.org/10.1016/j.matpr.2021.12.422]

[26] Gupta RH, Gurnule WB, Mishra P. Electrical Conductance Studies of o-Aminophenol - Melamine-Formaldehyde Copolymer-II. Journal of Bionano Frontier 2017; 1: 10.

[27] Gupta RH, Gupta PG, Gurnule WB. Electrical Conductance Properties Of Copolymer Derived From 2-Hydroxyacetophenone- Melamine- Formaldehyde International Journal of Current Engineering and Scientific Research 2019; 6(101): 6101-107.

[28] Rathod YU, Zanje SB, Gurnule WB. Hydroxyquinoline copolymers synthesis, characterization and thermal degradation studies. J Phys Conf Ser 2021; 1913(1): 012061. [http://dx.doi.org/10.1088/1742-6596/1913/1/012061]

[29] Bisen VR, Gurnule WB. Studies of Retension and Reusable Capacities of Semicarbazide Formaldehyde Based Copolymer Against Some Toxic Metal Ions by Batch Equilibrium Technique. Int J Curr Engg Sci Res 2019; 6: 1309-16.

[30] Gurnule WB, Gupta PG, Gupta RH, Rathod YU, Singh NB. Thermal degradation studies of 2-amino 6-nitrobenzothiazole-oxamide-formaldehyde copolymer and its composites. IOP Conf Ser Earth

Environ Sci 2023; 1281(1): 012026-32.
[http://dx.doi.org/10.1088/1755-1315/1281/1/012026]

[31] Tahoon, Mohamed A., Saifeldin M. Siddeeg, Norah Salem Alsaiari, Wissem Mnif, and Faouzi Ben Rebah. 2020. "Effective Heavy Metals Removal from Water Using Nanomaterials: A Review" Processes 8, no. 6: 645.
[http://dx.doi.org/10.3390/pr8060645]

[32] Azarudeen RS, Riswan Ahamed MA, Subha R, Burkanudeen AR. Heavy and toxic metal ion removal by a novel polymeric ion-exchanger: synthesis, characterization, kinetics and equilibrium studies. J Chem Technol Biotechnol 2015; 90(12): 2170-9.
[http://dx.doi.org/10.1002/jctb.4528]

[33] Azarudeen RS, Riswan Ahamed MA, Thirumarimurugan M, Prabu N, Jeyakumar D. Synthetic functionalized terpolymeric resin for the removal of hazardous metal ions: synthesis, characterization and batch separation analysis. Polym Adv Technol 2016; 27(2): 235-44.
[http://dx.doi.org/10.1002/pat.3626]

[34] Patle DB, Gurnule WB. An eco-friendly synthesis, characterization, morphology and ion exchange properties of terpolymer resin derived from p-hydroxybenzaldehyde. Arab J Chem 2016; 9: S648-58.
[http://dx.doi.org/10.1016/j.arabjc.2011.07.013]

[35] Mohurle R, Gurnule WB. Synthesis, Characterization and Ion-exchange properties of 2-Amino 6-Nitro benzothiazole-Adipamide-Formaldehyde Resin. International Journal in Physical and Applied Science 2018; 5(12): 33-47.

[36] Dengane R, Khobragade JV, Gurnule WB. Ion-exchange properties of copolymer resins derived from phthalic acid, thiosemicarbazide, and formaldehyde Int J of Researches in Biosciences, Agricultural and Technology 2023; II(XI): 256-66.

[37] Das NC, Gurnule WB. Electrical conductivity of newly synthesized copolymer resin-IV from 2,4-dihydroxypropiophenone, 1,5-diaminonapthalene and formaldehyde International J of Advanced Research in Science, Communication and Technology 2023; 3(2): 345-51.

[38] Rahangdale SS, Rahangadale MK, Gurnule WB. Chelation Ion exchange studies of copolymer resin from o-toluidine, biuret and formaldehyde. IJFANS UGC Care 2022; 11(13): 214-20.

CHAPTER 6

Biochar-Based Nanocomposites for Environmental Remediation

Tanisha Kathuria[1], Anjali Mehta[1], Sudhanshu Sharma[1,*] and Sudesh Kumar[2,*]

[1] *Department of Chemistry, Banasthali Vidyapith, Rajasthan 304022, India*

[2] *Department of Chemistry, National Council of Educational Research and Training (NCERT), New Delhi 110016, India*

Abstract: Biochar (BC) stands out as a remarkable material in the domain of environmental remediation. Waste biomass, like municipal solid waste, manure, wood chips, and agricultural residues, undergo pyrolysis in a controlled oxygen environment, producing BC with a carbonaceous composition ranging from 65% to 90%. Using metal nanoparticles (MNPs) to reinforce BC significantly increases its novelty. The synergistic advantages of BC, coupled with the enhanced catalytic activity of NPs, enhance physicochemical characteristics such as thermal stability, ideal pore size, surface area, and versatile functionalization. These attributes contribute to effectively addressing emerging environmental pollution challenges and their remediation. There are three major hazards in industrial wastewater: dyes, heavy metals, and pharmaceutical compounds. Thus, BC-based Nanocomposites (BNCs) are being investigated as a potential solution for wastewater pollution treatment that uses both adsorption and photocatalytic degradation. As a result of these composites, four integrated objectives can be achieved: the removal of pollutants, waste management, carbon sequestration, and energy production. It stands as a superior choice to conventional methods, marked by cost-effectiveness, sustainability, and environmental friendliness. This chapter provides a comprehensive insight into BC-based composite with precise preparation techniques, efficacy in eliminating pollutants, and underlying adsorption processes.

Keywords: Agricultural waste, Biomass, Biochar, Environmental remediation, Nanocomposites.

INTRODUCTION

Environmental pollution is an increasing global concern that requires innovative solutions to mitigate its impacts. One promising approach is the production of BC-based NCs (BNCs) for environmental remediation. The carbon-rich material known as BC, which is produced by pyrolyzing biomass, has drawn interest due

---

* **Corresponding author Sudesh Kumar:** Department of Chemistry, National Council of Educational Research and Training (NCERT), New Delhi 110016, India; E-mail: sudeshneyol@gmail.com

to its large porosity, surface area, and capacity to absorb contaminants. In BC production, biomass feedstock, such as lignocellulosic materials and agricultural residues, is pyrolyzed or gasified under controlled conditions without oxygen to provide combustion [1]. A diverse range of organic waste materials, including agricultural residues like cornstalks, cotton stalks, and wheat straws, garden waste such as leaves, grass, and wood, municipal waste like food waste, tire waste, and sludge, as well as different types of algae such as *Enteromorpha prolifera*, blue algae, and green algae, can be utilized as feedstock for BC production [2]. Its production can utilize a variety of thermochemical methods, including slow pyrolysis, fast pyrolysis, gasification, hydrothermal carbonization, torrefaction, and rectification, each offering distinct advantages and applications in sustainable biomass conversion [3]. Slow pyrolysis, conducted without oxygen, is a common method for producing BC, typically yielding about 35% [4] dry biomass and a minor amount of bio-oil [5]. While gasification is mostly utilized for the manufacture of syngas, which, as a result, produces heat and energy, fast pyrolysis is the favored method for producing biofuel [6]. Additionally, studies have shown that lignocellulosic materials yield a higher amount of BC compared to municipal solid waste [7]. Furthermore, its production stands out for its economic viability, environmental friendliness, and resource efficiency, requiring minimal energy input and achievable at temperatures below 700°C, making it a sustainable solution for waste utilization [8]. It is characterized by its significant specific surface area (SSA) and wide range of functional groups, such as hydroxyl, amino, and carboxyl groups. These properties contribute to its versatility and suitability for various applications across different industries [9]. There are multiple applications of BC across various industries, for example, wastewater treatment procedures, electrochemistry for energy storage devices, agriculture for soil amendment, catalysis for various chemical reactions, and so on, as illustrated in Fig. (**1**). Recent research has explored BC-based adsorbents for aqueous contaminant elimination, offering a synergistic approach to water pollution control and carbon sequestration [10].

The effectiveness of BC in water treatment is hindered by its restricted functionalities derived from the feedstock post-pyrolysis, which constrains its wider applications in this field. Furthermore, in aqueous solutions, powdered BC is difficult to separate because of its tiny particle size [11]. The development of "engineered BC" is an expanding area of research focused on enhancing BC's capabilities for multiple applications. The immobilization of materials at the nanoscale on BC is gaining significant scientific interest and attention. Functional NPs incorporated into BC can enhance its surface area, thermal stability, cation exchange capacity, and high-affinity adsorption sites. BC-based composite materials can be adapted to target many pollutant kinds, both positively and negatively charged. This involves selectively designing or producing composite materials by incorporating specific magnetic substances, functional materials, and NPs. Using these composite materials for environmental remediation can compensate for deficiencies found in pristine BCs in terms of functional groups [12, 13]. Engineered BC containing immobilized NPs has multiple potential applications, one of which is as a sorbent in environmental remediation.

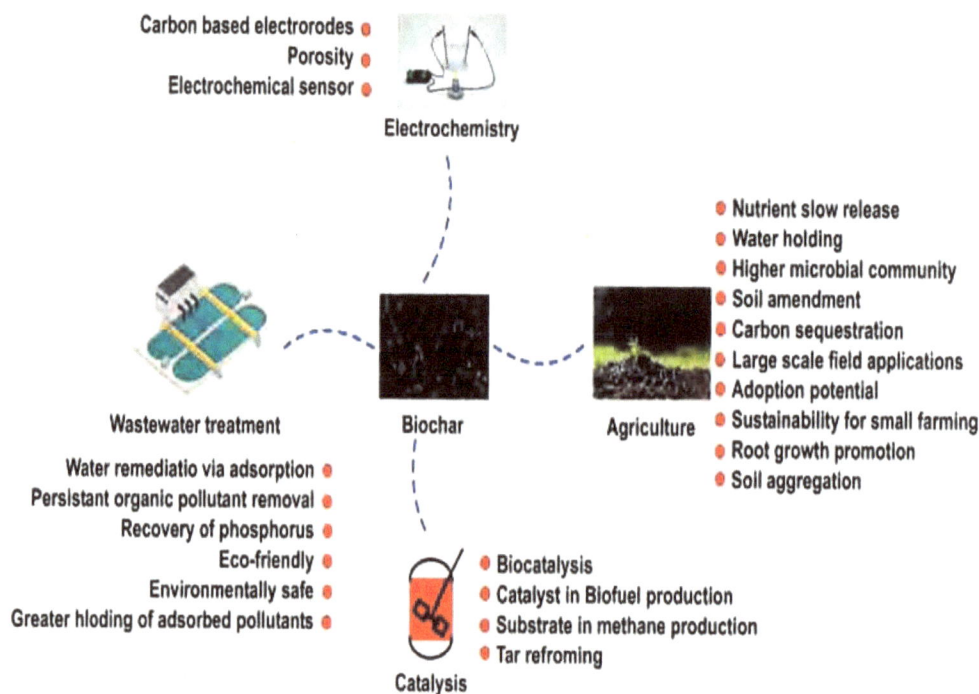

Fig. (1). Various Applications of BC [17].

Consequently, MNPs have been used as reductants, adsorbents, oxidants, and catalysts for the elimination of organic contaminants, diverse heavy metals, as well as other inorganic contaminants from aqueous solutions [14, 15]. However, MNPs tend to agglomerate and form larger particles due to their increased surface energy [16]. Activated carbon, BC, and silica porous supports have been employed to reduce the agglomeration of metal NPs. Thus, the immobilization of MNPs onto BC offers a novel approach to combine their advantages and overcome their drawbacks, creating a promising material called MNPs@BC for ecological restoration. Moreover, BC has demonstrated outstanding adsorption capabilities for various contaminants such as organic pollutants, dyes, heavy metals, pesticides, and volatile organic compounds (VOCs) in water. The chapter concludes with a concise summary and outlook on BC-based materials across diverse applications. In this study, BC-based materials are investigated for contamination and their application to pollution control.

BC COMPOSITES

Due to the constraints of using pure BC for environmental remediation, researchers have conducted numerous studies to develop BC composite materials. These investigations aim to enhance the physicochemical properties of BCs and create new composites that integrate the strengths of BCs with other materials [18]. A novel metal NPs incorporated BC (MNPs@BC) composite is created by immobilizing metal NPs on BC, synergizing their advantages and overcoming their limitations for environmental remediation. As a result of the synthesis of MNPs@BC, both BC and metal NPs are effectively mitigated [19]. It reduces the leaching, aggregation, and surface passivation of metal NPs by dispersing and stabilizing metal NPs at BC [20]. BC performs an essential function in dispersing and anchoring metal NPs, leading to reduced aggregation. Furthermore, enhancing BNC properties involves combining nanomaterial (NM) benefits with various functional groups found in pyrolyzed BC, such as -OH, amino acids, and -COOH. Such functional groups are vital for utilizing BC effectively, especially in the context of wastewater treatment and the removal of diverse pollutants [21]. Thirdly, MNPs@BC can speed up the oxidation of contaminants in the BC-metal interface by improving catalytic/redox performance. Additionally, MNPs@BC can enhance thermal stability and boost the yield of BC. As a result, these novel composites can be highly beneficial to environmental remediation. Based on the NMs incorporated into nano BC, they are classified into three main types:

a. Oxide/hydroxide BC NCs
b. Magnetic BC NCs
c. BC with functional nanoparticle coatings [8].

TYPES AND SYNTHESIS APPROACHES FOR BC NCS

Multiple approaches to the synthesis of BC NCs are depicted in Fig. (2) and described in following subsections.

Nanometal Oxide/Hydroxide-BC Composites

NMs (such as MgO and TiO_2) and BC produce novel composite materials that are capable of enhancing active sites, functional groups on the surface, and catalytic degradation efficiencies. Hence, the development of BC NCs stands as a promising method with extensive potential for use in environmental remediation. Metal salts are commonly selected as chemical reagents for biomass pretreatment. These BC NCs are primarily synthesized using three methods: i) Pyrolysis of the

BC rich in target metal elements, ii) Before pyrolysis treatment of biomass with metal salt, and iii) Immobilization of NPs on BC after pyrolysis. The first two processes primarily involve treating biomass by impregnating the desired metal into the biomass before undergoing pyrolysis. In the third technique, the BC produced by biomass pyrolysis is directly treated with nano-metal oxide or hydroxide [22].

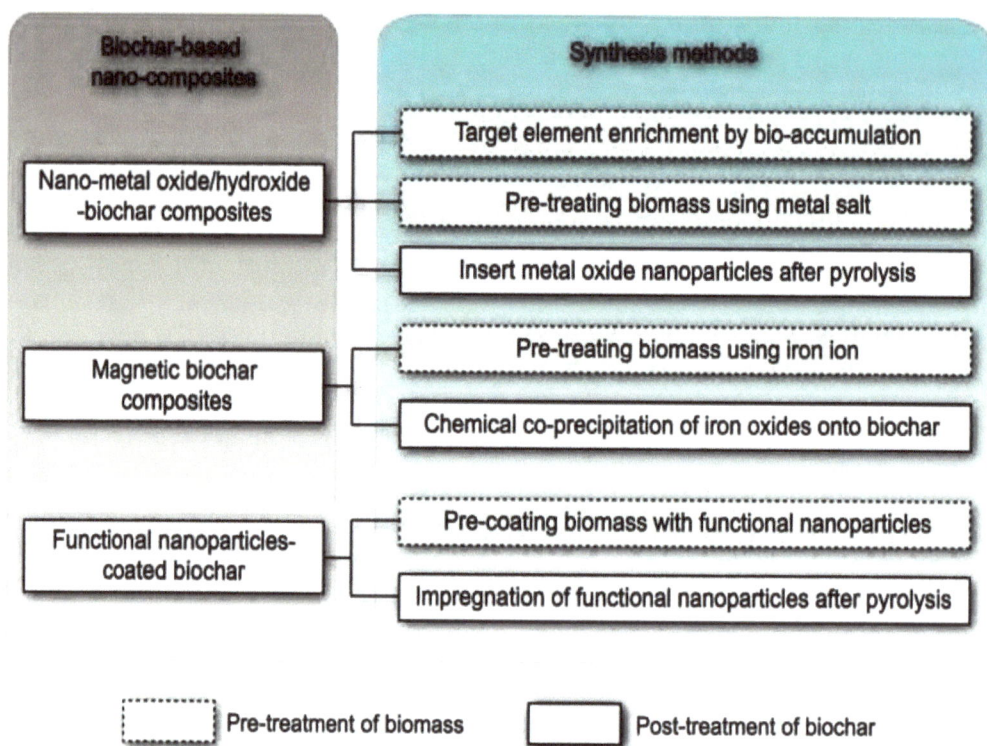

Fig. (2). Different methods for preparing BC-based NCs [22].

Treatment of Biomass with Metal Salt Before Pyrolysis

Using metal salts to pretreat biomass before pyrolysis is a method used to form MNPs@BC. The first step involves impregnating biomass with a range of metal salts, including $ZnCl_2$, $Fe(NO_3)_3$, $FeCl_3$, $MnCl_2$, $MgCl_2$, $AlCl_3$, $CaCl_2$, as well as sulfur-based and organic-based metal salt solutions, either with or without the application of an electric field [23]. Next, after drying the biomass, it is pyrolyzed at temperatures between 400 °C and 800 °C in an oxygen-limited or oxygen-free atmosphere. During pyrolysis, metal ions undergo conversion into metal oxide NPs (*e.g.*, Fe_2O_3, Al_2O_3, ZnO, MnO_2, MgO, CaO, *etc.*) that are deposited on the BC surface. Magnetic BC derived from Fe_3O_4 was produced by exposing $FeCl_3$-

treated corn straw to pyrolysis at 600°C for one hour in the presence of an electric field. The subsequent analysis revealed its remarkable lead adsorption capacity [24]. Similarly, MgO-BC NCs with high porosity were produced through the slow pyrolysis of MgCl$_2$-treated biomass. These NCs exhibited remarkable removal capabilities for phosphate and nitrate, attributed to the even distribution of MgO nano-flakes across the BC surface [25].

Immobilization of NPs on BC After Pyrolysis

BNCs can be synthesized by attaching metal oxide NPs onto the BC following the pyrolysis of biomass (illustrated in Fig. **2**). Various methods have been employed for this synthesis, including heat treatment [26] evaporative method [27], conventional wet impregnation method [28], and direct hydrolysis [29]. Initially, BC is immersed in solutions containing metal salts, allowing metal ions to be adsorbed onto the pores and surface of the BC. The composites are subsequently made by precipitating metal NPs onto the surface of the BC, pH modification of the metal salt solution, reduction procedures, and other methods [30]. An evaporative method utilizing Fe(NO$_3$)$_3$□9H$_2$O enables the enhancement of BC with iron oxides, leading to a significant rise in surface area of about 2.5 orders of magnitude [27]. Zero-valent iron magnetic BC composites (ZVI-MBC) were created by incorporating ZVI onto the surface of BC derived from paper mill sludge. This resulted in a higher surface area for the ZVI-MBC compared to the BC alone, with an increase of 1.5-fold. These composites were effective in removing pentachlorophenol from the effluent due to the combined adsorption capabilities of BC and ZVI particles, as demonstrated in a study [31]. The BC containing iron (Fe), generated by applying iron salt hydrolysis directly onto hickory BC, revealed that the primary sites for arsenic (As) absorption were the iron hydroxide particles located on the BC surface [29].

Pyrolysis the BC Rich in Target Metal Elements

The third approach documented for MNPs@BC synthesis entails the direct pyrolysis of biomass to enhance the concentration of the desired metal element [32]. Thermal pyrolysis can convert the metal element present in biomass into metal NPs. Magnesium-enriched BC was produced through the direct pyrolysis of tomato tissues rich in magnesium, carried out at 600°C for 2 hours within a nitrogen atmosphere [33]. The main sites for adsorbing phosphate on Mg-BC's surface were the MgO particles and Mg(OH)$_2$. There are advantages and disadvantages to each of the three techniques used to prepare MNPs@BC [34]. Before pyrolysis, the pretreatment of biomass with metal salt is advantageous due to its simplicity and cost-effectiveness, making it a practical and efficient method for large-scale production. However, it is difficult to achieve precise control over the composition, size, and variety of MNPs using this method. The process of depositing MNPs onto BC offers a convenient means of regulating the composition, shape, size, and specific types of MNPs. This method facilitates the synthesis of multilayer, double-layer, and monolayer MNPs on BC with ease. This approach is quite complex and involves higher costs. Pyrolyzing biomass directly to concentrate the target metal element is a straightforward and economical process. Despite this, there is a restricted supply of feedstocks that contain the specific metal element required for this purpose.

Magnetic BC NCs

BC's multi-functionality shows its effectiveness as a water sorbent for removing contaminants. While BC's small particle size and low density pose challenges for its recovery from large-scale sewage treatment, this issue could lead to secondary water pollution [35]. By using magnetic fields, BC-magnetic composites are easily reusable after being separated from water. Magnetic fields make it easy to separate and recycle BC-magnetic composites from water. There are two techniques for synthesizing BC-based magnetic composites: one involves pre-treating biomass with iron ions, while the other method entails loading magnetic material onto BC after its production. These techniques involve applying nanosized magnetic iron oxide particles, such as $CoFe_2O_4$ [36], Fe_3O_4 [37], and Fe_2O_3 [38], onto the BC's surface. This coating process helps induce magnetism and provides active sites for iron oxide, enhancing the efficiency of pollutant removal. Numerous studies have demonstrated that these composites exhibit excellent adsorption capabilities and can effectively adsorb a variety of contaminants from aqueous solutions [39]. BC's magnetic characteristics improve both its adsorption ability and ease of separation from solutions.

Biomass Pretreatment with Iron Ions

The process of biomass pretreatment involves the chemical coprecipitation of Fe(III)/Fe(II) onto biomass before pyrolysis or treating the biomass with Fe^{3+}/Fe^{2+}. Reddy and coworkers [36] produced magnetic BC through slow pyrolysis of pretreated biomass using $Co(NO_3)_2 \cdot 6H_2O$ and $Fe(NO_3)_3 \cdot 9H_2O$, which showed improved adsorption of Pb(II) and Cd(II) due to the spinel structure of cobalt ferrite incorporated onto the BC surface. Zhang and coworkers [40] utilized hydrothermal carbonization to pyrolyze a mix of biological and ferric sludge, resulting in a novel MBC with reduced diameter, a greater degree of carbonization, a larger SSA, and a smaller pore size degree compared to pure BC from the same process. Additionally, Li and coworkers [41] pointed out $BiFeO_3$ as another effective additive for making magnetic BC, noting that $BiFeO_3/BC$ exhibited reduced size, improved crystal morphology, exceptional ferromagnetic properties, and efficient degradation of organic pollutants and dyes in wastewater [42].

Iron Oxide Coating on BC via Pyrolysis

BC was first produced from the pyrolysis of biomass in this synthesis process. Then an aqueous ferrous/ferric solution was mixed with an aqueous charcoal suspension, and lastly, an aqueous NaOH treatment was performed [43]. Kulaksiz and coworkers [44] described the synthesis of a Fe-modified BC composite, referred to as MBC, using a co-precipitation process. In this investigation, pure BC was initially produced from a combination of paper sludge and wheat husks using standard pyrolysis at a temperature of 500 °C. Subsequently, a suspension was mixed with the dried BC containing $FeCl_3 \cdot 6H_2O$, ethanol, $FeCl_2 \cdot 4H_2O$, and HCl. The mixture was then stirred at ambient temperature and treated with an ammonia solution as the next step in the process. Shan and coworkers [45] used ball milling to create a hybrid material consisting of ultrafine magnetic BC and Fe_3O_4. The magnetic property of the BC/Fe_3O_4 composite demonstrated a saturation magnetization of 19.0 emu/g, allowing it to be readily separated from the solution using a permanent magnet.

Functional NPs Coated BC

The surface of BC has been proven to be more effective in eliminating a variety of pollutants when functional NPs are added. BC treated with NPs like graphene and its oxides, carbon nanotubes, and chitosan, among others, forms effective NM composites that are capable of efficiently eliminating various pollutants. The surface functional groups, thermal stability, porosity, and surface area of BC may all be significantly increased by these functional NPs, which will improve the material's capacity to remove contaminants.

Pre-treatment of Biomass with Functional NPs

Before pyrolysis, biomass was pretreated using several kinds of functional NPs. Typically, biomass can undergo conversion into functional BC composites through a dip-coating process. To prepare the biomass for pyrolysis, functional NPs are suspended in water. The biomass is mixed with the nanoparticle suspension, followed by separation and drying of the biomass before pyrolysis. The biomass and nanoparticle suspension are blended for a specified duration, and subsequently, the biomass is dried and separated to prepare for pyrolysis later. To enable the contaminants from aqueous solutions to be adsorbed onto the carbon surface of the BC matrix, these functional material particles were added. The resulting composites showcase enhanced functions and properties derived from the combination of both functional NPs and BC, highlighting their synergistic effects. Functional NPs coated on biomass feedstock may partially block BC pores, although this potential limitation can be overcome by leveraging the superior properties of these NPs [46]. BCs underwent a dip-coating process using different concentrations of carboxyl-functionalized carbon nanotube solutions before undergoing slow pyrolysis to produce multi-walled carbon nanotube-coated BCs [47]. For example, the wet impregnation technique is used to generate H_3PO_4-modified BC, which has increased surface area (1139.00 m^2/g) and mesopore volume (0.24 cm^3/g) and exhibits improved diuron sorption in water [48].

Post Pyrolysis Immobilization of Functional NPs

The technique of incorporating functional NPs onto untreated BC post-pyrolysis, leading to the formation of BC-based composites, synergizes the benefits of BC with the distinctive qualities of functional NPs (Fig. **2**). BC is frequently chemically modified by adding acids or bases. Both acidic and alkaline modifications have been found to enhance the physical and chemical properties of BC, as well as improve its removal efficiency. Chitosan coatings were employed to modify BC surfaces, leading to the formation of chitosan-modified BC [46]. This modification was found to integrate the strong chemical affinity of chitosan with the advantages offered by BC's surface area and porous structure. Andrew and coworkers [49] acidified BC using H_2SO_4, HNO_3, and a combination of both. The process of acidification increased the oxygen-to-carbon ratio and functional groups (nitro and carboxyl), showing that BC underwent chemical oxidation after acid treatment. However, acidification may lead to a reduction in the thermal stability of BC due to potential damage to its graphite structures. In this synthesis process, Mg/Al12 LDH [50], hydrogel [51], chitosan [46], ZnS nanocrystals [39], and zerovalent iron (ZVI) [31] were commonly utilized.

APPLICATION IN WASTEWATER TREATMENT

Dyes Removal from Aqueous Media

Hameed and coworkers [52] suggested that around 90% of produced dyes are expected to be employed in the textile sector, while the remaining portion will be distributed among other industries, including leather, paper, plastic, and chemicals. These contaminants exhibit features such as strong chemical stability, easy water solubility, intense coloration, and a complex aromatic structure, making them difficult to decompose naturally. The existence of textile dye colors not only physically damages water bodies [53] but also inhibits light transmission through water [54]. This obstruction results in reduced photosynthesis rates and lowered levels of dissolved oxygen, leading to a significant impact on aquatic biodiversity [54]. Furthermore, a significant number of organic dyestuffs, including azo dyes and anthraquinone dyes, have been identified as toxic or potentially cancerous to humans [55]. Therefore, it is crucial to treat wastewater containing dyes in a manner that is both cost-effective and environmentally friendly while also being efficient. BC (BC) has proven to be effective in adsorbing contaminants in water; however, its original form still has practical limitations that must be resolved for real-world use. As a result, there is increasing interest in developing BC-based hybrid materials to tackle these challenges.

These hybrid materials present opportunities to fulfill the need for cost-effective adsorbents with enhanced adsorption capacities by integrating the benefits of other materials with BC. Table **1** provides an overview of the studies on the elimination of dye from aqueous media using modified BC. Foroutan and coworkers [56] synthesized an $AC/CoFe_2O_4$ magnetic composite designed specifically for removing Nile blue (NB) dyes, methylene blue (MB), and methyl violet (MV) dyes from aqueous environments. The adsorption experiment was conducted using an ultrasonic device. The maximum sono-adsorption capacities were determined to be 87.48 mg/g for MB, 83.90 mg/g for MV, and 86.24 mg/g for NB. Lu and coworkers [57] utilized TiO_2-BC to eliminate methyl orange (MO), achieving a mineralization rate of 83.23 percent and a decolorization rate of 96.88 percent. Deng and coworkers [58] treated bamboo BC with $KMnO_4$ to modify it and utilized this modified BC as a very effective adsorbent to remove the malachite green dye. The adsorption mechanism of malachite green on potassium-modified bamboo BC involved pore filling, complexation with functional groups, and π–π interactions. Yıldırım and coworkers [59] engineered a composite consisting of BC and nano ZVI derived from organic waste, specifically mandarin peel, to remediate malachite green. A new magnetic BC was created by Thines and coworkers [60] from the peel of durian fruit. It has a large surface area of 820 m^2/g and an adsorption capacity of 87.32 mg/g, which allows it to remove congo red from an aqueous solution with 98% efficiency.

Table 1. Dye removal from water using modified BC (prepared in this work)

Biomass type	Pyrolytic Temperature	Strategies	Dye	Removal/adsorption capacity	Ref
Paper sludge and wheat husks	500°C	Co-precipitation	Malachite green	97.1%	[44]
Waste walnut shell	700°C	Pyrolysis of Tetrabuty lorthotitanate pretreated biomass	Methyl orange	96.88%	[57]
Wheat straw	500°C	CeO$_2$ hydrothermal method	Reactive Red 84	98.5%	[61]

Biomass type	Pyrolytic Temperature	Strategies	Dye	Removal/adsorption capacity	Ref
Sewage Sludge BC	450°C	Modified with magnetic Fe_3O_4 NPs	Acid orange 7 dye	110.27mg/g	[62]
Chestnut leaf	520°C	g-C_3N_4	Methylene blue	91%	[63]
Durian Rind	800°C	Treatment with iron (II) oxide, Fe_2O_3	Congo red	98%	[60]
Date Stones Of P. Dactylifera	450 °C	$CoFe_2O_4$	Nile Blue, methyl violet, and methylene blue	86.24 mg/g, 83.90 mg/g and 87.48 mg/g	[56]
Chicken feather	450°C	Titanate-treated biomass	Rhodamine B	90.91%	[64]
Sawdust Carbon		Loading of Fe_3O_4	Malachite green	89.22%	[65]
Walnut Shell	800°C	Copper-doped	Malachite green	2477 mg/g	[66]
Tapioca Peel Waste	800°C	Sulfur-doped	Malachite green and Rhodamine B dyes	30.18 and 33.10 mg/g	[67]

Heavy Metals Removal

Industries continue to release hazardous heavy metals, either directly or indirectly, into the environment. Even in trace amounts, metal ions can be highly toxic [68]. Heavy metals hurt agricultural productivity and human health since they are non-degradable and tend to collect in the environment [69]. Many studies have explored the use of BNCs as highly capable adsorbents. The aqueous phase adsorption of heavy metals by modified BC is summarized in Table **2**. Arsenic is a hazardous metal that is frequently found in drinking water and wastewater. Van Vinh and coworkers [70] used $Zn(NO_3)$ to surface-modify BC, increasing the adsorption capacity of As^{3+} from 5.7 µg/g to 7.0 µg/g.

Table 2. Modified BC studies on heavy metal removal (prepared in this work)

Biomass type	Pyrolytic Temperature	Strategies	Heavy metals	Removal/adsorption capacity	Ref
Hickory wood	600 °C	KMnO4 treatment	Cu(II), Cd(II) and Pb (II)	34.2 mg/g 28.1 mg/g 53.1 mg/g	[74]
Eucalyptus leaf residue	700 °C	Magnetic BC	Pb (II)	84.1%	[75]
Bamboo		$KMnO_4$ treatment	Pb^{2+}	123.47 mg/g	[58]

(Table 2) cont.....

Biomass type	Pyrolytic Temperature	Strategies	Heavy metals	Removal/adsorption capacity	Ref
Melia Azedarach Wood	400 °C	$Fe(NO_3)_3 \; 9H_2O$ treatment	Cr(VI)	99.8%	[76]
Corn stover	600 °C	Nano ZnS/ZnO modified BC	Cr(VI), Cu(II) and Pb(II)	91.2 mg/g, 24.5 mg/g, and 135.8 mg/g	[77]
Sawdust	600°C	MoS_2 treatment	Pb(II)	209 mg/g	[78]
Corn straw	600°C	Manganese oxide-modified BC	As(V)	14.36 mg/g	[79]
Corn straw (FBC)	600 °C	Iron (Fe)-impregnated BC	As(V)	Unmodified BC (0.017 mg/g), modified BC (6.80 mg/g),	[80]
Chestnut Shell	600°C	Fe	As(V)	45.8 mg/g	[81]
Corn Straw	800°C	α-FeOOH	Cu(II)	144.7 mg/g	[82]
Undaria Pinnatifida Roots	280°C	$MnFe_2O_4$	Pb(II)	175.4 mg/g	[83]
Rice hull	400°C	ZnS nanocrystals	Pb (II)	367.65 mg/g	[84]
Wetland reed	600°C	Iron NPs	Hexavalent chromium	58.82 mg/g, 82.2%	[85]
Peanut Shell	400°C	$MnSO_4 \cdot 4H_2O$	Cd(II)	22.3 mg/g	[86]
Rice Straw	700°C	$ZnCl_2/FeCl_3 \cdot 6H_2O$	Cu(II)	85.9 mg/g	[87]

Li and coworkers [71] extracted Cu(II), Pb(II), and Cr (VI) from the solution using nano-ZnO/ZnS modified BC that was produced by direct pyrolysis. The ZnO/ZnS-modified BC was found to have a greater porosity with total pore volumes of 0.43 cm^3/g and a higher specific surface area (SSA) of 397.4 m^2/g when compared to the pristine BC, which had an SSA of 102.9 m^2/g and total pore volumes of 0.2 cm^3/g. The carbon matrix was essential to the adsorption process because it allowed the heavy metals to combine and exchange cations with their heavy metal counterparts *via* hydroxyl groups on the nano-ZnO/ZnS surface [71].

BC made from walnut shells that have been copper-doped is capable of removing organic compounds from aqueous solutions, according to Shao and coworkers [72]. The surface complexation interaction between cationic dye Malachite green and Cu^0/Cu^{2+} led to effective adsorption. Modified BC has a better adsorption capacity when compared to activated carbon. Consequently, Wang and coworkers [73] reported that the maximum removal rate of Cr (VI) obtained with the Fe_3O_4-

BC composite was 99.44%. The strong attraction between the nanostructures in this composite and Cr (VI) promotes adsorption. Further, Fe_3O_4 contributes to Cr(VI) adsorption by reducing Cr(VI) through surface complexes with -CH, -OH, and –COOH.

Organic Compounds Removal

Organic pollutants are considered crucial environmental pollutants and are challenging for the ecosystem to break down naturally. Organic pollutants can deplete dissolved oxygen in water, posing risks to human health and aquatic ecosystems [88]. Therefore, managing pollutants wisely is extremely necessary. As a result, BC and modified BC have a significant capacity for adsorbing organic contaminants. By using BC in large-scale applications, removal can be effectively achieved without leaving behind environmentally harmful byproducts. BCs are cost-effective and efficient sorbents, removing organic compounds *via* hydrophobic interactions, electrostatic attraction, diffusion, pore filling, π–π interactions, and hydrogen bonding [89]. Using BC or modified BC, various organic materials such as pesticides, polychlorophenols, BPA, phenols, and pharmaceutical waste were eliminated [90].

Zhou and coworkers [91] introduced a novel Fe/Zn BC derived from sawdust and used it for effective tetracycline (TET) removal from aqueous solutions. There was an even distribution of crystal particles in the composite, resulting in a rough surface. The Fe/Zn BC displayed superior performance in removing TET compared to pure BC, Zn BC, and Fe BC. The enhanced performance was ascribed to the larger surface area and pore size, providing additional active sites for binding TET molecules through mechanisms such as π–π interactions, hydrogen bonding, and Fe(III)-TET coordination. It was estimated that this BC nanocomposite had a maximum adsorption capacity of 102.0 mg/g, demonstrating its relevance for practical TET treatment. Hu and coworkers [92] examined ZnO nanoparticle-modified BC derived from camphor leaf specifically for its efficacy in eliminating ciprofloxacin. The BC, derived from a $ZnCl_2$/BC mass ratio of 2 and subjected to calcination at 650°C, demonstrated typical microporous properties and an exceptional surface area measuring 915.0 m^2/g. The BC has a maximum adsorption capacity of 449.40 mg/L for ciprofloxacin. Even after completing three regeneration cycles, the adsorption capacity of the BC remained unchanged. Heo and coworkers [93] investigated the adsorption effectiveness of Bisphenol A (BPA) and sulfamethoxazole (SMX) using a $CuZnFe_2O_4$ BC composite. The primary mechanisms for adsorbing Bisphenol A and sulfamethoxazole included charge-assisted hydrogen bonding, π–π electron donor-acceptor interactions, and hydrophobic interactions. It exhibited significant adsorption capacities, facilitated simple magnetic separation, and demonstrated repeated recyclability. Table **3** outlines the adsorption of various organic pollutants by modified BC in aqueous solutions.

Table 3. Removal of organic contaminants using modified BC (prepared in this work)

Biomass type	Pyrolytic Temperature	Strategies	Organic Contaminants	Removal/adsorption capacity	Refs.
Bamboo Particles	500°C	Fe_3O_4	Polychlorinated Phenols	149.8 mg/g	[90]
Rice husk	800°C	Sulfide-modified nanoscale zerovalent iron	Nitrobenzene	550 mg/g,100%	[94]
Waste Douglas Fir	900- 1000°C	Fe_3O_4	4-nitroaniline, salicylic acid, benzoic acid, and phthalic acid	114, 109, 90, and 86 mg/g, respectively	[95]
Rice Straw TC	300°C	Alkali-acid combined, Magnetization	Tetracycline	98.33 mg/g	[96]
Sawdust	600°C	$ZnCl_2$- and $FeCl_3$ ·$6H_2O$-pretreated biomass	Tetracycline	94.1%	[91]
Reed	500°C	Zn- TiO_2	Sulfamethoxazole	81.21%	[97]
Eucalyptus globulus wood	380°C	Immobilization of NZVI onto BC	Chloramphenicol	70.5%	[98]
Camphor Leaf	650°C	ZnO	Ciprofloxacin	449.40 mg/g	[92]
Sawdust	600 ◦C	$ZnCl_2$- and $FeCl_2$-pretreated biomass	p-nitrophenol	170 mg/g	[99]
Vinasse Wastes	800°C	Fe/Mn	Levofloxacin	212 mg/g	[100]
Rice straw systems	450°C	$Co(NO_3)_2$ treated BCs	Ofloxacin	90.7%	[101]
Peanut Shells	450°C	$CuNO_3$	Doxycycline	93.22%	[102]
Pinus Taeda		NaOH	Tetracycline	274.8 mg/g	[103]

ADSORPTION MECHANISMS

The examination of adsorption mechanisms provides valuable insights into how different adsorbents perform in removing various adsorbates. Surface atoms or chemical groups on BC can attract adsorbate molecules, thus reducing surface energy. The driving force behind adsorption comprises multiple interactions, collectively contributing to the overall free energy of the adsorption process [104]. BNCs and modified BCs can be used to eliminate pollutants from aqueous solutions through a variety of methods that may work simultaneously, such as

a. Electrostatic attraction,
b. Ion exchange,
c. Hydrogen bonding,
d. Pore-filling, and
e. π–π electron donor-acceptor (EDA) interactions, *etc* [105].

Modified BC typically contains multiple functional groups like hydroxyl (OH), carboxyl (COOH), and carbonyl (CO), contributing to its exceptional performance in eliminating toxic metals or organic pollutants [106].

BC composites can have either a positively or negatively charged surface, depending on the medium's pH. Specifically, the adsorbent surface turns positively charged when the pH drops below the point of zero charge (pHpzc) and negatively charged when the pH rises above it [107]. Therefore, cationic or anionic contaminants are attracted to the BC surface by electrostatic attraction. According to Chen and coworkers [101], the enhanced electrostatic attraction between Tetracycline (TC) and BC is mainly responsible for the remarkable rise in modified BC's adsorption capacity for TC when the pH is raised from 5.0 to 9.0.

Ion exchange refers to a reversible process where ions in the solid phase and liquid phase interact. This phenomenon involves the adsorption of ions from the liquid phase by the solid substance, which then releases an equal number of ions back into the solution to preserve its electrical balance. An example of this process is seen in waste sludge BC modified with chitosan and Fe/S (BC Fe/S), which was developed for removing TC from water. The study found that key mechanisms involved in TC removal by BC included π–π interaction, hydrogen bonds, silicate bonds, electrostatic attraction, and pore filling. Additionally, chelation and ion exchange were identified as important adsorption mechanisms for Chitosan-Fe/S in TC removal [108].

An organic compound's chemical properties depend on its functional groups, which are groups of atoms that can react. Some functional groups can also create strong hydrogen bonds, which have high bond energy and are difficult to break apart. Oxygen-containing functional groups enable BC to adsorb toxic organic compounds through processes like complexation and hydrogen bonding. For example, using a hydrothermal technique, Heon and coworkers [93] developed a magnetic $CuZnFe_2O_4$ composite supported by BC to remove chemical pollutants such as Bisphenol A and sulfamethoxazole from aqueous solutions. Their research showed that π–π electron donor-acceptor (EDA) interactions, hydrogen bonding, and hydrophobic contacts were the primary processes behind this activity.

Through pore-filling processes, the adsorption of organic compounds is supported by the porous nature of BC surfaces. When BC undergoes modification, the application of modifying materials amplifies its surface area, thereby improving its capacity to adsorb a larger quantity of adsorbates [109]. The adsorption process may be hindered by the surface's functional groups and mineral components by blocking the pores on the BC surface [110]. Adsorption occasionally depends on intermolecular forces or physical bonding, such as dipole-dipole interactions and van der Waals forces.

Furthermore, the adsorption of organic pollutants may be impacted by the π–π electron donor-acceptor (EDA) interactions that occur between BC and aromatic groups in organic molecules. The adsorption processes of various carbonaceous materials on pollutants heavily depend on this specific adsorption mechanism. For example, potato stem and leaf BC resulted in a 2.3-fold increase in the highest adsorption capacity of Ciprofloxacin when modified with KOH. According to FTIR analysis, the primary interactions involved in the adsorption process were electrostatic, hydrogen bonding, and π–π interaction (EDA) [111]. Fig. (3) explains the different ways that different NMs remove contaminants, such as organic pollutants and heavy metals, by adsorbing them. This illustrates the range of uses that NMs have in environmental remediation.

CONCLUSION AND FUTURE RECOMMENDATIONS

In conclusion, the utilization of BNCs presents a promising avenue for environmental remediation, offering an efficient means of removing pollutants from water through adsorption processes. These NCs possess enhanced properties such as increased stability, catalytic ability, specific surface area, and porosity, which make them effective in regulating the absorption of contaminants. Research has demonstrated that incorporating nanostructured materials with BC enhances the sorption of organic contaminants, resulting in the high-capacity removal of impurities. Furthermore, the development of BNCs has shown significant potential in eliminating hazardous pollutants like dyes and heavy metals from industrial wastewater, underscoring their importance in sustainable environmental

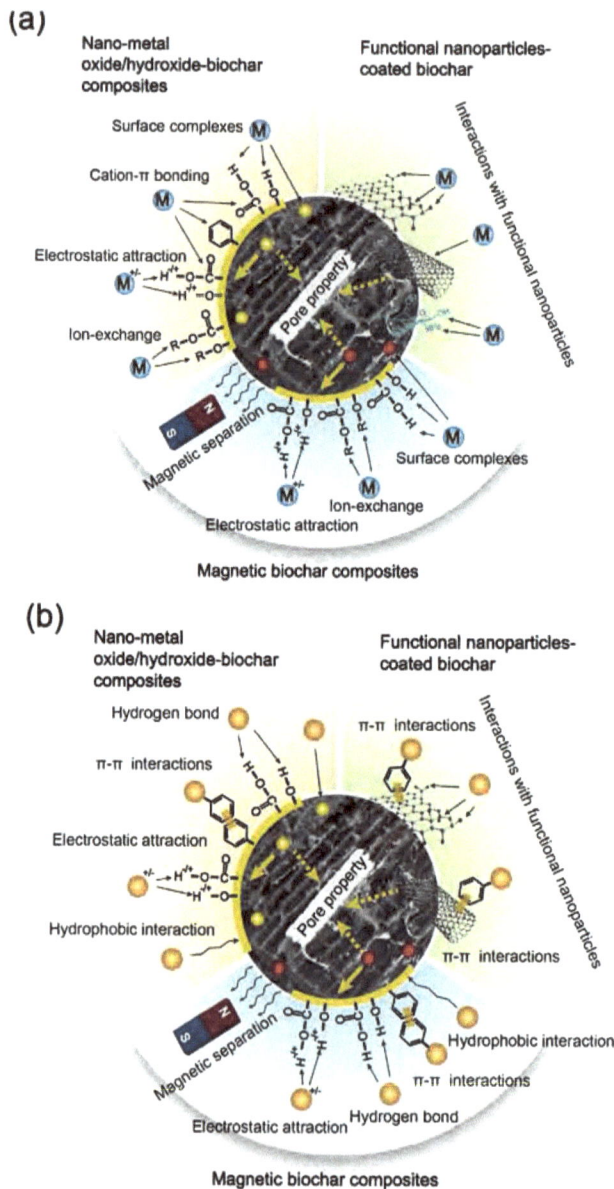

Fig. (3). Different NMs remove contaminants, such as a) heavy metals and b) organic pollutants [22].

remediation efforts. Additionally, the renewable nature of BC, derived from biomass sources, underscores the environmentally friendly aspect of these NCs, providing a greener alternative to traditional remediation methods. However, there remains a need for further research to fill the gap that exists between field applications and laboratory-scale research to fully harness the capabilities of

BNCs in industrial settings. Embracing BNCs represents a crucial step toward promoting cleaner and healthier ecosystems, contributing to a more sustainable future.

REFERENCES

[1]   Wang J, Wang S. Preparation, modification and environmental application of biochar: A review. J Clean Prod 2019; 227: 1002-22.
[http://dx.doi.org/10.1016/j.jclepro.2019.04.282]

[2]   Li L, Zou D, Xiao Z, *et al.* Biochar as a sorbent for emerging contaminants enables improvements in waste management and sustainable resource use. J Clean Prod 2019; 210: 1324-42.
[http://dx.doi.org/10.1016/j.jclepro.2018.11.087]

[3]   Meyer S, Glaser B, Quicker P. Technical, economical, and climate-related aspects of biochar production technologies: a literature review. Environ Sci Technol 2011; 45(22): 9473-83.
[http://dx.doi.org/10.1021/es201792c] [PMID: 21961528]

[4]   Tomczyk A, Sokołowska Z, Boguta P. Biochar physicochemical properties: pyrolysis temperature and feedstock kind effects. Rev Environ Sci Biotechnol 2020; 19(1): 191-215.
[http://dx.doi.org/10.1007/s11157-020-09523-3]

[5]   Ramanayaka S, Vithanage M, Alessi DS, Liu WJ, Jayasundera ACA, Ok YS. Nanobiochar: production, properties, and multifunctional applications. Environ Sci Nano 2020; 7(11): 3279-302.
[http://dx.doi.org/10.1039/D0EN00486C]

[6]   Cheah S, Jablonski WS, Olstad JL, *et al.* Effects of thermal pretreatment and catalyst on biomass gasification efficiency and syngas composition. Green Chem 2016; 18(23): 6291-304.
[http://dx.doi.org/10.1039/C6GC01661H]

[7]   Ashiq A, Adassooriya NM, Sarkar B, Rajapaksha AU, Ok YS, Vithanage M. Municipal solid waste biochar-bentonite composite for the removal of antibiotic ciprofloxacin from aqueous media. J Environ Manage 2019; 236: 428-35.
[http://dx.doi.org/10.1016/j.jenvman.2019.02.006] [PMID: 30769252]

[8]   Chausali N, Saxena J, Prasad R. NanoBC and BC based NCs: Advances and applications. J Agric Res (Lahore) 2021; 5: 100191.

[9]   Inyang MI, Gao B, Yao Y, *et al.* A review of biochar as a low-cost adsorbent for aqueous heavy metal removal. Crit Rev Environ Sci Technol 2016; 46(4): 406-33.
[http://dx.doi.org/10.1080/10643389.2015.1096880]

[10]   Ahmad M, Rajapaksha AU, Lim JE, *et al.* Biochar as a sorbent for contaminant management in soil and water: A review. Chemosphere 2014; 99: 19-33.
[http://dx.doi.org/10.1016/j.chemosphere.2013.10.071] [PMID: 24289982]

[11]   Chen B, Chen Z, Lv S. A novel magnetic biochar efficiently sorbs organic pollutants and phosphate. Bioresour Technol 2011; 102(2): 716-23.
[http://dx.doi.org/10.1016/j.biortech.2010.08.067] [PMID: 20863698]

[12]   Liu J, Jiang J, Meng Y, *et al.* Preparation, environmental application and prospect of biochar-supported metal nanoparticles: A review. J Hazard Mater 2020; 388: 122026.
[http://dx.doi.org/10.1016/j.jhazmat.2020.122026] [PMID: 31958612]

[13]   Liang L, Xi F, Tan W, Meng X, Hu B, Wang X. Review of organic and inorganic pollutants removal by BC and BC-based composites BC 2021; 3(1): 255-81.

[14]   Jiang D, Zeng G, Huang D, *et al.* Remediation of contaminated soils by enhanced nanoscale zero valent iron. Environ Res 2018; 163: 217-27.
[http://dx.doi.org/10.1016/j.envres.2018.01.030] [PMID: 29459304]

[15]   Siddiqui SI, Naushad M, Chaudhry SA. Promising prospects of nanomaterials for arsenic water remediation: A comprehensive review. Process Saf Environ Prot 2019; 126: 60-97.
[http://dx.doi.org/10.1016/j.psep.2019.03.037]

[16]   Rodriguez-Narvaez OM, Peralta-Hernandez JM, Goonetilleke A, Bandala ER. Biochar-supported nanomaterials for environmental applications. J Ind Eng Chem 2019; 78: 21-33.
[http://dx.doi.org/10.1016/j.jiec.2019.06.008]

[17]   Akhil D, Lakshmi D, Kartik A, Vo DVN, Arun J, Gopinath KP. Production, characterization, activation and environmental applications of engineered biochar: a review. Environ Chem Lett 2021; 19(3): 2261-97.
[http://dx.doi.org/10.1007/s10311-020-01167-7]

[18]   Huang Q, Song S, Chen Z, Hu B, Chen J, Wang X. BC-based materials and their applications in removal of organic contaminants from wastewater: state-of-the-art review BC 2019; 1(1): 45-73.

[19]   Li R, Wang JJ, Gaston LA, *et al.* An overview of carbothermal synthesis of metal–biochar composites for the removal of oxyanion contaminants from aqueous solution. Carbon 2018; 129: 674-87.
[http://dx.doi.org/10.1016/j.carbon.2017.12.070]

[20]   Ho SH, Zhu S, Chang JS. Recent advances in nanoscale-metal assisted biochar derived from waste biomass used for heavy metals removal. Bioresour Technol 2017; 246: 123-34.
[http://dx.doi.org/10.1016/j.biortech.2017.08.061] [PMID: 28893502]

[21]   Amdeha E. BC-based NCs for industrial wastewater treatment *via* adsorption and photocatalytic degradation and the parameterss affecting these processes. Biomass Convers Biorefin 2023; •••: 1-26.

[22]   Tan X, Liu Y, Gu Y, *et al.* Biochar-based nano-composites for the decontamination of wastewater: A review. Bioresour Technol 2016; 212: 318-33.
[http://dx.doi.org/10.1016/j.biortech.2016.04.093] [PMID: 27131871]

[23]   Jung KW, Ahn KH. Fabrication of porosity-enhanced MgO/biochar for removal of phosphate from aqueous solution: Application of a novel combined electrochemical modification method. Bioresour Technol 2016; 200: 1029-32.
[http://dx.doi.org/10.1016/j.biortech.2015.10.008] [PMID: 26476871]

[24]   Yang F, Zhang S, Sun Y, Du Q, Song J, Tsang DCW. A novel electrochemical modification combined with one-step pyrolysis for preparation of sustainable thorn-like iron-based biochar composites. Bioresour Technol 2019; 274: 379-85.
[http://dx.doi.org/10.1016/j.biortech.2018.10.042] [PMID: 30544043]

[25]   Zhang M, Gao B, Yao Y, Xue Y, Inyang M. Synthesis of porous MgO-BC NCs for removal of phosphate and nitrate from aqueous solutions. Chem Eng J 2012; 210: 26-32.
[http://dx.doi.org/10.1016/j.cej.2012.08.052]

[26]   Song Z, Lian F, Yu Z, Zhu L, Xing B, Qiu W. Synthesis and characterization of a novel MnOx-loaded biochar and its adsorption properties for Cu2+ in aqueous solution. Chem Eng J 2014; 242: 36-42.
[http://dx.doi.org/10.1016/j.cej.2013.12.061]

[27]   Cope CO, Webster DS, Sabatini DA. Arsenate adsorption onto iron oxide amended rice husk char. Sci Total Environ 2014; 488-489: 554-61.
[http://dx.doi.org/10.1016/j.scitotenv.2013.12.120] [PMID: 24529452]

[28]   Wang MC, Sheng GD, Qiu YP. A novel manganese-oxide/biochar composite for efficient removal of lead(II) from aqueous solutions. Int J Environ Sci Technol 2015; 12(5): 1719-26.
[http://dx.doi.org/10.1007/s13762-014-0538-7]

[29]   Hu X, Ding Z, Zimmerman AR, Wang S, Gao B. Batch and column sorption of arsenic onto iron-impregnated biochar synthesized through hydrolysis. Water Res 2015; 68: 206-16.
[http://dx.doi.org/10.1016/j.watres.2014.10.009] [PMID: 25462729]

[30]   Liao T, Li T, Su X, *et al.* La(OH)₃-modified magnetic pineapple biochar as novel adsorbents for

efficient phosphate removal. Bioresour Technol 2018; 263: 207-13.
[http://dx.doi.org/10.1016/j.biortech.2018.04.108] [PMID: 29747097]

[31]   Devi P, Saroha AK. Simultaneous adsorption and dechlorination of pentachlorophenol from effluent by Ni–ZVI magnetic biochar composites synthesized from paper mill sludge. Chem Eng J 2015; 271: 195-203.
[http://dx.doi.org/10.1016/j.cej.2015.02.087]

[32]   Mian MM, Liu G, Fu B. Conversion of sewage sludge into environmental catalyst and microbial fuel cell electrode material: A review. Sci Total Environ 2019; 666: 525-39.
[http://dx.doi.org/10.1016/j.scitotenv.2019.02.200] [PMID: 30802667]

[33]   Yao Y, Gao B, Chen J, *et al.* Engineered carbon (biochar) prepared by direct pyrolysis of Mg-accumulated tomato tissues: Characterization and phosphate removal potential. Bioresour Technol 2013; 138: 8-13.
[http://dx.doi.org/10.1016/j.biortech.2013.03.057] [PMID: 23612156]

[34]   Li P, Lin K, Fang Z, Wang K. Enhanced nitrate removal by novel bimetallic Fe/Ni nanoparticles supported on biochar. J Clean Prod 2017; 151: 21-33.
[http://dx.doi.org/10.1016/j.jclepro.2017.03.042]

[35]   Wei D, Li B, Huang H, *et al.* Biochar-based functional materials in the purification of agricultural wastewater: Fabrication, application and future research needs. Chemosphere 2018; 197: 165-80.
[http://dx.doi.org/10.1016/j.chemosphere.2017.12.193] [PMID: 29339275]

[36]   Harikishore Kumar Reddy D, Lee SM. Magnetic biochar composite: Facile synthesis, characterization, and application for heavy metal removal. Colloids Surf A Physicochem Eng Asp 2014; 454: 96-103.
[http://dx.doi.org/10.1016/j.colsurfa.2014.03.105]

[37]   Baig SA, Zhu J, Muhammad N, Sheng T, Xu X. Effect of synthesis methods on magnetic Kans grass biochar for enhanced As(III, V) adsorption from aqueous solutions. Biomass Bioenergy 2014; 71: 299-310.
[http://dx.doi.org/10.1016/j.biombioe.2014.09.027]

[38]   Wang S, Gao B, Zimmerman AR, *et al.* Removal of arsenic by magnetic biochar prepared from pinewood and natural hematite. Bioresour Technol 2015; 175: 391-5.
[http://dx.doi.org/10.1016/j.biortech.2014.10.104] [PMID: 25459847]

[39]   Yan L, Kong L, Qu Z, Li L, Shen G. Magnetic BC decorated with ZnS nanocrystals for Pb (II) removal. ACS Sustain Chem& Eng 2015; 3(1): 125-32.
[http://dx.doi.org/10.1021/sc500619r]

[40]   Zhang H, Xue G, Chen H, Li X. Magnetic biochar catalyst derived from biological sludge and ferric sludge using hydrothermal carbonization: Preparation, characterization and its circulation in Fenton process for dyeing wastewater treatment. Chemosphere 2018; 191: 64-71.
[http://dx.doi.org/10.1016/j.chemosphere.2017.10.026] [PMID: 29031054]

[41]   Li S, Wang P, Zheng H, Zheng Y, Zhang G. Adsorption and one-step degradation-regeneration of 4-amino-5-hydroxynaphthalene-2, 7-disulfonic acid using BC-based BiFeO3 NCs. Bioresour Technol 2017; 245: 1103-9.
[http://dx.doi.org/10.1016/j.biortech.2017.08.148] [PMID: 28950652]

[42]   Li S, Zhang G, Zhang W, *et al.* Microwave enhanced Fenton-like process for degradation of perfluorooctanoic acid (PFOA) using Pb-BiFeO3/rGO as heterogeneous catalyst. Chem Eng J 2017; 326: 756-64.
[http://dx.doi.org/10.1016/j.cej.2017.06.037]

[43]   Sun P, Hui C, Azim Khan R, Du J, Zhang Q, Zhao YH. Efficient removal of crystal violet using Fe3O4-coated biochar: the role of the Fe3O4 nanoparticles and modeling study their adsorption behavior. Sci Rep 2015; 5(1): 12638.
[http://dx.doi.org/10.1038/srep12638] [PMID: 26220603]

[44]   Kulaksiz E, Gözmen B, Kayan B, Kalderis D. Adsorption of Malachite Green on Fe-modified biochar: influencing factors and process optimization. Desalination Water Treat 2017; 74: 383-94. [http://dx.doi.org/10.5004/dwt.2017.20601]

[45]   Shan D, Deng S, Zhao T, *et al.* Preparation of ultrafine magnetic biochar and activated carbon for pharmaceutical adsorption and subsequent degradation by ball milling. J Hazard Mater 2016; 305: 156-63. [http://dx.doi.org/10.1016/j.jhazmat.2015.11.047] [PMID: 26685062]

[46]   Zhou Y, Gao B, Zimmerman AR, Fang J, Sun Y, Cao X. Sorption of heavy metals on chitosan-modified biochars and its biological effects. Chem Eng J 2013; 231: 512-8. [http://dx.doi.org/10.1016/j.cej.2013.07.036]

[47]   Inyang M, Gao B, Zimmerman A, Zhang M, Chen H. Synthesis, characterization, and dye sorption ability of carbon nanotube–BC NCs. Chem Eng J 2014; 236: 39-46. [http://dx.doi.org/10.1016/j.cej.2013.09.074]

[48]   Al Bahri M, Calvo L, Gilarranz MA, Rodríguez JJ. Activated carbon from grape seeds upon chemical activation with phosphoric acid: Application to the adsorption of diuron from water. Chem Eng J 2012; 203: 348-56. [http://dx.doi.org/10.1016/j.cej.2012.07.053]

[49]   Anstey A, Vivekanandhan S, Rodriguez-Uribe A, Misra M, Mohanty AK. Oxidative acid treatment and characterization of new biocarbon from sustainable Miscanthus biomass. Sci Total Environ 2016; 550: 241-7. [http://dx.doi.org/10.1016/j.scitotenv.2016.01.015] [PMID: 26820927]

[50]   Zhang M, Gao B, Yao Y, Inyang M. Phosphate removal ability of biochar/MgAl-LDH ultra-fine composites prepared by liquid-phase deposition. Chemosphere 2013; 92(8): 1042-7. [http://dx.doi.org/10.1016/j.chemosphere.2013.02.050] [PMID: 23545188]

[51]   Karakoyun N, Kubilay S, Aktas N, *et al.* Hydrogel–Biochar composites for effective organic contaminant removal from aqueous media. Desalination 2011; 280(1-3): 319-25. [http://dx.doi.org/10.1016/j.desal.2011.07.014]

[52]   Hameed BH, Ahmad AA, Aziz N. Isotherms, kinetics and thermodynamics of acid dye adsorption on activated palm ash. Chem Eng J 2007; 133(1-3): 195-203. [http://dx.doi.org/10.1016/j.cej.2007.01.032]

[53]   Setiadi T, Andriani Y, Erlania M. Treatment of textile wastewater by a combination of anaerobic and aerobic processes: A denim processing plant case. Proceedings of the Southeast Asian Water Environment. 1.

[54]   Hassan MM, Carr CM. A critical review on recent advancements of the removal of reactive dyes from dyehouse effluent by ion-exchange adsorbents. Chemosphere 2018; 209: 201-19. [http://dx.doi.org/10.1016/j.chemosphere.2018.06.043] [PMID: 29933158]

[55]   Shojaei S, Khammarnia S, Shojaei S, Sasani M. Removal of reactive red 198 by nanoparticle zero-valent iron in the presence of hydrogen peroxide J Water Environ Nanotechnol 2017; 1(2): 129-35.

[56]   Foroutan R, Mohammadi R, Ramavandi B. Elimination performance of methylene blue, methyl violet, and Nile blue from aqueous media using AC/CoFe$_2$O$_4$ as a recyclable magnetic composite. Environ Sci Pollut Res Int 2019; 26(19): 19523-39. [http://dx.doi.org/10.1007/s11356-019-05282-z] [PMID: 31077043]

[57]   Lu L, Shan R, Shi Y, Wang S, Yuan H. A novel TiO$_2$/biochar composite catalysts for photocatalytic degradation of methyl orange. Chemosphere 2019; 222: 391-8. [http://dx.doi.org/10.1016/j.chemosphere.2019.01.132] [PMID: 30711728]

[58]   Deng H, Zhang J, Huang R, *et al.* Adsorption of malachite green and Pb2+ by KMnO4-modified BC: insights and mechanisms. Sustainability (Basel) 2022; 14(4): 2040. [http://dx.doi.org/10.3390/su14042040]

[59]   Yıldırım GM, Bayrak B. The synthesis of biochar-supported nano zero-valent iron composite and its adsorption performance in removal of malachite green. Biomass Convers Biorefin 2022; 12(10): 4785-97.
[http://dx.doi.org/10.1007/s13399-021-01501-1]

[60]   Thines KR, Abdullah EC, Mubarak NM. Effect of process parameters for production of microporous magnetic biochar derived from agriculture waste biomass. Microporous Mesoporous Mater 2017; 253: 29-39.
[http://dx.doi.org/10.1016/j.micromeso.2017.06.031]

[61]   Khataee A, Gholami P, Kalderis D, Pachatouridou E, Konsolakis M. Preparation of novel CeO_2-biochar nanocomposite for sonocatalytic degradation of a textile dye. Ultrason Sonochem 2018; 41: 503-13.
[http://dx.doi.org/10.1016/j.ultsonch.2017.10.013] [PMID: 29137781]

[62]   Santhosh C, Daneshvar E, Tripathi KM, *et al.* Synthesis and characterization of magnetic BC adsorbents for the removal of Cr (VI) and Acid orange 7 dye from aqueous solution Environ Sci Pollut R 2020; 27(1): 32874-87.

[63]   Pi L, Jiang R, Zhou W, *et al.* g-C3N4 Modified biochar as an adsorptive and photocatalytic material for decontamination of aqueous organic pollutants. Appl Surf Sci 2015; 358: 231-9.
[http://dx.doi.org/10.1016/j.apsusc.2015.08.176]

[64]   Li H, Hu J, Zhou X, Li X, Wang X. An investigation of the biochar-based visible-light photocatalyst *via* a self-assembly strategy. J Environ Manage 2018; 217: 175-82.
[http://dx.doi.org/10.1016/j.jenvman.2018.03.083] [PMID: 29604411]

[65]   Bonyadi Z, Khatibi FS, Alipour F. Ultrasonic-assisted synthesis of Fe3O4 nanoparticles-loaded sawdust carbon for malachite green removal from aquatic solutions. Appl Water Sci 2022; 12(9): 221.
[http://dx.doi.org/10.1007/s13201-022-01745-w]

[66]   Shao Q, Li Y, Wang Q, Niu T, Li S, Shen W. Preparation of copper doped walnut shell-based biochar for efficiently removal of organic dyes from aqueous solutions. J Mol Liq 2021; 336: 116314.
[http://dx.doi.org/10.1016/j.molliq.2021.116314]

[67]   Vigneshwaran S, Sirajudheen P, Karthikeyan P, Meenakshi S. Fabrication of sulfur-doped biochar derived from tapioca peel waste with superior adsorption performance for the removal of Malachite green and Rhodamine B dyes. Surf Interfaces 2021; 23: 100920.
[http://dx.doi.org/10.1016/j.surfin.2020.100920]

[68]   Lai KC, Lee LY, Hiew BYZ, Thangalazhy-Gopakumar S, Gan S. Environmental application of three-dimensional graphene materials as adsorbents for dyes and heavy metals: Review on ice-templating method and adsorption mechanisms. J Environ Sci (China) 2019; 79: 174-99.
[http://dx.doi.org/10.1016/j.jes.2018.11.023] [PMID: 30784442]

[69]   Beidokhti MZ, Naeeni ST. AbdiGhahroudi MS. Biosorption of nickel (II) from aqueous solutions onto pistachio hull waste as a low-cost biosorbent. Civ Eng J 2019; 5(2): 447-57.
[http://dx.doi.org/10.28991/cej-2019-03091259]

[70]   Van Vinh N, Zafar M, Behera SK, Park HS. Arsenic(III) removal from aqueous solution by raw and zinc-loaded pine cone biochar: equilibrium, kinetics, and thermodynamics studies. Int J Environ Sci Technol 2015; 12(4): 1283-94.
[http://dx.doi.org/10.1007/s13762-014-0507-1]

[71]   Li C, Zhang L, Gao Y, Li A. Facile synthesis of nano ZnO/ZnS modified biochar by directly pyrolyzing of zinc contaminated corn stover for Pb(II), Cu(II) and Cr(VI) removals. Waste Manag 2018; 79: 625-37.
[http://dx.doi.org/10.1016/j.wasman.2018.08.035] [PMID: 30343795]

[72]   Goswami, Lalit, Anamika Kushwaha, Saroj Raj Kafle, and Beom-Soo Kim. 2022. "Surface Modification of Biochar for Dye Removal from Wastewater" Catalysts 12, no. 8: 817.

[http://dx.doi.org/10.3390/catal12080817]

[73]   Wang C, Tan H, Liu H, Wu B, Xu F, Xu H. A nanoscale ferroferric oxide coated biochar derived from mushroom waste to rapidly remove Cr(VI) and mechanism study. Bioresour Technol Rep 2019; 7: 100253.
[http://dx.doi.org/10.1016/j.biteb.2019.100253]

[74]   Wang H, Gao B, Wang S, Fang J, Xue Y, Yang K. Removal of Pb(II), Cu(II), and Cd(II) from aqueous solutions by biochar derived from KMnO4 treated hickory wood. Bioresour Technol 2015; 197: 356-62.
[http://dx.doi.org/10.1016/j.biortech.2015.08.132] [PMID: 26344243]

[75]   Wang S, Tang Y, Chen C, *et al.* Regeneration of magnetic biochar derived from eucalyptus leaf residue for lead(II) removal. Bioresour Technol 2015; 186: 360-4.
[http://dx.doi.org/10.1016/j.biortech.2015.03.139] [PMID: 25857768]

[76]   Zhang X, Lv L, Qin Y, Xu M, Jia X, Chen Z. Removal of aqueous Cr(VI) by a magnetic biochar derived from Melia azedarach wood. Bioresour Technol 2018; 256: 1-10.
[http://dx.doi.org/10.1016/j.biortech.2018.01.145] [PMID: 29427861]

[77]   Bingbing Qiu, Xuedong Tao, Hao Wang, Wenke Li, Xiang Ding, Huaqiang Chu, Biochar as a low-cost adsorbent for aqueous heavy metal removal: A review, Journal of Analytical and Applied Pyrolysis,Volume 155, 2021, 105081, ISSN 0165-2370.
[http://dx.doi.org/10.1016/j.jaap.2021.105081]

[78]   Zhu H, Tan X, Tan L, *et al.* BC derived from sawdust embedded with molybdenum disulfide for highly selective removal of Pb2+. ACS Appl Nano Mater 2018; 1(6): 2689-98.
[http://dx.doi.org/10.1021/acsanm.8b00388]

[79]   Yu Z, Zhou L, Huang Y, Song Z, Qiu W. Effects of a manganese oxide-modified biochar composite on adsorption of arsenic in red soil. J Environ Manage 2015; 163: 155-62.
[http://dx.doi.org/10.1016/j.jenvman.2015.08.020] [PMID: 26320008]

[80]   He R, Peng Z, Lyu H, Huang H, Nan Q, Tang J. Synthesis and characterization of an iron-impregnated biochar for aqueous arsenic removal. Sci Total Environ 2018; 612: 1177-86.
[http://dx.doi.org/10.1016/j.scitotenv.2017.09.016] [PMID: 28892862]

[81]   Zhou Z, Liu Y, Liu S, *et al.* Sorption performance and mechanisms of arsenic(V) removal by magnetic gelatin-modified biochar. Chem Eng J 2017; 314: 223-31.
[http://dx.doi.org/10.1016/j.cej.2016.12.113]

[82]   Yang F, Zhang S, Li H, *et al.* Corn straw-derived biochar impregnated with α-FeOOH nanorods for highly effective copper removal. Chem Eng J 2018; 348: 191-201.
[http://dx.doi.org/10.1016/j.cej.2018.04.161]

[83]   Jung KW, Lee SY, Lee YJ. Facile one-pot hydrothermal synthesis of cubic spinel-type manganese ferrite/biochar composites for environmental remediation of heavy metals from aqueous solutions. Bioresour Technol 2018; 261: 1-9.
[http://dx.doi.org/10.1016/j.biortech.2018.04.003] [PMID: 29635102]

[84]   Pare, Brijesh & Joshi, Roshni & Mehta, Sanika & Solanki, Vijendra & Gupta, Rupesh & Agarwal, Neha & Yadav, Virendra. (2024). Preparation and characterisation of BiOCl nano photocatalyst for the remediation of wastewater under LED light. International Journal of Environmental Analytical Chemistry. 1-25.
[http://dx.doi.org/10.1080/03067319.2024.2442086]

[85]   Zhu S, Huang X, Wang D, Wang L, Ma F. Enhanced hexavalent chromium removal performance and stabilization by magnetic iron nanoparticles assisted biochar in aqueous solution: Mechanisms and application potential. Chemosphere 2018; 207: 50-9.
[http://dx.doi.org/10.1016/j.chemosphere.2018.05.046] [PMID: 29772424]

[86]   Wan S, Wu J, Zhou S, Wang R, Gao B, He F. Enhanced lead and cadmium removal using biochar-

supported hydrated manganese oxide (HMO) nanoparticles: Behavior and mechanism. Sci Total Environ 2018; 616-617: 1298-306.
[http://dx.doi.org/10.1016/j.scitotenv.2017.10.188] [PMID: 29103653]

[87]   Yin Z, Liu Y, Liu S, *et al.* Activated magnetic biochar by one-step synthesis: Enhanced adsorption and coadsorption for 17β-estradiol and copper. Sci Total Environ 2018; 639: 1530-42.
[http://dx.doi.org/10.1016/j.scitotenv.2018.05.130] [PMID: 29929316]

[88]   Ahmed MB, Zhou JL, Ngo HH, Guo W, Chen M. Progress in the preparation and application of modified biochar for improved contaminant removal from water and wastewater. Bioresour Technol 2016; 214: 836-51.
[http://dx.doi.org/10.1016/j.biortech.2016.05.057] [PMID: 27241534]

[89]   Reguyal F, Sarmah AK. Adsorption of sulfamethoxazole by magnetic biochar: Effects of pH, ionic strength, natural organic matter and 17α-ethinylestradiol. Sci Total Environ 2018; 628-629: 722-30.
[http://dx.doi.org/10.1016/j.scitotenv.2018.01.323] [PMID: 29454212]

[90]   Jun BM, Kim Y, Han J, Yoon Y, Kim J, Park CM. Preparation of activated BC-supported magnetite composite for adsorption of polychlorinated phenols from aqueous solutions. Water 2019; 11(9): 1899.
[http://dx.doi.org/10.3390/w11091899]

[91]   Zhou Y, Liu X, Xiang Y, *et al.* Modification of biochar derived from sawdust and its application in removal of tetracycline and copper from aqueous solution: Adsorption mechanism and modelling. Bioresour Technol 2017; 245(Pt A): 266-73.
[http://dx.doi.org/10.1016/j.biortech.2017.08.178] [PMID: 28892700]

[92]   Hu Y, Zhu Y, Zhang Y, *et al.* An efficient adsorbent: Simultaneous activated and magnetic ZnO doped biochar derived from camphor leaves for ciprofloxacin adsorption. Bioresour Technol 2019; 288: 121511.
[http://dx.doi.org/10.1016/j.biortech.2019.121511] [PMID: 31132594]

[93]   Heo J, Yoon Y, Lee G, Kim Y, Han J, Park CM. Enhanced adsorption of bisphenol A and sulfamethoxazole by a novel magnetic $CuZnFe_2O_4$–biochar composite. Bioresour Technol 2019; 281: 179-87.
[http://dx.doi.org/10.1016/j.biortech.2019.02.091] [PMID: 30822638]

[94]   Zhang D, Li Y, Tong S, *et al.* Biochar supported sulfide-modified nanoscale zero-valent iron for the reduction of nitrobenzene. RSC Advances 2018; 8(39): 22161-8.
[http://dx.doi.org/10.1039/C8RA04314K] [PMID: 35541698]

[95]   Karunanayake AG, Todd OA, Crowley ML, *et al.* Rapid removal of salicylic acid, 4-nitroaniline, benzoic acid and phthalic acid from wastewater using magnetized fast pyrolysis biochar from waste Douglas fir. Chem Eng J 2017; 319: 75-88.
[http://dx.doi.org/10.1016/j.cej.2017.02.116]

[96]   Dai J, Meng X, Zhang Y, Huang Y. Effects of modification and magnetization of rice straw derived biochar on adsorption of tetracycline from water. Bioresour Technol 2020; 311: 123455.
[http://dx.doi.org/10.1016/j.biortech.2020.123455] [PMID: 32413637]

[97]   Xie X, Li S, Zhang H, Wang Z, Huang H. Promoting charge separation of biochar-based $Zn-TiO_2$/pBC in the presence of ZnO for efficient sulfamethoxazole photodegradation under visible light irradiation. Sci Total Environ 2019; 659: 529-39.
[http://dx.doi.org/10.1016/j.scitotenv.2018.12.401] [PMID: 31096382]

[98]   Ahmed MB, Zhou JL, Ngo HH, *et al.* Nano-Fe 0 immobilized onto functionalized biochar gaining excellent stability during sorption and reduction of chloramphenicol *via* transforming to reusable magnetic composite. Chem Eng J 2017; 322: 571-81.
[http://dx.doi.org/10.1016/j.cej.2017.04.063]

[99]   Wang P, Tang L, Wei X, *et al.* Synthesis and application of iron and zinc doped biochar for removal of p-nitrophenol in wastewater and assessment of the influence of co-existed Pb(II). Appl Surf Sci 2017; 392: 391-401.

[http://dx.doi.org/10.1016/j.apsusc.2016.09.052]

[100] Xiang Y, Xu Z, Zhou Y, *et al.* A sustainable ferromanganese biochar adsorbent for effective levofloxacin removal from aqueous medium. Chemosphere 2019; 237: 124464.
[http://dx.doi.org/10.1016/j.chemosphere.2019.124464] [PMID: 31394454]

[101] Chen L, Yang S, Zuo X, Huang Y, Cai T, Ding D. Biochar modification significantly promotes the activity of Co3O4 towards heterogeneous activation of peroxymonosulfate. Chem Eng J 2018; 354: 856-65.
[http://dx.doi.org/10.1016/j.cej.2018.08.098]

[102] Liu S, Xu W, Liu Y, *et al.* Facile synthesis of Cu(II) impregnated biochar with enhanced adsorption activity for the removal of doxycycline hydrochloride from water. Sci Total Environ 2017; 592: 546-53.
[http://dx.doi.org/10.1016/j.scitotenv.2017.03.087] [PMID: 28318694]

[103] Jang HM, Yoo S, Choi YK, Park S, Kan E. Adsorption isotherm, kinetic modeling and mechanism of tetracycline on Pinus taeda-derived activated biochar. Bioresour Technol 2018; 259: 24-31.
[http://dx.doi.org/10.1016/j.biortech.2018.03.013] [PMID: 29536870]

[104] Zhang R, Somasundaran P. Advances in adsorption of surfactants and their mixtures at solid/solution interfaces. Adv Colloid Interface Sci 2006; 123-126: 213-29.
[http://dx.doi.org/10.1016/j.cis.2006.07.004] [PMID: 17052678]

[105] Peng B, Chen L, Que C, *et al.* Adsorption of antibiotics on graphene and BC in aqueous solutions induced by π-π interactions. Sci Rep 2016; 6(1): 1-0.
[PMID: 28442746]

[106] Wu J, Lu J, Zhang C, Zhang Z, Min X. Adsorptive removal of tetracyclines and fluoroquinolones using yak dung BC. Bull Environ Contam Toxicol 2019; 102(3): 407-12.
[http://dx.doi.org/10.1007/s00128-018-2516-0] [PMID: 30552439]

[107] Bazrafshan AA, Hajati S, Ghaedi M. Synthesis of regenerable Zn(OH) 2 nanoparticle-loaded activated carbon for the ultrasound-assisted removal of malachite green: optimization, isotherm and kinetics. RSC Advances 2015; 5(96): 79119-28.
[http://dx.doi.org/10.1039/C5RA11742A]

[108] Liu J, Zhou B, Zhang H, Ma J, Mu B, Zhang W. A novel Biochar modified by Chitosan-Fe/S for tetracycline adsorption and studies on site energy distribution. Bioresour Technol 2019; 294: 122152.
[http://dx.doi.org/10.1016/j.biortech.2019.122152] [PMID: 31557651]

[109] Li Y, Wang Z, Xie X, Zhu J, Li R, Qin T. Removal of Norfloxacin from aqueous solution by clay-biochar composite prepared from potato stem and natural attapulgite. Colloids Surf A Physicochem Eng Asp 2017; 514: 126-36.
[http://dx.doi.org/10.1016/j.colsurfa.2016.11.064]

[110] Nguyen TH, Cho HH, Poster DL, Ball WP. Evidence for a pore-filling mechanism in the adsorption of aromatic hydrocarbons to a natural wood char. Environ Sci Technol 2007; 41(4): 1212-7.
[http://dx.doi.org/10.1021/es0617845] [PMID: 17593721]

[111] Li R, Wang Z, Guo J, *et al.* Enhanced adsorption of ciprofloxacin by KOH modified biochar derived from potato stems and leaves. Water Sci Technol 2018; 77(4): 1127-36.
[http://dx.doi.org/10.2166/wst.2017.636] [PMID: 29488976]

CHAPTER 7

Bionanomaterials: Harnessing Transformative Approaches for Environmental Remediation

Amrit Krishna Mitra[1,*]

¹ Department of Chemistry, Government General Degree College Singur, Hooghly, West Bengal, India

Abstract: Numerous biotic life forms on earth are being negatively impacted by the rising amounts of environmental pollutants caused by human activity. Heavy metals and certain organic pollutants are widely recognized as significant environmental contaminants globally because of their hazardous ability to persist in the environment. Contaminants present in various forms in the environment pose a challenge for eradication, as conventional technologies encounter difficulties in effectively eliminating them. Contemporary research primarily aims to devise cost-effective solutions for eliminating environmental contaminants. The latest investigation into minimizing environmental contaminants with minimal ecological impact involves leveraging the adsorption principles from traditional technologies alongside modified nanoscale adsorbents. In the past decade, the untapped prospective of biological resources enabling the biofabrication of nanomaterials (NMs) has spurred extensive investigation for benign pollution remediation. Processes such as surface active site interactions, electrostatic contact, photo and enzymatic catalysis, and other distinctive phenomena associated with biofabricated NMs play essential roles in detoxifying various contaminants.

In light of this context, the present chapter concentrates on the mechanism of environmental remediation by emerging biofabricated nano-based adsorbent while also addressing the remediation of persistent organic pollutants (POPs). Every category has been demonstrated with appropriate examples, basic mechanisms as well as societal applications. Last but not least, the long-term development of environmentally benign biofabricated NM-based adsorbents is highlighted.

Keywords: Adsorbents, Bioremediation, Biofabricated NMs, Heavy metals, Persistent organic pollutants, Wastewater treatment.

---

* **Corresponding author Amrit Krishna Mitra:** Department of Chemistry, Government General Degree College Singur, Hooghly, West Bengal, India; E-mail: ambrosia12june@gmail.com

Neha Agarwal, Vijendra Singh Solanki, Neetu Singh & Maulin P. Shah (Eds.)

INTRODUCTION

Researchers have shown a great deal of interest for the past few decades in materials with nanoscale dimensions owing to their exponential potential [1]. Nanotechnology has played a crucial role in fostering notable progress in bioengineering and medicine, influencing the development of biomaterials and the treatment of various ailments [2, 3]. Biological NMs encompass an interdisciplinary field, finding applications in materials science, quantum technology, biotechnology, chemistry, engineering, and physics [4]. Diverse nanoparticles (NPs) have been harnessed for specific applications, particularly contributing to the rapid expansion of research related to biomaterials [5]. This growth is propelled by the surface modification of biomaterials using NPs and the integration of NMs with medications, photosensitizers, and genes to create advanced delivery systems. The primary drivers for employing NMs include their exceptional biocompatibility, extensive surface area, and distinctive optical properties [6].

Biomaterials possess distinct physical features, dimensions, and improved stability and bio-efficacy due to their narrower size range of 1–100 nm. Science now has more options because of the skill of modifying materials at the nanoscale and using their characteristics for the benefit of society. Recent advances in bioanalytical instruments for accurate atom and molecular probing have brought this field of study to the forefront of scientific attention. Comprehending these distinct characteristics has facilitated the creation of novel and enhanced products worldwide through green process technology [7].

Through the reduction of industrial processes and materials to the nanoscale, nanobiotechnology has enabled the full utilization of surface and quantum phenomena. Biofabrication of NMs has become a popular area of study in nanobiotechnology in recent years because of the ease of production and the abundance of biological sources that include metabolites with a variety of properties [8]. In the search for safer alternatives to conventional NPs with potential hazards in biomedical applications, bio-nanomaterials (BioNMs) have emerged as an ideal choice. BioNMs are either created using biomolecules or can be employed to encapsulate or immobilize traditional NMs. Over the past fifteen years, attempts have been made to create NPs from biological resources, including plant extracts [9 - 11], bacteria [12], actinomycetes [13], basidiomycetes [14], and fungi [15]. These bioNMs exhibit enhanced biocompatibility, bioavailability, and bioreactivity, with minimal toxicity to humans, other living creatures, and the environment. Such special qualities make these bioNMs useful for a range of applications, including controlled drug delivery systems, electronics, tissue engineering, agriculture, biosensing, biolabeling, electronics, and agriculture [16].

Bioactive chemicals that serve as stabilizing agents for NPs, inhibiting their aggregation over time and providing them with further stabilization, would mediate the biocompatibility [17]. These bioactive substances can include carbohydrates, fats, carotenoids, proteins, vitamins, and other secondary metabolites with a variety of biological functions. These organisms function as bionanofactories, able to synthesize biochemicals necessary for the production of very stable NPs [18]. Due to their rich content of secondary metabolites, high effectiveness, widespread availability, and cost-effectiveness as reducing agents in environmentally friendly NP synthesis, plants are often referred to as chemical factories [19]. Biogenic NPs, synthesized using plants, are considered more biocompatible compared to chemical methods, thus rendering them more appropriate for a wide range of applications [20].

The objective of this chapter is to offer a thorough introduction to bioNMs, encompassing their definitions, origins, types, and characteristics. Additionally, the chapter presents advancements in the application of bio-NPs for environmental remediation.

WHAT ARE BIONMS

The terminology associated with NMs is rapidly evolving due to recent breakthroughs, leading to various definitions for these materials. Before delving into a detailed explanation of bioNMs, it is essential to outline the definitions of these terms. The science of the process by which NMs or NPs are created is known as nanoscience. In broad terms, nanoscience entails the exploration of matter at the nanoscale, with a specific emphasis on its size and structural characteristics that distinguish it from atoms, molecules, or bulk materials. NPs are commonly characterized as particles with at least one dimension spanning between 1 and 1000 nanometers. When particles reach a size between 1 and 100 nanometers, their properties alter significantly [21 - 23].

In the realm of nanotechnology, the term 'bioNM' is relatively recent compared to NPs. The key distinction between these two types of NMs lies in their origins: bioNMs are created using biological entities, while NMs are generated through physical and chemical methods. Consequently, bioNMs refer to materials at the nanoscale produced using biomolecules—such as enzymes, proteins, and amino acids—sourced from plants, animals, agricultural waste, or microorganisms. These biomolecules are derived from various sources, such as microorganisms, marine species, plants, agricultural wastes, insects, and certain mammals. BioNMs exhibit enhanced biocompatibility, bioavailability, and bioreactivity, with low or negligible toxicity towards humans, other species, and the environment. Additionally, the category of bioNMs encompasses NPs that have been fabricated, coated, or immobilized with biomolecules. It is important to highlight that discerning bioNMs from regular NMs requires careful consideration of their sources.

NMs and bioNMs are categorized according to five primary criteria: uniformity, morphology, agglomeration, geometry, and composition. Moreover, NMs can be classified into 1D, 2D, or 3D based on the shape of the NPs [22]. BioNMs may comprise a single material or a mixture of different materials, such as ceramics, metals, alloys, and polymers. Engineered NMs, which are pure single-composition NPs, can be produced through various techniques like gas-phase and mechanical processes, co-precipitation techniques, sol-gel method, vapor deposition synthesis, and others. In contrast, bioNMs produced through natural processes often result in multicomposite agglomerations [23].

SOURCES OF BIONMS

Based on the origin, nanosized materials are categorized into three classes:

- **Incidental NMs:** These are inadvertently produced as by-products of industrial processes. Examples include NPs found in vehicle exhaust, combustion emissions, fumes from welding processes, and certain natural phenomena like forest fires.
- **Engineered NMs:** Such NMs are intentionally created by humans, with properties specifically tailored for particular applications.
- **Naturally Occurring NMs:** Such NMs are present in microbes and macroorganisms naturally, without human intervention.

It is worth noting that incidental NMs may sometimes be regarded as a subset of natural NMs. Common biological sources used to create bioNMs include bacteria, viruses, plant extracts, polysaccharides, and proteins from plants or animals [24 - 27].

Microorganisms for the Production of NPs

Microorganisms are widespread in the environment and can be found in water, soil, and air. While some microbes play crucial roles in maintaining human health, others can be harmful and cause infections. Over the past few decades, there have been numerous reports on the utilization of microorganisms for synthesizing bioNMs. These materials have been applied across various fields, including antimicrobial and anticancer treatments, medical diagnosis, biosensors, mosquito larvicidal activities, catalysis, and antibiofouling [28].

The production of NMs through microbial biosynthesis has a number of noteworthy benefits and characteristics. These include the well-defined shape and chemical composition of the NMs, their ease of handling and microbial cell cultivation, scalability, and excellent environmental adaptability [29]. However, it should be noted that certain properties of biosynthesized NMs, such as pH and temperature, may differ from those of the original NM. For example, research by Fabrega *et al.* suggests that the particle sizes of silver increase with the rising pH of *Pseudomonas fluorescens* [30].

The production of NMs using microbial biosynthesis offers several notable benefits and characteristics. Two important categories for the microbial synthesis of NMs are extracellular and intracellular (Fig. **1**). Metal ion internalization, which can occur within microbial cells (intracellular method) or adhere to the cell surface (extracellular approach), usually starts the microbial-mediated metal NP manufacturing pathway. Studies have shown that oxidoreductase enzymes play a significant role in both intracellular and extracellular techniques. Microbial enzymes can specifically catalyze the reduction of metal ions within microbial cells or on the cell surface, resulting in the formation of metallic NPs. In recent decades, significant progress has been made in a variety of disciplines employing microbial-based NMs.

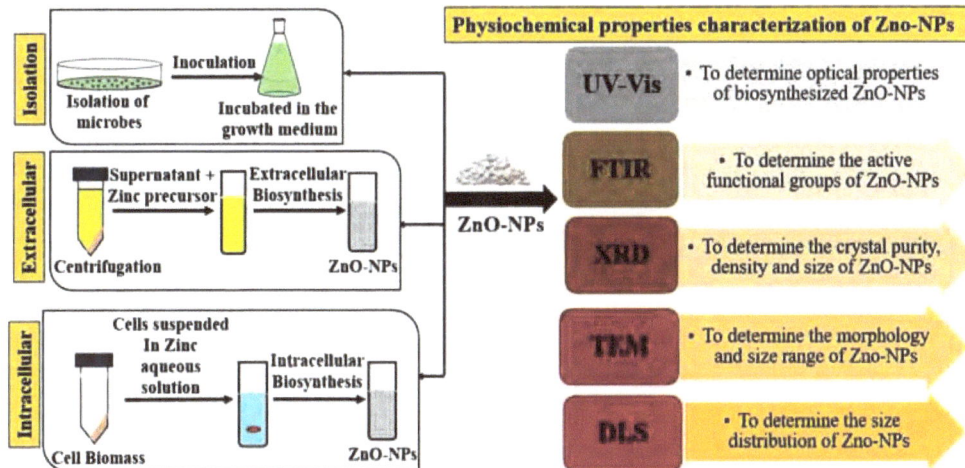

Fig. (1). Schematic illustration of microbial-mediated production and the physicochemical features of zinc oxide NPs.

In particular, using the mycelia-free culture filtrate of *Penicillium aculeatum*, Barabadi *et al.* produced biogenic gold NPs and effectively demonstrated that these particles might be used as a possible scolicidal therapy to treat cystic hydatid illness [31]. The microbial-based NPs' potential was further highlighted in a study by Pugazhendhi *et al.*, where they successfully inhibited bacterial growth, demonstrating bactericidal activity against gram-negative and gram-positive biofilm-forming pathogens. Marine red *Gelidium amansii* algae were used to generate silver NPs, and these NPs demonstrated exceptional antifouling capabilities [32].

Virus for the Production of NPs

Our ability to prevent, detect, and treat diseases is being revolutionized by nanoscale engineering. Due to their ability to act as prefabricated nanoscaffolds with distinct features and ease of modification, viruses have been particularly important in these advancements [33]. Viruses are noncellular, microscopic parasites that propagate by transferring their genetic material, which can be single- or double-stranded segments, to infect the host cells present in plants, fungi, animals, and bacteria. Viruses are often categorized based on their hosts, such as bacteriophages for bacteria, mycophages for fungi, or phytophages for plants. Typically, viral capsids have diameters ranging from 20 to 500 nm, making them natural NPs. Due to their nanoscale nature, viruses are considered ready-made NPs and find applications in the fields of nanoscience and nanoengineering. Virus particles can be engineered to present large and small molecules in precisely arranged arrays on their exteriors while simultaneously enclosing and protecting sensitive chemicals within their interiors. Viruses have been developed as chiefly targeted drug delivery methods that complement and enhance existing pharmaceutical choices because of these characteristics as well as their inherent biocompatibility.

Developments in the fields of synthetic biology and chemistry have made it possible to fabricate nanoscale devices with ever-more-controllable architectures, which have had an impact on numerous fields of study, business, and medicine. However, the preparation of structurally uniform populations of particles is still a challenge in the large-scale manufacture of such materials. On the other hand, the manufacture of millions of identical NPs by templated assembly in live cells is made possible by bioNMs derived from viruses. The environment is full of viruses, and those that infect plants, animals, or bacteria have all been utilized to create virus-based NPs (V).

Plant viruses are one of the most prevalent phytophages. Over the years, these viruses have become a promising platform for the synthesis of bioNMs. This is chiefly due to several advantages like small size, lack of envelope, ease of functionalization, excellent stability in a variety of environmental conditions, capacity for self-assembly, and structural symmetry. Plant viruses have a variety of morphologies, although they are primarily helical and icosahedral. *Tobacco mosaic virus* (TMV), a rod-shaped virus, is one of the most widely employed viruses in nanotechnological applications, whereas *cowpea mosaic virus* (CPMV), and *brome mosaic virus* (BMV) are examples of frequently used icosahedral viruses [34, 35]. Furthermore, fermentation and the molecular farming method can generate vast amounts of plant viruses. The majority of plant viruses offer three-dimensional components that can be assembled to create multi-dimensional nanostructures. For instance, gold NPs were guided to self-assemble into a complex 3D binary superlattice using CPMV. This complex structure can be used as a biocompatible substrate to aid in the creation of biosensors [36, 37].

Plant viruses have the capability to produce three main types of proteins by utilizing their host's machinery:

- **Replication Proteins:** These proteins are essential for the synthesis of nucleic acids, facilitating the replication of the viral genome.
- **Movement Proteins:** Movement proteins enable the virus to travel within the host, particularly through structures like plasmodesmata, which are channels connecting plant cells.
- **Structural Proteins:** These proteins are in possession of forming the virus's capsid, the protective outer shell that encloses its genetic material.

The proteins present in plant viral capsids exhibit remarkable properties, providing three distinct surfaces— the interior, the exterior, and the interface between subunits—that can be exploited for genetic or chemical modifications. This versatility allows for the creation of materials with controlled nanostructures. For example, Millán *et al.* employed a hierarchical self-assembly approach to develop CCMV protein cage NPs [38]. These NPs incorporated optically active and paramagnetic micelles, including chelated Gd^{3+} and Zn^{2+} phthalocyanine dye, with the aim of applications in multimodal imaging and therapy. The nucleation and oxidation of a Fe(II) cargo were aided by an electrostatically designed CCMV, which resulted in the creation of spatially confined iron oxide nanocrystals appropriate for applications such as magnetic resonance imaging (MRI) and hyperthermia treatment. Materials often necessitate encapsulation by inducing capsid formation around a cargo, allowing certain substances to diffuse through the capsid into the interior cavity. Once inside, they can be permanently attached to handles through bioconjugation or persuaded to stay inside by noncovalent interactions with nucleic acids or internally projecting side chains of amino acids. This method has been used to load fluorescent dyes for optical imaging, Gd^{3+} ions for MRI, and tiny medicinal compounds. A potent method for modifying V entails employing traditional chemical methods to modify specific amino acid side chains, such as phenol groups present on tyrosine residues, carboxylate groups found on glutamic acid and aspartic acid residues, sulfhydryl groups located on cysteine residues, and reactive amines present on lysine residues [39]. These groups can be introduced in order to link certain molecules to one another or adjusted to include functional groups required for more complex conjugation methods [40]. Tyrosine side chain phenol rings exposed to solvents can be changed by reacting them with the diazonium salts of a specific conjugate. Copper-catalyzed azide-alkyne cycloaddition (CuAAC), which produces the biocompatible 1,4-substituted triazole derivative irreversibly in the presence of Cu(I) between azides and alkynes, is a common click chemistry technique utilized with V [41]. Furthermore, payloads can be added to V *via* azo-coupling. Generally, azo-coupling to a bifunctional linker is the initial step in introducing an aldehyde into the VNP surface. Following that, the aldehyde can be employed in hydrazone or oxime condensation processes. Pyridoxal 5-phosphate can also be used to specifically oxidize the main *N*-terminal amine.

Plant extracts for the Production of NPs

Plant extracts stand out as a reliable biological source for bioNM synthesis due to their wide availability, environmental friendliness, ability to provide refined dominance of NP morphology, and do not require the use or production of harmful compounds [42, 43]. Due to their remarkable advantages, plant extracts, encompassing amino acids, vitamins, polysaccharides, carbonyls, proteins, phenolics, tannins, and ketones, have been extensively used in the synthesis of different NMs, particularly metal NPs (Fig. **2**). These functional groups facilitate the biogenesis of metal NPs by serving as a basis of a reductant, effectively reducing metal ions into atoms. The production of diverse metal NPs has led to a re-evaluation of plants' probable innate potential to reduce metal ions *via* neutral atoms without the usage of harmful and dangerous compounds. The design of NPs comprises the usage of plants. Recently, an enormous amount of study has been done on metal NPs, including those made of platinum, silver, zinc sulfide, gold, barium titanate, copper, and palladium. Agricultural by-products have been utilized to construct bioNPs and experiments involving different natural reducing agents remain an underexplored area in biomaterial synthesis. Plant-mediated NMs can also alleviate a wide range of symptoms, including those brought on by cancer, hepatitis, HIV, malaria, and other acute disorders.

In 2020, Kambale *et al.* synthesized leaf extracts from three species of Congolese plant—*Brillantaisia patula*, *Crossopteryx febrifuga*, and *Senna siamea*—to generate globular silver NPs that ranged in size from 45 nm to 110 nm [45]. The research findings indicated that the biogenic designing of silver NPs displays significant bactericidal characteristics against *Pseudomonas aeruginosa*, *Escherichia coli*, and *Staphylococcus aureus*. These qualities could have significant applications in the management of microbial disorders, including infectious skin diseases. Again, *Eucalyptus globulus* aqueous leaf extracts were used to create metal oxide nanorods by Jeevanandam *et al.* in 2018 [46]. These nanorods displayed potential uses in biosensing and medicine. Furthermore, a contemporary analysis by Abbas *et al.* in 2020 revealed that the production of both silver and copper NPs from *Aloe barbedensis*, *Azadirachta indica*, and *Coriandrum sativum* for wastewater remediation is feasible in wastewater treatment applications [47]. The study also highlighted the possibility of using the produced plant-mediated metal NPs as possible adsorbents for the elimination of naphthalene from aqueous environments.

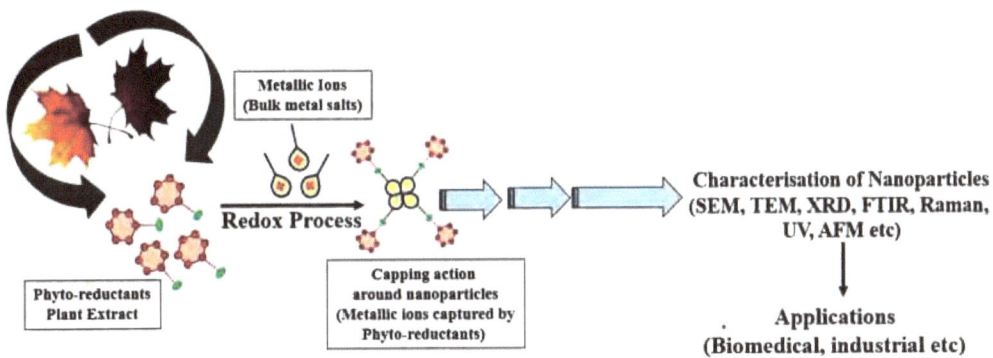

Fig. (2). Schematic representation of the *in vitro* green synthesis of plant-based metal NPs [44].

Polysaccharides and Proteins for the Production of NPs

Polysaccharides, natural biopolymers, offer excellent biocompatibility and biodegradability, making them valuable for creating bio-based NMs with specialized functions [48]. Polysaccharides derived from botanical sources such as wood, plants, and algae like cellulose and starch, owing to their abundance and budget-friendly and uncomplicated processing, have garnered significant attention. Cellulose, an abundant polysaccharide derived from plant fiber wastes, has garnered significant attention for its synthesized forms of NMs, including cellulose nanofibers, cellulose nanocrystals, and bacterial nanocellulose [49]. These materials offer excellent biocompatibility, impressive strength, biodegradability, and functionalizability for enhanced utility. In a recent study, a 3D nanoporous starch-based material was created to compress clove essential oil (CEO) as a natural food preservative, revealing improved antimicrobial action against *Bacillus subtilis*, *S. aureus*, and *E. coli* compared to pure CEO [50]. Ahmad *et al.* utilized spherical starch NPs from sago (*Metroxylon sagu*) starch granules as nanofillers in a composite film, resulting in increased tensile strength and improved water vapor permeability [51]. This indicates possible applications in green composite materials and medication delivery carriers.

Proteins, essential in biological systems and sourced from animals or plants (*e.g.*, silk, elastin, resilin, collagen, and keratin), provide diverse chemical groups in amino acids [52]. This diversity makes them exceptional building blocks for synthesizing novel functional NMs. Biomedical uses for protein-based NMs, which are generally categorized as NPs and nanofibers, are numerous and include delivery of drugs, biosensing, and diagnostics [53]. Zein, which comes from corn, is a commonly used bio-template for the synthesis of NMs. Using laccase-gold NPs crosslinked with zein-based nanofibers, Chen *et al.* created a novel laccase-based biosensor with a strong electrochemical reaction toward catechol, enabling the measurement of catechol in real solution samples [53]. Additionally, animal silk demonstrates notable potential for crafting functional NMs, including membranes, hydrogels, and fibers, with hierarchical self-assembled silk proteins.

PARADIGM SHIFT FROM TRADITIONAL NMS TO BIONMS IN ENVIRONMENTAL MITIGATION

In the realm of environmental mitigation, NMs emerge as a highly efficient tool for eliminating biological contaminants and pollutants, including toxic gases such as SO_2, CO, and NO_x, contaminated chemicals such as arsenic, iron, manganese, nitrate, heavy metals, toxic organic pollutants and biological entities such as viruses, bacteria, parasites, and antibiotics. Employing composites of NMs with diverse shapes and morphologies, ranging from NPs to tubes, wires, and fibers, has multiple applications. The heightened performance of NMs in environmental cleanup is attributed to their substantial surface area (expressed as the surface-t--volume ratio) and the resulting elevated reactivity, surpassing the capabilities of conventional approaches [1 - 7]. Various types of NMs show promise in environmental cleanup due to their unique properties. Some of these NMs include:

- **Nanocatalysts:** Nanocatalysts can be used to accelerate chemical reactions that take part in the degradation of pollutants. They can be employed in processes such as catalytic oxidation to break down contaminants.
- **Nanofibers:** Nanofibers, particularly those made from materials like carbon nanotubes or polymers, can be used in filtration systems to capture and remove pollutants, including particulate matter and contaminants in water.
- **NPs:** Metal and metal oxide NPs, such as titanium dioxide or iron NPs, can be used in the disintegration of organic pollutants in water and soil through photocatalysis or other chemical reactions.
- **Nanocomposites:** Materials composed of nanoscale components, such as nanocomposite membranes, can be employed in water purification processes to selectively remove contaminants while allowing the passage of clean water.

- **Nanospheres and Nanocapsules:** These structures can be designed to encapsulate and deliver specific agents for targeted remediation, releasing them at the site of contamination to enhance efficiency.
- **Nanoremediation agents:** Engineered NMs can be designed for specific drives of remediation, such as the removal of heavy metals or organic pollutants. These agents can be applied directly to contaminated sites.
- **Nanostructured Materials:** Materials with nanoscale features, such as zeolites or clays, can be utilized for their broad surface area and capacity for ion exchange in soil and water remediation.
- **Quantum Dots:** These semiconductor nanocrystals can be employed in environmental sensing and imaging for the detection and monitoring of pollutants.

All forms of organic and hazardous wastes are treated using conventional methods through the processes of adsorption, chemical oxidation, biological oxidation, and burning. The environmental applications of NMs have garnered significant attention, driven by the rapid advancement of nanotechnology. In the challenging task of environmental remediation, NMs play a crucial role in the removal of air and water contaminants. Their remarkable reactivity and extensive specific surface area make them highly effective as catalysts, adsorbents, and sensors. NPs heighten the adsorption efficiency of sorbent materials owing to their high surface area per mass. Furthermore, because of their small size and great mobility in solution, modest amounts of NMs can be used to quickly scan the entire volume. These special qualities can be used to scavenge and break down air and water contaminants. By using a gentle gravitational or magnetic (in the case of magnetic NPs) force, the particles adsorbed onto the NMs can be removed. Water and air quality in the natural environment are significantly impacted by NMs in a variety of sizes, morphologies, and forms. Because magnetic nanoadsorbents are so easy to extract and retain from treated water, they are particularly appealing. Additionally, the distribution of disordered surface areas and reactive surface sites varies amongst NMs [1 - 3]. Indeed, both artificial and natural NMs exhibit potent antibacterial qualities. Some examples include:

- **Carbon Nanotubes (CNTs):** These cylindrical structures possess strong antibacterial attributes and can be utilized in diverse applications, including antimicrobial coatings and filters.
- **Silver NPs (nAg):** Silver NPs are well-known for their antimicrobial features and are employed in different applications, ranging from medical devices to water purification systems.
- **Photocatalytic TiO$_2$:** Titanium dioxide (TiO$_2$) NPs, especially in their photocatalytic form, demonstrate antimicrobial properties. They are used in self-

cleaning surfaces and water treatment applications.
• **Chitosan:** Derived from chitin, chitosan NPs exhibit antimicrobial properties and are used in various biomedical and environmental applications.

In addition to antibacterial qualities, nanotechnology has proven effective in detecting pesticides and heavy metals. Nanoscale materials can be engineered for sensing applications to detect and quantify the presence of contaminants, including cadmium, copper, lead, mercury, arsenic, *etc*. Moreover, NMs exhibit enhanced photocatalytic and redox capabilities, making them valuable in environmental applications. These capabilities can be harnessed for pollutant degradation, water treatment, and the rectification of contaminated sites. The multifunctionality of NMs makes them versatile tools in addressing various environmental challenges, but it is crucial to consider their potential environmental and health impacts during their application.

When utilized in bioremediation, NMs show a quantum effect, requiring a reduced amount of activation energy to enable some chemical processes. Furthermore, the surface plasmon resonance characteristic that nanoscale matter exhibits is used to identify hazardous substances. The ability of NMs to enter contaminated areas and exhibit increased reactivity towards redox pollutants allows them to be employed in various forms and sizes for environmental rehabilitation. As a result, the use of NMs to mitigate pollution is starting to gain traction, which should help the environment in the decades to come. In order to address the shortcomings of current approaches, it is imperative to consider alternative approaches. A variety of cost-effective treatments with few or no adverse consequences are made possible by nanotechnology.

As environmental conditions change, traditional engineering environmental NMs (EENM) may struggle to adjust or forfeit their original functionalities because they were made to fulfill relatively straightforward and specific jobs. The proactive nature of smart environmental bio-NMs (SEBN), which can self-regulate their characteristics for improved performance in dynamic environmental conditions, is a result of their intentional design. The design and synthesis of SEBN have the potential to introduce disruptive technologies that outperform traditional methods in advancing environmental engineering reform. Recently developed bioinspired self-healing NMs, a significant subset of smart materials, have been made to autonomously repair structural impairment or surface activities in standard environments without requiring external energy input, aligning them more closely with natural processes. The traditional approaches to synthesizing NMs rely on chemical and physical principles, which can pose environmental risks and lead to toxicity for living organisms. To address these concerns, alternative strategies involve the utilization of various biological techniques derived from flora and fauna. The synthesis of bioNMs from biological sources is comparatively less complicated and offers the advantage of generating recyclable waste products, contributing to environmental pollution reduction. Green bio-based nanofillers, produced through biological synthesis, have potential applications in food packaging and storage, given their renewable and biodegradable nature. Moreover, waste from food and agricultural goods can be turned into NPs, which helps to reduce waste materials and pollution. This chapter offers a summary of NP toxicity and its biological production. It also discusses the usage of bioNMs to mitigate the negative impacts of environmental contamination [55].

BIONMS FOR THE PURPOSE OF ENVIRONMENTAL REMEDIATION

Nanoscale biomaterials have fostered interdisciplinary connections and brought about substantial advantages in the realms of biotechnology and biomedical fields. Different methods, such as physical, chemical, and biological approaches, can be employed to create bioNMs. Among these, biological methods have gained greater popularity due to their simplicity, practicality, environmental friendliness, and reduced reliance on harmful chemicals compared to other methods [56].

The sustainability concerns raised by traditional synthesis methods have drawn a lot of interest to green synthesis procedures for NMs in recent years. Nevertheless, recent literature lacks a comprehensive portrayal of biogenic NMs. The synthesis of inorganic and organic NMs can be intricately controlled by living organisms through internal processes that depend on physiological factors and growth conditions. These conditions include pH levels, temperature, culture duration, and concentrations of metal ions. The interaction between inorganic materials and biological entities leads to the formation of NMs through various intracellular and extracellular mechanisms. These nanoscale materials act as a link between larger substances and atomic or molecular frameworks. The biosynthesized metallic materials hold positive prospects in diverse fields such as packaging, cosmetics, electronics, coatings, and biotechnology. The use of toxic solvents, the possibility of adulteration from antecedent chemicals, and the creation of unsafe by-products are some of the drawbacks of the chemical processes used in the synthesis of NMs [57]. Therefore, there is an increasing need for highly effective, non-toxic, economical, and ecologically acceptable processes for the production of materials at the nanoscale.

Biological approaches are becoming more and more popular in modern applications because they are easy to use, convenient, environmentally friendly, lesser dependent on harmful chemicals and more economical [58]. This desire also carries over to the biomedical field, where green-synthesized bioNMs are highly valued for their non-toxicity and biocompatibility. Various living species, such as bacteria, fungi, yeast, plants, mammals, *etc.*, are included in biological approaches. Furthermore, a variety of natural ingredients is used, including starch, glucose, pectin, honey, and so forth. Compared to other green-based synthesis techniques, the use of plant extracts—also referred to as "green synthesis of bioNMs"—holds particular appeal. This is attributed to plants' immunity to the accumulation of heavy metal ions, eliminating the need for maintaining microorganism culture media. Furthermore, the synthesis rate of plants surpasses that of microorganisms.

In the plant-based synthesis of bioNMs, a crucial factor is the selection of the right plant and its components. This is due to the variations in enzyme activities, phytochemical compositions, biochemical processes, *etc.*, among different plant species. Plant-based synthesis can occur through two distinct methods: *in vitro* synthesis involves utilizing the plant itself, while *in vivo* synthesis involves the creation of NPs inside the plant. Plant extracts are integral to the designing of NMs as they naturally act as both reductants and stabilizers. The typical process for plant-based bioNM synthesis involves cleaning, compressing, and filtering the plant extract. Subsequently, salt (metal or non-metal) is introduced, resulting in a noticeable color change. BioNMs, owing to their unique properties, have gained widespread use in research and have led to the development of markets for products containing nano-objects. Various characterization techniques are employed to comprehend the physical and chemical characteristics of environmentally friendly green-synthesized bioNMs. When producing bioNMs in an environmentally friendly manner, considerations include the solvent medium, as well as the stabilizing and reducing agent. The primary objective of characterization techniques is to analyze and enhance the unique dimensions, configurations, molecular masses, purity levels, solubilities, chemical compositions, and stabilities of bioNMs synthesized using green methods. These properties are crucial as they determine the materials' applications in diverse fields such as biomedical, sensing, agriculture, and environmental applications. In light of their catalytic qualities, metal NMs are utilized in environmental remediation processes. A recent technique involves precipitating transition metals, like iron, gold, and palladium, at the exterior of bacteria to create bio-NMs or nanoparticulate catalysts (bionano-Met).

A novel avenue for investigating biological mechanisms leading to the creation of zero-valent elements, bi/multi-elemental quantum dots, and metal-containing NMs/ has been opened up by the biotransformation of metals by microbes. In a specific application, the reduction of azo dyes was achieved using reduced Pd(0) NM on *C. pasteurianum* [59]. The generation of bio-Pd includes the precipitation of palladium onto a bacterium. Specific bacterial species are capable of reducing

Pd(II), leading to the formation of Pd(0) NPs that occurs through precipitation, both on the cell wall and within the periplasmic space, when hydrogen is supplied. This represents a novel biologically inspired method for generating a nanopalladium catalyst. Two main Gram-negative model organisms employed in this process are the sulfate-reducing bacterium *Desulfovibrio desulfuricans* and the metal-respiring bacterium *Shewanella oneidensis* [60 - 62]. It has been observed that Pd and bioPd NPs exhibit reactivity towards various halogenated groundwater and soil contaminants, including hexavalent chromium, polychlorobiphenyls, and chlorinated solvents. Additionally, they serve as significant catalysts in chemical synthesis. However, all these methods are two-step processes involving the formation of Pd NPs in a separate reaction before pollutant treatment, which reduces their effectiveness in treating subterranean contaminants. The necessity of applying hydrogen *in situ* poses both financial and technological challenges, serving as an additional limiting factor in these procedures.

In 2010, Chidambaram *et al.* generated and implemented a successful bioremediation technique based on the *in situ* creation of hydrogen and bio-Pd NPs [63]. Pd(II) ions were reduced by *C. pasteurianum* BC1, an anaerobic hydrogen-producing member of the Clostridium group, to create Pd NPs (bio-Pd), which mostly precipitated in the cytoplasm and on the cell wall. Following the loading of bio-Pd NPs into *C. pasteurianum* BC1 cells in the presence of glucose, hydrogen was produced fermentatively, and soluble Cr(VI) was efficiently removed by reductive transformation into insoluble Cr(III) species. In both batch and aquifer microcosm tests, effective reductive removal of Cr(VI) was demonstrated using *C. pasteurianum* BC1 cells loaded with bio-Pd. In contrast, control experiments using viable or deceased bacterial cultures lacking Pd did not exhibit reductive removal of Cr(VI). Seven highly dangerous polychlorinated biphenyls (PCBs) could only be reduced by 50 mg l^{-1} of bio-Pd(0) to 27% of their initial level, according to a 2009 study by White *et al.* In contrast, 500 mg l^{-1} of commercial Pd(0) powder could eliminate the same concentration of PCBs. Sulfate-reducing bacteria synthesized iron sulfide (FeS) NPs with magnetic properties on their surface. These NPs were subsequently isolated from the solution through the application of a high-gradient magnetic field, creating an adsorbent capable of capturing several heavy metals and certain anions [64]. According to Singh *et al.*, the combination of *Pseudomonas aeruginosa* and nanoscale zerovalent iron (NZVI) particles contributed to the reduction of heavy metals, achieving percentages of 72.97% and 87.63% for Cr(VI) and Cd(II), respectively [65].

Another study in 2010 reported that copper (II) ions were adsorbed from aqueous solutions using saccharomyces cerevisiae immobilized on the surface of chitosan-coated magnetic NPs. (SICCM), a newly developed magnetic adsorbent [66]. Characterization of the produced magnetic adsorbent was conducted using TEM, XRD, and FTIR. No conglomeration was observed, and TEM images demonstrated the successful immobilization of *S. cerevisiae* on the surface of chitosan-coated magnetic NPs (CCM). XRD results indicated that the immobilization process did not induce a phase change in Fe_3O_4, confirming that the Fe_3O_4 NPs maintained a pure spinel structure. Optimal Cu(II) absorption occurred at a pH of 4.5. The highest removal efficiency, reaching 96.8%, was achieved when the initial concentration of Cu(II) was 60 mg L^{-1}, and the adsorption capacity increased with higher initial concentrations of Cu(II). Notably, SICCM exhibited exceptional efficiency in the rapid adsorption of Cu(II) in the initial 10 minutes, leading to the attainment of adsorption equilibrium in less than one hour.

The amine and carboxyl groups present in the cell walls of Bacillus subtilis were individually chemically modified to neutralize their electrochemical charge, aiming to determine their respective contributions to the metal uptake process [67]. Elevation of the metal interaction with phosphodiester bonds was made possible by the removal of approximately 94% of the ingredient teichoic acid (represented as inorganic phosphorus) by mild alkali treatment. If teichoic acid was extracted, the levels decreased stoichiometrically, but chemical alterations of amine functions did not lower the metal absorption values in comparison to native walls. However, for the majority of the metals examined, changes to carboxyl groups significantly restricted metal deposition. Here, electron microscopy and X-ray diffraction revealed that the metal deposit's structure and form might differ from native walls. These observations indicate that carboxyl groups play a more substantial role in metal deposition within the *B. subtilis* wall compared to amine groups.

According to Nair *et al.* in 2002, when exposed to the precursor ions, lactobacillus strains, which are prevalent in buttermilk, aid in the formation of submicron-sized crystals of gold, silver, and gold-silver alloys [68]. There are several distinct crystal morphologies seen. Tens of crystals are present within the bacterial outline, where crystal formation is attributed to the amalgamation of clusters. The development of crystals has no effect on the bacteria's capacity to survive. Nanoclusters are produced inside the bacteria and are likely the means by which crystals are nucleated. It is possible to fully harvest the biomass that contains crystals. The findings suggest possible uses in metal ion recovery, nanotechnology, analytical chemistry, and medicine. The surface area of the crystal appears to be decreased through coalescence, providing it with an efficient defense against biological harm.

In 2007, Husseiny *et al*. utilized Pseudomonas aeruginosa for the extracellular production of gold NPs [69]. The bacterial strain *P. aeruginosa* ATCC 90271 was employed, and its cell supernatant successfully reduced gold ions, leading to the formation of gold. UV–vis and fluorescence spectra were obtained for both the bacterial and chemically produced gold. Transmission electron microscopy (TEM) micrographs revealed the development of evenly distributed gold NPs in the 15–30 nm range. The extracellular nature of the reduction process suggests the potential for developing a straightforward bioprocess for Au production.

Superoxide dismutase (SOD) enzymes were effectively employed as solid supports on layered double hydroxide (LDH) NPs by Pavlovic *et al*. in 2017 [70]. The structural features of the resulting materials were examined using XRD, spectroscopic techniques (IR, UV-Vis, and fluorescence), and TEM. The colloidal stability of the materials was evaluated using electrophoresis and light scattering in aqueous dispersions. The SOD was quantitatively adsorbed onto the LDH *via* electrostatic and hydrophobic interactions, preserving its structural integrity following restriction. To enhance the colloidal stability of the system and mitigate salt-induced aggregation, heparin polyelectrolyte was employed with the composite material exhibiting a moderate level of resistance to aggregation in dispersions. With an appropriate loading of polyelectrolyte, heparin, characterized by a highly negative line charge density, exhibited strong adsorption on the oppositely charged hybrid particles. This led to charge neutralization and overcharging of the particles.

In 2012, Majumder *et al*. synthesized copper NPs () using *Fusarium oxysporum*, *Pseudomonas sp.*, and *Lantana camara* leaf extract for electronic waste bioremediation [71]. Again, microorganisms such as *Bacillus subtilis*, *Escherichia coli*, *Lactobacillus*, *Pseudomonas aeruginosa*, and *Rhodopseudomonas capsulata* were used to generate gold NPs [72]. Numerous microbes accumulate intracellular gold NPs to withstand the elevated metal ion concentration. Studies indicated that by exposing the bacterial cells to gold chloride when they were at room temperature and pressure, *Bacillus subtilis* was able to decrease Au^{3+} ions to form octahedral gold particles of nanoscale dimensions. Fe(III)-lowering microorganism, *Shewanella alga,*has the ability to decrease Au(III) ions in anaerobic settings. The Au ions were totally reduced in the presence of *S. alga* and hydrogen gas, resulting in the creation of 10–20 nm gold NPs. By forming platinum NPs, S. algae also decreased the amount of platinum metal ions. After supplying lactate as the electron donor at room temperature and neutral pH, dormant S. algae cells converted $[PtCl6]^{2-}$ into elemental platinum within the periplasm in less than 60 minutes. Two chalcogenide oxyanion species, selenite/nate and tellurite/rate, were found to be reduced to elemental selenium and tellurium by two anaerobic bacteria, *Bacillus selenireducens* and

Sulfurospirillum barnesii, respectively, during the synthesis of selenium NMs [73, 74].

Because of their non-toxic, biodegradable, and biocompatible characteristics, biopolymer-based NMs are very adaptable in a range of applications. For the manufacture of biopolymer-based NMs, biopolymers such as chitosan, carboxymethyl cellulose (CMC), polyhydroxyalkanoates (PHA), poly lactic acid (PLA), *etc.*, can be utilized. Samrot *et al.* produced chitosan laden with curcumin from crab shells for pharmaceutical delivery [75]. Sodium tripolyphosphate and barium chloride were used as chelators during a demineralization process to extract chitosan NPs from the shells of Metacarcinus magister for this investigation. The production of chitosan NPs was reported to be non-toxic and free of organic solvents by Sathiyabama *et al.* in 2016. Using anionic proteins derived from *Penicillium oxalicum*, chitosan NPs were produced in a biological manner. Chloramphenicol and ketoconazole were shown to be released under regulated conditions *in vitro* by chitosan NPs that were produced using leaf extract from *Catharanthus roseus*. The study found that by maximizing the ratio of chitosan and leaf extract in a ratio of 3:1 with a 30-minute crosslinking period at pH~3, chitosan NPs with a minimal diameter (45 nm) could be formed. In numerous microorganisms under nutrient-imbalanced conditions, polyesters of hydroxyalkanoate monomers build up to form polyhydroxyalkanoates (PHAs). Using the *Bacillus subtilis* NCDC0671 strain, Umesh *et al.* developed an orange-peel-based medium for the economically viable biogenic production of PHA [77]. Similar to this, gram-positive bacteria spontaneously create poly-γ-glutamic acid (γ PLA), a biopolymer made of repeating units of the amino acid glutamic acid. This bio-based chemical has been successfully used in the fields of heavy metal environmental remediation due to its non-toxic and biodegradable qualities. A variety of substances can be flocculated by poly-γ-glutamic acid (γ PLA). A commercially available and ecologically benign substance called carboxymethyl cellulose (CMC) is used to produce supports for iron and bimetallic NPs () that are more chemically reactive and physically stable. CMC has been used as a stabilizer for iron because it provides steric stabilization and inhibits aggregation by applying stronger repulsion forces than electrostatic repulsion. Cr(VI) toxicity was successfully decreased by biopolymer-based carboxymethyl cellulose-stabilized iron NPs (), highlighting the enhanced performance attained by adding stabilizers such as CMC.

The distinct physico-chemical properties of NMs/ from various sources offer new pathways for developing innovative biological sensors. *Acacia nilotica*, also known as wild twig bark, synthesized Ag used for detecting 4-nitrophenol. As reported by Karuppiah *et al.*, the detection limit of the method was 15 nM, with a broader linear range spanning from 100 nM to 350 µM. In another study, gold

(Au) and silver (Ag) NPs were synthesized using the stem extract of Breynia rhamnoides. The rapid reduction of Au^{3+} ions to Au facilitated the effective reduction of 4-nitrophenol. The green synthesis of Au was demonstrated by Emmanuel *et al.* using an extract of *A. nilotica*'s twig bark. The green-synthesized Au-modified electrode showed superior electrochemical detection capability [78].

Harmful heavy metal ions such as Cu^{2+}, Cr^{6+}, Zn^{2+}, Cd^{2+}, Pb^{2+}, Cr^{3+}, Hg^{2+}, and Mn^{2+} serve as inorganic pollutants alongside toxic organic contaminants in the environment. NMs are extensively used as affordable and highly sensitive colorimetric sensors have been developed for detecting these heavy metal ions. Silver (Ag) and gold (Au) NPs are well known for their tunable size and distance-dependent optical properties, particularly exhibit strong attenuation coefficients in the visible range, making them valuable for metal sensing applications in environmental monitoring. To achieve colorimetric metal ion sensing, biosynthesized Au NPs require the addition of a chelating agent to their surface. *Citrus paradisi* extract was used by Silva-De Hoyos *et al.* as a capping and reducing agent in a quick and efficient biogenic approach for the synthesis of Au with adjustable size and optical characteristics [79]. Using plasmonic and fluorescence sensing techniques, biosynthesized Au showed metal-sensing capability for Pb^{2+}, Cu^{2+}, Hg^{2+}, Zn^{2+}, and Ca^{2+}. Additionally, Nag *et al.* evaluated the catalytic activity of these Au using the intracellular protein extract (IPE) of the bacterial strain *Staphylococcus warneri* [80]. More surface-enhanced Raman scattering (SERS) activity was observed for the detection of toxic compounds by these synthesized Au. Additionally, they were able to completely degrade toxic nitro aromatic pollutants, such as 4-nitroaniline, 4-nitrophenol, 2-nitroaniline, and 2-nitrophenol, with three times the catalytic activity's recyclability.

A metal ion biosensor mediated by green synthesis that detects Hg^{2+}, Pb^{2+}, and Mn^{2+} ions in an aqueous medium was demonstrated by the use of L-tyrosine as a capping and reducing agent. Using lotus root (LR) microwave treatment, Gu *et al.* produced nitrogen-doped carbon nanodots (CDs) [81]. The produced LR-CDs showed an 18.7 nM detection limit and selective sensitivity to Hg^{2+}. In order to sense a variety of organic and inorganic environmental toxins, more NMs are being biosynthesized and functionalized.

MECHANISM OF BIONMS FOR THE MITIGATION OF ENVIRONMENTAL CONTAMINANTS

Microorganisms have been instrumental in synthesizing diverse inorganic and organic NMs/NPs (NMs/) with varying chemical compositions, sizes, and morphologies. These naturally occurring NMs/ are essential for environmental remediation and innovative applications. Three main processes—adsorption, transformation, and catalysis—are used by NMs to remove contaminants from the environment (Fig. **3**). Biologically produced NMs/ have proven useful in mitigating organic and hazardous metal pollution *via* photocatalysis, adsorption, transformation, and catalytic reduction, among other mechanisms (Table **1**). An overview of the various techniques used by biologically generated nanoscale materials to remediate organic and inorganic contaminants in the environment is given in this section.

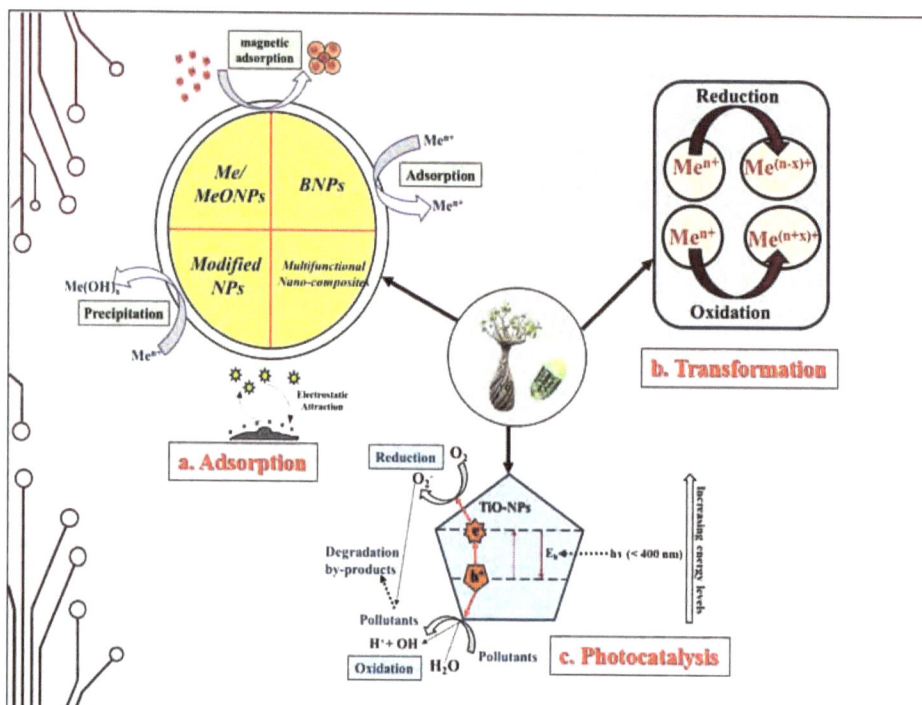

Fig. (3). Schematic representation of the mechanisms of the organic and inorganic pollutant elimination using biologically produced NMs through a) adsorption, b) transformation, and c) photocatalysis.

Table 1. Nanoscale materials developed biologically to eliminate both organic and inorganic contaminants

Biologically Produced NMs	Pollutants	Biological Source	Mode of Action	Refs.
Ca-Alginate	Cr(VI)	Honey	Sorption	[82]
CuO-	Cr(VI)	Plant (*Citrus limon*)	Adsorption	[83]
Co_3O_4-	Lithium	*Bacillus subtilis*	Transformation	[84]
NiO-	Ni(II)	*Microbacterium* sp.	Transformation	[85]
α-Fe_2O_3	As (V)	Plant (*Aloe vera*)	Adsorption	[86]
CdS-	Cd(II)	*Pseudomonas aeruginosa* JP-11	Adsorption	[87]

(Table 1) cont.....

Biologically Produced NMs	Pollutants	Biological Source	Mode of Action	Refs.
Au and Ag-	Organics (hexadecane)	Fungus (*Gordonia amicalis* HS-11)	Catalysis	[88]
Pd-	Nitroarenes	Plant (fruit extract)	Catalysis	[89]
Cu-	Organics (methylene blue)	Plant (*Citrus grandis*)	Photocatalysis	[90]
Ag-	Organics (phenolic compound)	Plant (tea polyphenols)	Catalysis	[91]
Ag-	Organics	Plant (spinach)	Electrocatalysis	[92]
Biomatrixed Au	Organics (methylene blue, nitrophenol)	Fungus (*Flammulina velutipes*)	Catalytic reduction	[93]
Zv-I	Ibuprofen	Plant (grape marc, black tea, vine leaves)	Adsorption	[94]
SnO_2-	Organics (nitrophenol)	Microwave	Catalyst	[95]
CuO-	Organics	Plant (*Thymus vulgaris*)	Catalyst	[96]
Ag-	Organics (methylene blue)	Fungus (*Penicillium oxalicum*)	Catalytic reduction	[97]

Adsorption

Contaminants attach themselves to the surface of an adsorbent, a process known as adsorption. Ionic and surface interactions are what propel this exothermic reaction. The heat produced when one mole of adsorbate binds to an adsorbent is known as the enthalpy of adsorption, and it is at all times negative. This negative results from the adsorbate's limited mobility during adsorption, which lowers entropy. At constant pressure and temperature, adsorption happens spontaneously and helps to lower Gibb's free energy. Prokaryotes, eukaryotes, and other biological entities act as valuable factories for synthesizing nanoscale adsorbents, crucial for combating pollution due to their economic and environmentally friendly properties. These organisms possess the necessary elements for nanoscale adsorbent synthesis, and various biologically produced nanoadsorbents are currently deployed to eliminate environmental contaminants from polluted areas [98].

Metal/oxide NPs

Metal/oxide NPs (Me/MeO) can efficiently adsorb heavy metals and eliminate organic impurities [99]. In particular, magnetic NPs play a key role in the effective treatment of huge volumes of wastewater by easy magnetic separation.

Due to their high adsorption capacity, iron oxide NPs are cost-effective and efficient types of adsorbents. Salvadori *et al.* investigated the use of *R. mucilaginosa* yeast in the bioremediation of wastewater containing copper by the creation of copper NPs [100]. They demonstrated that CuO- is efficient in removing Pb(II) and organic dyes from the environment by using copper oxide (CuO) nanostructures for adsorption. In environmental remediation, the biosynthesis of zero-valent silver nanoadsorbents using Phyllanthus emblica leaf extract showed a remarkable adsorption capacity, reaching 312 mg g^{-1} against Hg(II) ions.

Bimetallic NPs

An environmentally friendly and biocompatible method for creating NMs is to use biologically produced bimetallic NPs, which are made in microorganisms, namely fungi, bacteria, yeast, and plants at ambient temperature. Iron NPs that have had noble metals deposited on them react more quickly and with increased reactivity. When it comes to environmental remediation, biologically produced NPs (B) provide an economical and sustainable way to improve the decontamination of both organic and inorganic pollutants. One example is the synthesis of bimetallic Au/Ag NPs utilizing *Spirulina platensis* single-cell proteins [101].

Utilizing the functional groups in the reductants to decrease and cap the particles, plants contribute to the ecologically friendly synthesis of bimetallic NPs. Different from chemical reduction, this non-toxic method is frequently used, and substances found in plant extracts, such as flavonoids and phenolics, aid in the reduction and stabilization of bimetallic NPs. For example, pomegranate fruit extract's phenolic hydroxyls and proteins are essential for stabilizing and reducing bimetallic Au–Ag NPs. Bioreduction has been used to successfully synthesize Ti/Ni and Au/Pd bimetallic NPs in addition to Au–Ag and Ag/Se bimetallic NPs. Bimetallic NPs (B) produced biologically have a high catalytic potential, making them useful for cleaning up environmental pollutants. More investigation is required to examine the green synthesis of B for efficient contamination remediation, even if there are many publications on the biosynthesis and use of B in environmental remediation.

Multifunctional nano-composite

The goal of researching multifunctional properties of NMs/ is to develop smart materials with a range of functions. In order to create a nanocomposite encapsulated in a polyvinyl alcohol membrane that is patented for the treatment of industrial effluents, Nehru *et al.* biologically produced Ag using the extract from *Amaranthus tristis* [102]. Furthermore, Alsabagh *et al.* reported on the effectiveness of a multifunctional nanocomposite that contained carbon nanotubes, Ag, Cu, and chitosan for treating water contaminated with harmful metals such as Pb(II), Cd(II), and Cu(II) [103]. With the capacity to renew, these nanocomposites removed metal ions almost entirely and quickly in less than ten minutes. A reusable nanocomposite made of nanocellulose and silver NPs embedded in pebbles was created by Suman *et al.* to provide a thorough method of eliminating heavy metals, dyes, and microbiological burdens from water [104].

Transformation

Emerging toxins build up in the environment as a result of anthropogenic activities. Organic and inorganic contaminants can be effectively removed from contaminated locations by transformation through oxidation or reduction. Metal speciation and a decrease in metal toxicity may result from these interactions. In addition, modifications to the site and/or state of toxicity result from the occurrence of electron-transfer processes to steady oxidation states. Nano zero-valent iron (NZVI) has been used to explore the transformations of heavy metals, including Cd, Cr, Ni, Zn, and Pb, in great detail [105]. Numerous bench- and field-scale investigations have looked into the decrease of Cr(VI) by NZVI particles. NZVI acts as an electron donor in Cr(VI) reduction, releasing ferrous iron, atomically active hydrogen, and molecular hydrogen through NZVI corrosion. These by-products aid in the reduction process. In a separate study, NZVI particles efficiently removed 85% of Zn^{2+} from an aqueous solution in the presence of dissolved oxygen (DO) [106]. Corrosion by DO led to the development of an iron (oxy) hydroxide shell on the NZVI particle surface, enhancing Zn^{2+} adsorption and co-precipitation. The redox cycling between metal ions and NMs may occur, and in the context of environmental remediation, "transformation" is commonly used to describe the removal, transport, and toxicity of both organic and inorganic chemical pollutants.

Catalysis

To create efficient catalysts, utilizing cost-effective and environmentally friendly raw materials is crucial for generating high-value products. Achieving an active catalyst involves reducing particle size, increasing the number of sites, and expanding surface area. The biological synthesis of NMs presents an affordable and eco-friendly option for active catalysis. In the domain of wastewater treatment, magnetic nanoscale materials, particularly iron oxide magnetic NMs/, dominate both bench-scale research and field applications. These materials stand out as promising solutions for heavy metal treatment due to their ease of magnetic separation and effective treatment of large wastewater volumes [107].

Photocatalysis refers to the acceleration of a photoreaction in the presence of a catalyst, applicable to the photodegradation of both organic and inorganic pollutants. A catalyst composed of semiconductor matter is first made sensitive to visible or ultraviolet light *via* photocatalysis, leading to the formation of electron-hole pairs. When incident light energy matches or exceeds the bandgap energy of the semiconductor, electrons are excited from the valence band to the conduction band, creating holes in the valence band. These holes induce oxidative breakdown of organic contaminants by splitting water molecules into hydrogen gas and hydroxyl radicals. Redox reactions are triggered by the reaction between the dissolved oxygen and the conductive band electrons, which produces superoxide anions. In the presence of light, these holes and electrons proceed through a series of oxidation and reduction processes that finally break down organic contaminants. These days, there is a lot of interest in the biological synthesis of metal oxide NPs, including zinc, titanium, and silver, with the purpose of using photocatalysis to remove pollutants from the environment. Photocatalytic degradation is highly effective due to its rapid oxidation process, excellent stability, and cost-effectiveness. The stem, bark, and root of *Helicteres isora* extracts were used by Bhakya *et al.* to biosynthesize silver NPs, investigating their role in degrading various chemical dyes [108]. The catalytic degradation of methylene blue dye was also studied using biosynthesized gold and silver NPs. Additionally, Shi *et al.* demonstrated the successful synthesis of Au- using *Pycnoporus sanguineus* intracellular protein extract (IPE) and its efficient catalytic destruction of 4-nitroaniline [109]. TiO_2 NPs, which may be produced naturally by bacteria and plants, have been shown to function as nanocatalysts in the breakdown of both organic and inorganic pollutants and as disinfectants.

Hydrogen peroxide (H_2O_2) oxidizes the Fe^{2+} ion to Fe^{3+} in the heterogeneous Fenton reaction, producing a hydroxide ion (OH^-) and a hydroxyl radical (·OH). An additional H_2O_2 molecule catalyzes the breakdown of Fe^{3+} to create Fe^{2+} and produces a proton and a hydroperoxyl radical (·OOH). These free radicals are strong oxidizing agents that aid in the oxidation of organic contaminants. Li *et al.* found that white rot fungi enhance hematite NP-mediated bisphenol A degradation through a Fenton-like reaction [110]. Green-synthesized Fe displayed the Fenton reaction for nitrate removal from wastewater. Garcia-Segura *et al.* extensively assessed the fluidized-bed Fenton (FBF) technique and the Fenton reaction. Biologically produced Fe NPs are emerging as a highly favorable method for reducing both organic and inorganic contaminants in the environment. The recent progress in the FBF technique encompasses the treatment of both real wastewater effluents and synthetic wastewater.

EXPLORING THE TOXICITY PROFILES OF BIONMS: CHALLENGES AND CONSIDERATIONS

NMs produced through physical and chemical methods are typically viewed as toxic due to the use of harmful chemicals or the presence of toxic byproducts and functional groups. In an effort to advance their biocompatibility, bioreactivity, and bioavailability, bioNMs are suggested as a replacement for traditional hazardous NMs in biomedical applications. Gradually, a new area of nanotechnology-based remediation has evolved as a result of the unearthing of multiple biologically generated NMs/, which have been successfully employed in the remediation of several organic and inorganic contaminants. Recent research, however, suggests that these bioNMs might react adversely to human cells and animal models, especially when used in high doses or concentrations. Saha *et al.* [111] utilized aqueous extracts from *Swertia chirata* leaves to synthesize spherical-shaped silver NPs with a size of 20 nm. They then exposed these NPs to *Allium cepa* to evaluate their toxicity. The study revealed that *A. cepa* exhibited adverse effects even at low doses of the bioNM, leading to reduced mitotic and meiotic indices and various chromosomal abnormalities with increasing nanomaterial concentration [111]. Similarly, Sharma *et al.* investigated the toxicity of silica NPs derived from rice husk ash biomass [112]. The study indicated that the bioNM exhibited concentration-dependent high toxicity against pathogenic bacterial strains *S. aureus* and *E. coli*, along with enhanced bioadhesion properties. Khan *et al.* showed how to use *Rumex acetosa* to synthesize spherical silver NPs that are 20–25 nm in size. The toxicity of these NPs was seen in both human cell lines and zebrafish models, which can result in oxidative stress, apoptotic cell death, and cell senescence [112].

CONCLUSION

Novel bioNMs have been developed through the latest progress in nanoscience, offering a substitute for traditional counterparts in various biomedical applications. The nomenclature and toxicity assessment techniques for bioNMs vary across national borders, making it exceedingly difficult to standardize and apply them in industrial settings. Furthermore, it is yet unclear whether a particular biomolecule reduces the stabilization of nanosized material, making the mechanism of bioNM production an issue of intense discussion. The most frequently proposed mechanism for the generation of NMs is the collaborative effect of biomolecules found in microbial or plant extract; nevertheless, toxicity studies will benefit from the identification of specific biomolecules. As of now, no bioNM has received approval for use in biomedicine, and bioNMs are subject to the same rules governing NMs.

Despite challenges like instability in biological fluids and morphological control issues, ongoing research aims to enhance these materials. Notably, bioNMs exhibit lower and generally negligible toxicity compared to conventional NPs, making them a more favorable choice for biological applications. Further research in this field promises the creation of innovative bioNMs, understanding their distinctive mechanisms, and obtaining regulatory approval for future commercial applications.

ACKNOWLEDGEMENT

The author is thankful for the financial assistance provided by the Department of Science & Technology and Biotechnology, Government of West Bengal, India (Memo No. 860(Sanc.)/STBT-11012(25)/5/2019-ST SEC dated 03/11/2023).

REFERENCES

[1]   Gajanan K, Tijare SN. Applications of nanomaterials. Mater Today Proc 2018; 5(1): 1093-6.
[http://dx.doi.org/10.1016/j.matpr.2017.11.187]

[2]   Prakash Sharma V, Sharma U, Chattopadhyay M, Shukla VN. Advance applications of NMs: a review. Mater Today Proc 2018; 5(2): 6376-80.
[http://dx.doi.org/10.1016/j.matpr.2017.12.248]

[3]   Kolahalam LA, Kasi Viswanath IV, Diwakar BS, Govindh B, Reddy V, Murthy YLN. Review on nanomaterials: Synthesis and applications. Mater Today Proc 2019; 18: 2182-90.
[http://dx.doi.org/10.1016/j.matpr.2019.07.371]

[4]   Wagner S, Gondikas A, Neubauer E, Hofmann T, von der Kammer F. Spot the difference: engineered and natural nanoparticles in the environment--release, behavior, and fate. Angew Chem Int Ed 2014; 53(46): 12398-419.
[http://dx.doi.org/10.1002/anie.201405050] [PMID: 25348500]

[5]   Sahayaraj K, Rajesh S. BioNPs: synthesis and antimicrobial applications Science Against Microbial Pathogens: Communicating Current Research and Technological Advances 2011; 23: 228-44.

[6]   Wei G, Ed. Self-Assembled Bio-NMs: Synthesis. Characterization, and Applications 2020.

[7]   Mishra RK, Ha SK, Verma K, Tiwari SK. Recent progress in selected bio-nanomaterials and their engineering applications: An overview. J Sci Adv Mater Devices 2018; 3(3): 263-88.
[http://dx.doi.org/10.1016/j.jsamd.2018.05.003]

[8]   Karanassios V. Brief introduction to nanoscience and nanotechnology. Nanoscience Journal 2018; 1(1): 1-6.

[8]   Aravinthan A, Govarthanan M, Selvam K, *et al.* Sunroot mediated synthesis and characterization of silver nanoparticles and evaluation of its antibacterial and rat splenocyte cytotoxic effects. Int J Nanomedicine 2015; 10: 1977-83.
[PMID: 25792831]

[9]   Chinnappan S, Kandasamy S, Arumugam S, Seralathan KK, Thangaswamy S, Muthusamy G. Biomimetic synthesis of silver nanoparticles using flower extract of Bauhinia purpurea and its antibacterial activity against clinical pathogens. Environ Sci Pollut Res Int 2018; 25(1): 963-9.
[http://dx.doi.org/10.1007/s11356-017-0841-1] [PMID: 29218578]

[10]   Ameen F, Srinivasan P, Selvankumar T, *et al.* Phytosynthesis of silver nanoparticles using Mangifera indica flower extract as bioreductant and their broad-spectrum antibacterial activity. Bioorg Chem 2019; 88: 102970.

[http://dx.doi.org/10.1016/j.bioorg.2019.102970] [PMID: 31174009]

[11]   Ameen F, AlYahya S, Govarthanan M, *et al.* Soil bacteria Cupriavidus sp. mediates the extracellular synthesis of antibacterial silver nanoparticles. J Mol Struct 2020; 1202: 127233.
[http://dx.doi.org/10.1016/j.molstruc.2019.127233]

[12]   Golinska P, Wypij M, Ingle AP, Gupta I, Dahm H, Rai M. Biogenic synthesis of metal nanoparticles from actinomycetes: biomedical applications and cytotoxicity. Appl Microbiol Biotechnol 2014; 98(19): 8083-97.
[http://dx.doi.org/10.1007/s00253-014-5953-7] [PMID: 25158833]

[13]   Jogaiah S, Kurjogi M, Abdelrahman M, Hanumanthappa N, Tran LSP. Ganoderma applanatum-mediated green synthesis of silver nanoparticles: Structural characterization, and *in vitro* and *in vivo* biomedical and agrochemical properties. Arab J Chem 2019; 12(7): 1108-20.
[http://dx.doi.org/10.1016/j.arabjc.2017.12.002]

[14]   Widikdo W, Sugita P, Khotib M, Restu W K, Rahmayeni R, Agusnar H. Optimizing the Synthesis of Silver NPs Using Five Curcuma Species under UV Irradiation as a Potent Sunscreen Substrate Tropical Journal of Natural Product Research 2023; 7(5): 100-20.

[15]   Tian M, Ticer T, Wang Q, *et al.* Adipose☐derived biogenic NPs for suppression of inflammation. Small 2020; 16(10): 1904064.
[http://dx.doi.org/10.1002/smll.201904064] [PMID: 32067382]

[16]   Bhavya G, Belorkar SA, Mythili R, *et al.* Remediation of emerging environmental pollutants: A review based on advances in the uses of eco-friendly biofabricated nanomaterials. Chemosphere 2021; 275: 129975.
[http://dx.doi.org/10.1016/j.chemosphere.2021.129975] [PMID: 33631403]

[17]   Moise S, Céspedes E, Soukup D, Byrne JM, El Haj AJ, Telling ND. The cellular magnetic response and biocompatibility of biogenic zinc- and cobalt-doped magnetite nanoparticles. Sci Rep 2017; 7(1): 39922.
[http://dx.doi.org/10.1038/srep39922] [PMID: 28045082]

[18]   Geetha N, Bhavya G, Abhijith P, Shekhar R, Dayananda K, Jogaiah S. Insights into nanomycoremediation: Secretomics and mycogenic biopolymer nanocomposites for heavy metal detoxification. J Hazard Mater 2021; 409: 124541.
[http://dx.doi.org/10.1016/j.jhazmat.2020.124541] [PMID: 33223321]

[19]   Karatutlu A, Barhoum A, Sapelkin A. Theories of nanoparticle and nanostructure formation in liquid phase. Emerging Applications of NPs and Architecture Nanostructures. Elsevier 2018; pp. 597-619.
[http://dx.doi.org/10.1016/B978-0-323-51254-1.00020-8]

[20]   Mulvaney P. Nanoscience vs nanotechnology--defining the field. ACS Nano 2015; 9(3): 2215-7.
[http://dx.doi.org/10.1021/acsnano.5b01418] [PMID: 25802086]

[21]   Dolez PI. NMs definitions, classifications, and applications. Nanoengineering. Elsevier 2015; pp. 3-40.
[http://dx.doi.org/10.1016/B978-0-444-62747-6.00001-4]

[22]   Oksman K, Mathew AP, Bismarck A, Rojas O, Sain M, Eds. Handbook of green materials: processing technologies, properties and applications (in 4 volumes). World Scientific 2014; 5.
[http://dx.doi.org/10.1142/8975]

[23]   Barhoum A, Jeevanandam J, Rastogi A, *et al.* Plant celluloses, hemicelluloses, lignins, and volatile oils for the synthesis of nanoparticles and nanostructured materials. Nanoscale 2020; 12(45): 22845-90.
[http://dx.doi.org/10.1039/D0NR04795C] [PMID: 33185217]

[24]   Sharma VK, Agrawal MK. A historical perspective of liposomes-a bio nanomaterial. Mater Today Proc 2021; 45: 2963-6.
[http://dx.doi.org/10.1016/j.matpr.2020.11.952]

[25]   Jeevanandam J, San Chan Y, Jing Wong Y, Siang Hii Y. Biogenic synthesis of magnesium oxide

nanoparticles using Aloe barbadensis leaf latex extract. IOP Conf Series Mater Sci Eng 2020; 943(1): 012030. []. IOP Publishing.].
[http://dx.doi.org/10.1088/1757-899X/943/1/012030]

[26] Jonoobi M, Mathew AP, Oksman K. Natural resources and residues for production of bioNMs. In HANDBOOK OF GREEN MATERIALS: 1 BioNMs: separation processes, characterization and properties. 2014; pp. 19-33.

[27] Salunke BK, Sawant SS, Lee SI, Kim BS. Microorganisms as efficient biosystem for the synthesis of metal nanoparticles: current scenario and future possibilities. World J Microbiol Biotechnol 2016; 32(5): 88.
[http://dx.doi.org/10.1007/s11274-016-2044-1] [PMID: 27038958]

[28] Jacob JM, Ravindran R, Narayanan M, Samuel SM, Pugazhendhi A, Kumar G. Microalgae: A prospective low cost green alternative for nanoparticle synthesis. Curr Opin Environ Sci Health 2021; 20: 100163.
[http://dx.doi.org/10.1016/j.coesh.2019.12.005]

[29] Fabrega J, Fawcett SR, Renshaw JC, Lead JR. Silver nanoparticle impact on bacterial growth: effect of pH, concentration, and organic matter. Environ Sci Technol 2009; 43(19): 7285-90.
[http://dx.doi.org/10.1021/es803259g] [PMID: 19848135]

[30] Barabadi H, Honary S, Ali Mohammadi M, *et al.* Green chemical synthesis of gold nanoparticles by using Penicillium aculeatum and their scolicidal activity against hydatid cyst protoscolices of Echinococcus granulosus. Environ Sci Pollut Res Int 2017; 24(6): 5800-10.
[http://dx.doi.org/10.1007/s11356-016-8291-8] [PMID: 28054267]

[31] Pugazhendhi A, Prabakar D, Jacob JM, Karuppusamy I, Saratale RG. Synthesis and characterization of silver nanoparticles using Gelidium amansii and its antimicrobial property against various pathogenic bacteria. Microb Pathog 2018; 114: 41-5.
[http://dx.doi.org/10.1016/j.micpath.2017.11.013] [PMID: 29146498]

[32] Soto CM, Ratna BR. Virus hybrids as nanomaterials for biotechnology. Curr Opin Biotechnol 2010; 21(4): 426-38.
[http://dx.doi.org/10.1016/j.copbio.2010.07.004] [PMID: 20688511]

[33] Zhang Y, Dong Y, Zhou J, Li X, Wang F. Application of plant viruses as a biotemplate for nanomaterial fabrication. Molecules 2018; 23(9): 2311.
[http://dx.doi.org/10.3390/molecules23092311] [PMID: 30208562]

[34] Wen AM, Steinmetz NF. Design of virus-based nanomaterials for medicine, biotechnology, and energy. Chem Soc Rev 2016; 45(15): 4074-126.
[http://dx.doi.org/10.1039/C5CS00287G] [PMID: 27152673]

[35] Culver JN, Brown AD, Zang F, Gnerlich M, Gerasopoulos K, Ghodssi R. Plant virus directed fabrication of nanoscale materials and devices. Virology 2015; 479-480: 200-12.
[http://dx.doi.org/10.1016/j.virol.2015.03.008] [PMID: 25816763]

[36] Kostiainen MA, Hiekkataipale P, Laiho A, *et al.* Electrostatic assembly of binary nanoparticle superlattices using protein cages. Nat Nanotechnol 2013; 8(1): 52-6.
[http://dx.doi.org/10.1038/nnano.2012.220] [PMID: 23241655]

[37] Millán JG, Brasch M, Anaya-Plaza E, *et al.* Self-assembly triggered by self-assembly: Optically active, paramagnetic micelles encapsulated in protein cage nanoparticles. J Inorg Biochem 2014; 136: 140-6.
[http://dx.doi.org/10.1016/j.jinorgbio.2014.01.004] [PMID: 24513535]

[38] Steele JFC, Peyret H, Saunders K, *et al.* Synthetic plant virology for nanobiotechnology and nanomedicine. Wiley Interdiscip Rev Nanomed Nanobiotechnol 2017; 9(4): e1447.
[http://dx.doi.org/10.1002/wnan.1447] [PMID: 28078770]

[39] Gunnoo SB, Madder A. Chemical protein modification through cysteine. ChemBioChem 2016; 17(7):

529-53.
[http://dx.doi.org/10.1002/cbic.201500667] [PMID: 26789551]

[40]   Thirumurugan P, Matosiuk D, Jozwiak K. Click chemistry for drug development and diverse chemical-biology applications. Chem Rev 2013; 113(7): 4905-79.
[http://dx.doi.org/10.1021/cr200409f] [PMID: 23531040]

[41]   Seena S, Rai A, Kumar S, Eds. NPs and Plant-Microbe Interactions: An Environmental Perspective. Elsevier 2023.

[42]   Vijayaraghavan K, Ashokkumar T. Plant-mediated biosynthesis of metallic nanoparticles: A review of literature, factors affecting synthesis, characterization techniques and applications. J Environ Chem Eng 2017; 5(5): 4866-83.
[http://dx.doi.org/10.1016/j.jece.2017.09.026]

[43]   Marslin G, Siram K, Maqbool Q, *et al.* Secondary metabolites in the green synthesis of metallic NPs. Materials (Basel) 2018; 11(6): 940.
[http://dx.doi.org/10.3390/ma11060940] [PMID: 29865278]

[44]   Kambale EK, Nkanga CI, Mutonkole BPI, *et al.* Green synthesis of antimicrobial silver nanoparticles using aqueous leaf extracts from three Congolese plant species (Brillantaisia patula, Crossopteryx febrifuga and Senna siamea). Heliyon 2020; 6(8): e04493.
[http://dx.doi.org/10.1016/j.heliyon.2020.e04493] [PMID: 32793824]

[45]   Jeevanandam J, Pal K, Danquah MK. Virus-like nanoparticles as a novel delivery tool in gene therapy. Biochimie 2019; 157: 38-47.
[http://dx.doi.org/10.1016/j.biochi.2018.11.001] [PMID: 30408502]

[46]   Abbas S, Nasreen S, Haroon A, Ashraf MA. Synhesis of silver and copper NPs from plants and application as adsorbents for naphthalene decontamination. Saudi J Biol Sci 2020; 27(4): 1016-23.
[http://dx.doi.org/10.1016/j.sjbs.2020.02.011] [PMID: 32256162]

[47]   Bertolino V, Cavallaro G, Milioto S, Lazzara G. Polysaccharides/Halloysite nanotubes for smart bionanocomposite materials. Carbohydr Polym 2020; 245: 116502.
[http://dx.doi.org/10.1016/j.carbpol.2020.116502] [PMID: 32718613]

[48]   Zhou R, Zhao L, Wang Y, *et al.* Recent advances in food-derived nanomaterials applied to biosensing. Trends Analyt Chem 2020; 127: 115884.
[http://dx.doi.org/10.1016/j.trac.2020.115884]

[49]   Fang Y, Fu J, Liu P, Cu B. Morphology and characteristics of 3D nanonetwork porous starch-based nanomaterial *via* a simple sacrifice template approach for clove essential oil encapsulation. Ind Crops Prod 2020; 143: 111939.
[http://dx.doi.org/10.1016/j.indcrop.2019.111939]

[50]   Ahmad AN, Lim SA, Navaranjan N, Hsu YI, Uyama H. Green sago starch nanoparticles as reinforcing material for green composites. Polymer (Guildf) 2020; 202: 122646.
[http://dx.doi.org/10.1016/j.polymer.2020.122646]

[51]   Abascal NC, Regan L. The past, present and future of protein-based materials. Open Biol 2018; 8(10): 180113.
[http://dx.doi.org/10.1098/rsob.180113] [PMID: 30381364]

[52]   Freeman A. Protein-mediated biotemplating on the nanoscale. Biomimetics (Basel) 2017; 2(3): 14.
[http://dx.doi.org/10.3390/biomimetics2030014] [PMID: 31105177]

[53]   Chen X, Li D, Li G, *et al.* Facile fabrication of gold nanoparticle on zein ultrafine fibers and their application for catechol biosensor. Appl Surf Sci 2015; 328: 444-52.
[http://dx.doi.org/10.1016/j.apsusc.2014.12.070]

[54]   Das S, Chakraborty J, Chatterjee S, Kumar H. Prospects of biosynthesized nanomaterials for the remediation of organic and inorganic environmental contaminants. Environ Sci Nano 2018; 5(12): 2784-808.

[http://dx.doi.org/10.1039/C8EN00799C]

[55]   Jeevanandam J, Ling JKU, Barhoum A, San Chan Y, Danquah MK. BioNMs: Definitions, sources, types, properties, toxicity, and regulations. Fundamentals of BioNMs. Elsevier 2022; pp. 1-29.

[56]   Thakkar KN, Mhatre SS, Parikh RY. Biological synthesis of metallic nanoparticles. Nanomedicine 2010; 6(2): 257-62.
       [http://dx.doi.org/10.1016/j.nano.2009.07.002] [PMID: 19616126]

[57]   Sastry M, Ahmad A, Khan MI, Kumar R. Biosynthesis of metal NPs using fungi and actinomycete. Curr Sci 2003; •••: 162-70.

[58]   Johnson A, Merilis G, Hastings J, Elizabeth Palmer M, Fitts JP, Chidambaram D. Reductive degradation of organic compounds using microbial nanotechnology. J Electrochem Soc 2013; 160(1): G27-31.
       [http://dx.doi.org/10.1149/2.053301jes]

[59]   Lloyd JR, Yong P, Macaskie LE. Enzymatic recovery of elemental palladium by using sulfate-reducing bacteria. Appl Environ Microbiol 1998; 64(11): 4607-9.
       [http://dx.doi.org/10.1128/AEM.64.11.4607-4609.1998] [PMID: 9797331]

[60]   Yong P, Rowson NA, Farr JPG, Harris IR, Macaskie LE. Bioreduction and biocrystallization of palladium by *Desulfovibrio desulfuricans* NCIMB 8307. Biotechnol Bioeng 2002; 80(4): 369-79.
       [http://dx.doi.org/10.1002/bit.10369] [PMID: 12325145]

[61]   White BR, Stackhouse BT, Holcombe JA. Magnetic γ-Fe2O3 nanoparticles coated with poly--cysteine for chelation of As(III), Cu(II), Cd(II), Ni(II), Pb(II) and Zn(II). J Hazard Mater 2009; 161(2-3): 848-53.
       [http://dx.doi.org/10.1016/j.jhazmat.2008.04.105] [PMID: 18571848]

[62]   Chidambaram D, Hennebel T, Taghavi S, *et al.* Concomitant microbial generation of palladium nanoparticles and hydrogen to immobilize chromate. Environ Sci Technol 2010; 44(19): 7635-40.
       [http://dx.doi.org/10.1021/es101559r] [PMID: 20822130]

[63]   Watson JHP, Ellwood DC, Soper AK, Charnock J. Nanosized strongly-magnetic bacterially-produced iron sulfide materials. J Magn Magn Mater 1999; 203(1-3): 69-72.
       [http://dx.doi.org/10.1016/S0304-8853(99)00191-2]

[64]   Singh S, Barick KC, Bahadur D. Surface engineered magnetic nanoparticles for removal of toxic metal ions and bacterial pathogens. J Hazard Mater 2011; 192(3): 1539-47.
       [http://dx.doi.org/10.1016/j.jhazmat.2011.06.074] [PMID: 21784580]

[65]   Peng Q, Liu Y, Zeng G, Xu W, Yang C, Zhang J. Biosorption of copper(II) by immobilizing Saccharomyces cerevisiae on the surface of chitosan-coated magnetic nanoparticles from aqueous solution. J Hazard Mater 2010; 177(1-3): 676-82.
       [http://dx.doi.org/10.1016/j.jhazmat.2009.12.084] [PMID: 20060211]

[66]   Fang L, Cai P, Chen W, Liang W, Hong Z, Huang Q. Impact of cell wall structure on the behavior of bacterial cells in the binding of copper and cadmium. Colloids Surf A Physicochem Eng Asp 2009; 347(1-3): 50-5.
       [http://dx.doi.org/10.1016/j.colsurfa.2008.11.041]

[67]   Nair B, Pradeep T. Coalescence of nanoclusters and formation of submicron crystallites assisted by Lactobacillus strains. Cryst Growth Des 2002; 2(4): 293-8.
       [http://dx.doi.org/10.1021/cg0255164]

[68]   Husseiny MI, El-Aziz MA, Badr Y, Mahmoud MA. Biosynthesis of gold nanoparticles using Pseudomonas aeruginosa. Spectrochim Acta A Mol Biomol Spectrosc 2007; 67(3-4): 1003-6.
       [http://dx.doi.org/10.1016/j.saa.2006.09.028] [PMID: 17084659]

[69]   Pavlovic M, Rouster P, Szilagyi I. Synthesis and formulation of functional bionanomaterials with superoxide dismutase activity. Nanoscale 2017; 9(1): 369-79.
       [http://dx.doi.org/10.1039/C6NR07672F] [PMID: 27924343]

[70] Majumder DR. Bioremediation: copper NPs from electronic-waste. Int J Eng Sci Technol 2012; 4(10)

[71] Ahmed S, Annu , Ikram S, Yudha S S. Biosynthesis of gold nanoparticles: A green approach. J Photochem Photobiol B 2016; 161: 141-53.
[http://dx.doi.org/10.1016/j.jphotobiol.2016.04.034] [PMID: 27236049]

[72] Oremland RS, Herbel MJ, Blum JS, *et al.* Structural and spectral features of selenium nanospheres produced by Se-respiring bacteria. Appl Environ Microbiol 2004; 70(1): 52-60.
[http://dx.doi.org/10.1128/AEM.70.1.52-60.2004] [PMID: 14711625]

[73] Baesman SM, Bullen TD, Dewald J, *et al.* Formation of tellurium nanocrystals during anaerobic growth of bacteria that use Te oxyanions as respiratory electron acceptors. Appl Environ Microbiol 2007; 73(7): 2135-43.
[http://dx.doi.org/10.1128/AEM.02558-06] [PMID: 17277198]

[74] Samrot AV, Burman U, Philip SA, N S, Chandrasekaran K. Synthesis of curcumin loaded polymeric nanoparticles from crab shell derived chitosan for drug delivery. Inform Med Unlocked 2018; 10: 159-82.
[http://dx.doi.org/10.1016/j.imu.2017.12.010]

[75] Sathiyabama M, Parthasarathy R. Biological preparation of chitosan nanoparticles and its *in vitro* antifungal efficacy against some phytopathogenic fungi. Carbohydr Polym 2016; 151: 321-5.
[http://dx.doi.org/10.1016/j.carbpol.2016.05.033] [PMID: 27474573]

[76] Umesh M, Priyanka K, Thazeem B, Preethi K. Biogenic PHA nanoparticle synthesis and characterization from *Bacillus subtilis* NCDC0671 using orange peel medium. Int J Polym Mater 2018; 67(17): 996-1004.
[http://dx.doi.org/10.1080/00914037.2017.1417284]

[77] Emmanuel R, Karuppiah C, Chen SM, Palanisamy S, Padmavathy S, Prakash P. Green synthesis of gold nanoparticles for trace level detection of a hazardous pollutant (nitrobenzene) causing Methemoglobinaemia. J Hazard Mater 2014; 279: 117-24.
[http://dx.doi.org/10.1016/j.jhazmat.2014.06.066] [PMID: 25048622]

[78] Silva-De Hoyos LE, Sánchez-Mendieta V, Camacho-López MA, Trujillo-Reyes J, Vilchis-Nestor AR. Plasmonic and fluorescent sensors of metal ions in water based on biogenic gold nanoparticles. Arab J Chem 2020; 13(1): 1975-85.
[http://dx.doi.org/10.1016/j.arabjc.2018.02.016]

[79] Nag S, Pramanik A, Chattopadhyay D, Bhattacharyya M. Green-fabrication of gold nanomaterials using Staphylococcus warneri from Sundarbans estuary: an effective recyclable nanocatalyst for degrading nitro aromatic pollutants. Environ Sci Pollut Res Int 2018; 25(3): 2331-49.
[http://dx.doi.org/10.1007/s11356-017-0617-7] [PMID: 29124636]

[80] Gu D, Shang S, Yu Q, Shen J. Green synthesis of nitrogen-doped carbon dots from lotus root for Hg(II) ions detection and cell imaging. Appl Surf Sci 2016; 390: 38-42.
[http://dx.doi.org/10.1016/j.apsusc.2016.08.012]

[81] Geetha P, Latha MS, Pillai SS, Deepa B, Santhosh Kumar K, Koshy M. Green synthesis and characterization of alginate nanoparticles and its role as a biosorbent for Cr(VI) ions. J Mol Struct 2016; 1105: 54-60.
[http://dx.doi.org/10.1016/j.molstruc.2015.10.022]

[82] Mohan S, Singh Y, Verma DK, Hasan SH. Synthesis of CuO nanoparticles through green route using Citrus limon juice and its application as nanosorbent for Cr(VI) remediation: Process optimization with RSM and ANN-GA based model. Process Saf Environ Prot 2015; 96: 156-66.
[http://dx.doi.org/10.1016/j.psep.2015.05.005]

[83] Shim HW, Jin YH, Seo SD, Lee SH, Kim DW. Highly reversible lithium storage in Bacillus subtilis - directed porous Co_3O_4 nanostructures. ACS Nano 2011; 5(1): 443-9.
[http://dx.doi.org/10.1021/nn1021605] [PMID: 21155558]

[84] Sathyavathi S, Manjula A, Rajendhran J, Gunasekaran P. Extracellular synthesis and characterization of nickel oxide nanoparticles from Microbacterium sp. MRS-1 towards bioremediation of nickel electroplating industrial effluent. Bioresour Technol 2014; 165: 270-3.
[http://dx.doi.org/10.1016/j.biortech.2014.03.031] [PMID: 24685513]

[85] Mukherjee D, Ghosh S, Majumdar S, Annapurna K. Green synthesis of α-Fe 2 O 3 nanoparticles for arsenic(V) remediation with a novel aspect for sludge management. J Environ Chem Eng 2016; 4(1): 639-50.
[http://dx.doi.org/10.1016/j.jece.2015.12.010]

[86] Raj R, Dalei K, Chakraborty J, Das S. Extracellular polymeric substances of a marine bacterium mediated synthesis of CdS nanoparticles for removal of cadmium from aqueous solution. J Colloid Interface Sci 2016; 462: 166-75.
[http://dx.doi.org/10.1016/j.jcis.2015.10.004] [PMID: 26454375]

[87] Sowani H, Mohite P, Munot H, *et al.* Green synthesis of gold and silver nanoparticles by an actinomycete Gordonia amicalis HS-11: Mechanistic aspects and biological application. Process Biochem 2016; 51(3): 374-83.
[http://dx.doi.org/10.1016/j.procbio.2015.12.013]

[88] Nasrollahzadeh M, Mohammad Sajadi S, Rostami-Vartooni A, Alizadeh M, Bagherzadeh M. Green synthesis of the Pd nanoparticles supported on reduced graphene oxide using barberry fruit extract and its application as a recyclable and heterogeneous catalyst for the reduction of nitroarenes. J Colloid Interface Sci 2016; 466: 360-8.
[http://dx.doi.org/10.1016/j.jcis.2015.12.036] [PMID: 26752431]

[89] Sinha T, Ahmaruzzaman M. Biogenic synthesis of Cu nanoparticles and its degradation behavior for methyl red. Mater Lett 2015; 159: 168-71.
[http://dx.doi.org/10.1016/j.matlet.2015.06.099]

[90] Wang Z, Xu C, Li X, Liu Z. in situ green synthesis of Ag nanoparticles on tea polyphenols-modified graphene and their catalytic reduction activity of 4-nitrophenol. Colloids Surf A Physicochem Eng Asp 2015; 485: 102-10.
[http://dx.doi.org/10.1016/j.colsurfa.2015.09.015]

[91] Megarajan S, Ayaz Ahmed KB, Rajendra Kumar Reddy G, Suresh Kumar P, Anbazhagan V. Phytoproteins in green leaves as building blocks for photosynthesis of gold nanoparticles: An efficient electrocatalyst towards the oxidation of ascorbic acid and the reduction of hydrogen peroxide. J Photochem Photobiol B 2016; 155: 7-12.
[http://dx.doi.org/10.1016/j.jphotobiol.2015.12.009] [PMID: 26722997]

[92] Narayanan KB, Park HH, Han SS. Synthesis and characterization of biomatrixed-gold nanoparticles by the mushroom Flammulina velutipes and its heterogeneous catalytic potential. Chemosphere 2015; 141: 169-75.
[http://dx.doi.org/10.1016/j.chemosphere.2015.06.101] [PMID: 26207976]

[93] Machado S, Grosso JP, Nouws HPA, Albergaria JT, Delerue-Matos C. Utilization of food industry wastes for the production of zero-valent iron nanoparticles. Sci Total Environ 2014; 496: 233-40.
[http://dx.doi.org/10.1016/j.scitotenv.2014.07.058] [PMID: 25089685]

[93] Bhattacharjee A, Ahmaruzzaman M. A green approach for the synthesis of SnO2 nanoparticles and its application in the reduction of p-nitrophenol. Mater Lett 2015; 157: 260-4.
[http://dx.doi.org/10.1016/j.matlet.2015.05.053]

[94] Nasrollahzadeh M, Sajadi SM, Rostami-Vartooni A, Hussin SM. Green synthesis of CuO nanoparticles using aqueous extract of Thymus vulgaris L. leaves and their catalytic performance for N-arylation of indoles and amines. J Colloid Interface Sci 2016; 466: 113-9.
[http://dx.doi.org/10.1016/j.jcis.2015.12.018] [PMID: 26707778]

[95] Du L, Xu Q, Huang M, Xian L, Feng JX. Synthesis of small silver nanoparticles under light radiation by fungus Penicillium oxalicum and its application for the catalytic reduction of methylene blue.

Mater Chem Phys 2015; 160: 40-7.
[http://dx.doi.org/10.1016/j.matchemphys.2015.04.003]

[96] Ruthven DM. Principles of adsorption and adsorption processes. John Wiley & Sons 1984.

[97] Hu H, Wang Z, Pan L. Synthesis of monodisperse Fe_3O_4@silica core–shell microspheres and their application for removal of heavy metal ions from water. J Alloys Compd 2010; 492(1-2): 656-61.
[http://dx.doi.org/10.1016/j.jallcom.2009.11.204]

[98] Salvadori MR, Ando RA, Oller do Nascimento CA, Corrêa B. Intracellular biosynthesis and removal of copper nanoparticles by dead biomass of yeast isolated from the wastewater of a mine in the Brazilian Amazonia. PLoS One 2014; 9(1): e87968.
[http://dx.doi.org/10.1371/journal.pone.0087968] [PMID: 24489975]

[99] Govindaraju K, Basha SK, Kumar VG, Singaravelu G. Silver, gold and bimetallic nanoparticles production using single-cell protein (Spirulina platensis) Geitler. J Mater Sci 2008; 43(15): 5115-22.
[http://dx.doi.org/10.1007/s10853-008-2745-4]

[100] Nehru K, Sivakumar M. Biosynthesis of Ag NPs using Amaranthus tristis extract for the fabrication of nanoparticle embedded PVA membrane. Curr Nanosci 2012; 8(5): 703-8.
[http://dx.doi.org/10.2174/157341312802884436]

[101] Alsabagh AM, Fathy M, Morsi RE. Preparation and characterization of chitosan/silver nanoparticle/copper nanoparticle/carbon nanotube multifunctional nano-composite for water treatment: heavy metals removal; kinetics, isotherms and competitive studies. RSC Advances 2015; 5(69): 55774-83.
[http://dx.doi.org/10.1039/C5RA07477K]

[102] Suman K, Kardam A, Gera M, Jain VK. A novel reusable nanocomposite for complete removal of dyes, heavy metals and microbial load from water based on nanocellulose and silver nano-embedded pebbles. Environ Technol 2015; 36(6): 706-14.
[http://dx.doi.org/10.1080/09593330.2014.959066] [PMID: 25243917]

[103] Singh R, Misra V, Singh RP. Removal of hexavalent chromium from contaminated ground water using zero-valent iron nanoparticles. Environ Monit Assess 2012; 184(6): 3643-51.
[http://dx.doi.org/10.1007/s10661-011-2213-5] [PMID: 21769560]

[104] Liang W, Dai C, Zhou X, Zhang Y. Application of zero-valent iron nanoparticles for the removal of aqueous zinc ions under various experimental conditions. PLoS One 2014; 9(1): e85686.
[http://dx.doi.org/10.1371/journal.pone.0085686] [PMID: 24416439]

[105] Neyaz N, Siddiqui WA, Nair KK. Application of surface functionalized iron oxide NMs as a nanosorbents in extraction of toxic heavy metals from ground water: a review. Int J Environ Sci 2014; 4(4): 472-83.

[106] Bhakya S, Muthukrishnan S, Sukumaran M, Muthukumar M, Kumar ST, Rao MV. Catalytic degradation of organic dyes using synthesized silver NPs: a green approach. J Bioremediat Biodegrad 2015; 6(5): 1.

[107] Shi C, Zhu N, Cao Y, Wu P. Biosynthesis of gold nanoparticles assisted by the intracellular protein extract of Pycnoporus sanguineus and its catalysis in degradation of 4-nitroaniline. Nanoscale Res Lett 2015; 10(1): 147.
[http://dx.doi.org/10.1186/s11671-015-0856-9]

[108] Li M, Zhang C. γ-Fe2O3 nanoparticle-facilitated bisphenol A degradation by white rot fungus. Sci Bull (Beijing) 2016; 61(6): 468-72.
[http://dx.doi.org/10.1007/s11434-016-1021-2]

[109] Saha N, Dutta Gupta S. Low-dose toxicity of biogenic silver nanoparticles fabricated by Swertia chirata on root tips and flower buds of Allium cepa. J Hazard Mater 2017; 330: 18-28.
[http://dx.doi.org/10.1016/j.jhazmat.2017.01.021] [PMID: 28208089]

[110] Sharma SK, Sharma AR, Pamidimarri SDVN, *et al.* Bacterial compatibility/toxicity of biogenic silica

(b-SiO2) NPs synthesized from biomass rice husk ash. Nanomaterials (Basel) 2019; 9(10): 1440.
[http://dx.doi.org/10.3390/nano9101440] [PMID: 31614501]

[111]   Khan I, Bahuguna A, Krishnan M, *et al.* The effect of biogenic manufactured silver nanoparticles on human endothelial cells and zebrafish model. Sci Total Environ 2019; 679: 365-77.
[http://dx.doi.org/10.1016/j.scitotenv.2019.05.045] [PMID: 31085416]

[112]   Tan KX, Barhoum A, Pan S, Danquah MK. Risks and toxicity of NPs and nanostructured materials. Emerging applications of NPs and architecture nanostructures. Elsevier 2018; pp. 121-39.
[http://dx.doi.org/10.1016/B978-0-323-51254-1.00005-1]

Nanobioremediation and Phytonanotechnology for Remediation of Various Categories of Pollutants

Nilesh Gupta[1,*], Hemant Khambete[1], Sourab Billore[1], Sanjay Jain[1], Kamal Kant Sharma[2] and Sapna A. Kondalkar[3]

[1] *Faculty of Pharmacy, Medicaps University, Rau, Indore, India*

[2] *Gurukula kangri (Deemed to be university), Haridwar, Uttarakhand, India*

[3] *Regional Ayurved Research Institute, Gwalior, Amkhoo, India*

Abstract: In the present scenario, the most serious threat to the environment and worldwide food safety is the anthropological incursion due to rapid development that has led to severe pollution. Pollutants such as dyes, heavy metals, pesticides, and polycyclic aromatic hydrocarbons can merge into nature in a number of different modes, both naturally and through human activities. These pollutants majorly contaminate soil, water, and air through solubilization, precipitation, and accumulation processes. Various traditional methods such as zeolite adsorption, photocatalysis, electro kinetics, electrochemical advanced oxidation processes, advanced oxidation process, electro-coagulation, ozonation, classical Fenton process, and biological processes are used to overcome the harmful effects of pollutants from the ecosystem, but they have some limitations due to the generation of hazardous compounds, high costs, ineffective clean-up methods, and significant capital needs. Hence, presently, more attention is on alternative methods such as nanobioremediation and phytonanotechnology due to their more effectiveness and eco-friendly nature to achieve better outcomes. A relatively new area of nanotechnology called phytonanotechnology combines nanotechnology and plant biotechnology and aims to produce nanoparticles (NPs) from natural sources by employing the main accessible synthesis methods, using fungal mycelial surfaces, plant bacterial culture, and secondary metabolite extracts. Therefore, it is very crucial to understand these remediation techniques that avoid the production of harmful by-products during the synthesis process. This chapter gives a detailed account of the great efficiency of these methods in environmental remediation.

Keywords: Bioremediation, Environment, Nanoparticles, Pollutant, Phytonanotechnology.

* **Corresponding author Nilesh Gupta:** Faculty of Pharmacy, Medicaps University, Rau, Indore, India;
E-mail: nilesh.gupta@medicaps.ac.in

Neha Agarwal, Vijendra Singh Solanki, Neetu Singh & Maulin P. Shah (Eds.)

INTRODUCTION

Urbanization, as well as the Industrial Revolution, has brought great economical and technological advancements worldwide. These advancements are responsible for increasing environmental pollution through the process of excessive resource extraction, product dissemination, and greater disposal of waste materials into the environment without diligence. Pollutants such as dyes, heavy metals, pesticides, and poly aromatic hydrocarbons (PAHs) contaminate soil, water, and air through solubilization, precipitation, and accumulation processes. Various traditional and advanced methods such as zeolite adsorption, photocatalysis, electro kinetics, electrochemical advanced oxidation processes, electrocoagulation, ozonation, classical Fenton process, and biological processes are used to mitigate the harmful effects of pollutants on the environment, but they have some limitations due to the generation of hazardous by-products, high costs, inefficiency, and significant capital needs. Hence, more attention is being paid to alternative methods such as nanobioremediation and phytonanotechnology due to their better efficiency and eco-friendly nature [1, 2]. Nanobioremediation is a largely accepted naturally occurring waste treatment method because it is more affordable than other remediation techniques for the remediation of pollutants from coastal regions, estuaries, and marine environments [3, 4].

It is an environment-friendly and highly efficient remediation technique that avoids the production of harmful by-products during the synthesis process [5, 6]. The main purpose of this chapter is to discuss the principles of bioremediation assisted by NPs and their interaction with different environmental matrices. The international regulatory frameworks that are applicable to these technologies and how they might contribute to sustainability are also discussed.

DIFFERENT CATEGORIES OF POLLUTANTS

The toxic substances that are released naturally or *via* anthropogenic activities that change the equilibrium of the environment, causing the contamination of soil, water, and air, are known as pollutants. Understanding different types of pollutants is crucial for addressing environmental issues and developing strategies for pollution prevention and remediation. A detailed account of various types of pollutants is discussed here [7].

Organic pollutants

Organic pollutants may be produced by plants or animals, or they can be synthetic chemicals created by human activities. Organic pollutants can have diverse chemical structures and properties. These organic pollutants can have adverse effects on human health, wildlife, and ecosystems. They may persist in the environment for extended periods, leading to long-term impacts. Efforts to manage and reduce the presence of organic pollutants often involve regulatory measures, pollution prevention strategies, and the development of cleaner technologies and practices. Various categories of organic pollutants, their sources, and their impacts are summarized in Table **1**.

Table 1. Different categories of organic pollutants and their impacts.

Categories of Organic Pollutants	Source	Impacts	Example
Polycyclic Aromatic Hydrocarbons (PAHs)	Combustion of fossil fuels, industrial processes, and incomplete combustion of organic matter.	Carcinogenic and mutagenic properties, bioaccumulation in the environment	Benzo-[α]-pyrene, naphthalene [8]
Volatile Organic Compounds (VOCs):	Evaporation of fuels, industrial processes, and use of certain products like paints and solvents.	Contribution to ground-level ozone formation, respiratory and neurological effects, and indoor air pollution.	Benzene, toluene, xylene, *etc* [9].
Pesticides	Agricultural activities, pest control.	Harmful to non-target organisms, bioaccumulation, and potential health risks to humans.	Organophosphates, organochlorines, and carbamates [10].
Chlorinated Solvents (CSs)	Industrial processes, dry cleaning, and use as cleaning agents.	Groundwater contamination, potential health risks	Trichloroethylene (TCE) and perchloroethylene (PCE) [11].
Pharmaceuticals and Personal Care Products (PPCPs)	Disposal of medications and personal care products	Potential ecological impacts, antibiotic resistance concerns	Antibiotics, hormones, personal care product ingredients [12].
Dioxins	Combustion of organic materials, certain industrial processes	Carcinogenic and toxic effects, bioaccumulation	Polychlorinated dibenzo--dioxins (PCDDs) [13].
Polybrominated Flame Retardants (PBDEs)	Flame retardants in electronics, textiles, and furniture	Persistent in the environment, potential endocrine-disrupting properties	Deca-bromo-diphenyl ether (DecaBDE) [14 - 16].
Herbicides	Agricultural and landscaping applications.	Runoff into water bodies, impact on aquatic ecosystems	Atrazine, glyphosate [17].
BTEX Compounds	Combustion of fossil fuels, industrial activities	Ground-level ozone formation, respiratory and neurological effects	Benzene, toluene, ethylbenzene, xylene

Categories of Organic Pollutants	Source	Impacts	Example
Per- and Polyfluoroalkyl Substances (PFAS)	Industrial processes, firefighting foam, and various consumer products	Persistent in the environment, potential health concerns, bioaccumulation	Perfluorooctanoic acid (PFOA), perfluorooctanesulfonic acid (PFOS).

Inorganic Pollutants

Inorganic pollutants are substances that do not contain carbon-hydrogen (C-H) bonds and are typically derived from minerals, metals, and various industrial processes. These pollutants can enter the environment through natural sources as well as human activities. Inorganic pollutants encompass a diverse range of elements and compounds, and their environmental impact depends on various factors such as concentration, chemical form, and persistence. Some common types of inorganic pollutants and their impacts are discussed in Table **2**. Inorganic pollutants can have significant environmental and health impacts, depending on their concentrations and the ecosystems they affect. Addressing inorganic pollution often involves implementing pollution control technologies, regulating industrial emissions, improving waste management practices, and promoting sustainable agricultural and industrial practices.

Table 2. Different categories of inorganic pollutants and their impacts.

Categories	Source	Effect	Example
Heavy Metals	**Natural**: Weathering of rocks and minerals. **Industrial**: Mining, smelting, industrial processes, and the use of heavy metal-containing products (*e.g.*, batteries, paints).	Bind to the body's essential compounds such as oxygen, sulfur, and nitrogen, interrupt the activity of enzymes, and disrupt the synthesis of essential compounds.	**Mercury (Hg):** Released from coal combustion, industrial processes, and certain manufacturing activities [18 - 20]. **Lead (Pb):** From lead-acid batteries, lead-based paints, and industrial activities [20].
Arsenic (As)	**Natural Sources:** Weathering of rocks and minerals. **Human Activities:** Mining, industrial processes, and agricultural applications.	Increase the growth and spread of viral, bacterial, and fungal infections.	**Arsenic Compounds:** Found in certain pesticides and wood preservatives [20 - 24].

(Table 2) cont.....

Categories	Source	Effect	Example
Nitrogen Compounds	**Agricultural Activities:** Use of nitrogen-based fertilizers. **Industrial Processes:** Combustion of fossil fuels, industrial discharges. **Transportation:** Vehicle emissions, especially nitrogen oxides (NOx).		**Nitrogen Oxides (NOx):** Nitric oxide (NO) and nitrogen dioxide (NO2) from combustion processes. **Ammonia (NH₃):** Released from agricultural practices and livestock waste [25, 26].
Phosphorus Compounds	**Agricultural Runoff:** Phosphorus-containing fertilizers and animal manure. **Wastewater Discharges:** Effluents from sewage treatment plants.		**Phosphates (PO₄)²⁻:** Commonly found in fertilizers, detergents, and wastewater.
Sulfur Compounds	**Fossil Fuel Combustion:** Release of sulfur dioxide (SO₂) from burning coal and oil. **Industrial Processes:** Smelting, refining, and other industrial activities.		**Sulfur Dioxide (SO₂):** A common air pollutant from combustion processes. **Hydrogen Sulfide (H₂S):** Emitted from certain industrial processes and natural sources.
Fluorides	**Industrial Emissions:** Aluminum smelting, phosphate fertilizer production. **Natural Sources:** Volcanic activity and weathering of rocks.	Paralyzing bone diseases, blemished teeth, stooped backs, contorted hands and legs, and blindness.	**Hydrogen Fluoride (HF):** Emitted from industrial processes, such as aluminum manufacturing [27].
Chlorides	**Road Salting:** Application of de-icing salts on roads. **Industrial Processes:** Chemical manufacturing.		**Sodium Chloride (NaCl):** Commonly used as road salt.
Cyanides	**Industrial Processes:** Metal plating, mining, and chemical manufacturing. **Combustion:** Some combustion processes produce small amounts of cyanides.	Bind with iron in the blood by creating complexes and its tendency to suffocate animals; cyanide is thought to be an extremely dangerous toxin.	**Hydrogen Cyanide (HCN):** Released from various industrial activities [24].

TRADITIONAL TECHNIQUES OF REMEDIATION OF POLLUTANTS

Remediation of pollutants involves the application of various techniques and strategies to mitigate or eliminate the adverse effects of contaminants on the environment. The choice of remediation method depends on the type of pollutant, the environmental setting, and the specific goals of the cleanup. There are various traditional methods like coagulation, adsorption, and advanced oxidation processes combined with electrochemical and biological processes, *etc.*, have been used, with certain limitations, such as inefficiency in scarping metal ions, high-energy input, and production of no reusable chemicals, before the revealment of nanobiotechnology. Common traditional remediation techniques used for different types of pollutants and their applications are discussed in Table 3.

It is important to note that a combination of remediation techniques, known as a *"remediation train"*, may be employed to address complex contamination scenarios. Additionally, the choice of remediation method should align with sustainability goals and minimize environmental disruption.

Table 3. Various traditional techniques of remediation of pollutants.

S. no.	Traditional Remediation Techniques	Type of Pollutants	Description	Applications
1.	**Bioremediation**	Biological	Bioremediation utilizes living organisms (microbes, plants) to break down or transform pollutants into less harmful substances.	**Microbial Bioremediation:** Bacteria and fungi are used to degrade organic pollutants such as hydrocarbons and pesticides. **Phytoremediation:** Plants absorb, accumulate, or transform pollutants in soil or water. They are used for heavy metal and organic pollutant removal.
2.	**Chemical precipitation**	Chemical	Chemicals are added to a contaminated solution to form insoluble precipitates, removing the contaminants from the water.	**Heavy Metal Removal:** Addition of chemicals like lime or ferrous sulfate to precipitate and remove heavy metals.
3.	**Pump and treat**	Physical/Chemical	Groundwater is pumped to the surface, treated to remove contaminants, and then either discharged or re-injected into the ground.	**Groundwater Contamination:** Commonly used for the removal of pollutants like solvents and petroleum hydrocarbons.
4.	**Soil Vapor Extraction (SVE)**	Physical	Volatile contaminants in the soil are vaporized and removed by extraction, often using vacuum systems.	**Contaminated Soil:** Effective for removing volatile organic compounds (VOCs).

(Table 3) cont.....

S. no.	Traditional Remediation Techniques	Type of Pollutants	Description	Applications
5.	*In situ Chemical Oxidation (ISCO)*	Chemical	Chemical agents (*e.g.*, hydrogen peroxide, permanganate ion) are injected into the subsurface to chemically oxidize and break down contaminants.	**Organic Contaminants:** Useful for treating groundwater contaminated with hydrocarbons or other organic compounds.
6.	*Ex situ* **Thermal Treatment**	Physical	Contaminated soil or sediment is excavated and subjected to high temperatures to volatilize or destroy pollutants.	**Highly Contaminated Sites:** Effective for treating soils contaminated with persistent organic pollutants.
7.	**Activated Carbon Adsorption**	Physical	Pollutants are adsorbed onto activated carbon particles, removing them from air or water.	**Water Treatment:** Effective for removing a wide range of contaminants, including organic pollutants and certain heavy metals.
8.	**Landfill Gas Collection**	Physical	Collection systems capture and extract gases produced by decomposing waste in landfills, preventing their release into the atmosphere.	**Methane Emissions:** Reduces greenhouse gas emissions from landfills.
9.	**Phytostabilization**	Biological	Plants are used to immobilize contaminants in the soil, preventing their migration.	**Heavy Metal Contamination:** Suitable for stabilizing areas with elevated levels of heavy metals.
10.	**Electrokinetic Remediation**	Physical/Chemical	An electric field is applied to the soil, promoting the movement of contaminants toward collection electrodes.	**Soil Remediation:** Effective for treating soils contaminated with heavy metals and certain organic compounds [28].

NANOBIOREMEDIATION AND PHYTONANOTECHNOLOGY FOR REMEDIATION OF POLLUTANTS

Nanotechnology

Nanotechnology is a field of science and engineering focused on creating materials at the scale of atoms and molecules. It holds promise for significantly contributing to the development of cleaner, more sustainable technologies with significant environmental and health benefits. Researchers are exploring nanotechnology techniques for their potential to address pollution and enhance traditional environmental cleanup methods. For instance, in recent years, NMs have been used to remediate contaminated soil and groundwater at hazardous waste sites, such as those affected by chlorinated solvents or oil spills. One commonly employed treatment method involves immobilization or adsorption, which offers effectiveness, affordability, and environmentally friendly removal of metal pollutants from soils [29]. Similarly, carbon-based NMs, including carbon nanotubes, metal oxides (such as ferric oxide and titanium oxide), and various nano-composites, have been employed to immobilize soil pollutants. For instance, ferric oxide NPs demonstrate exceptional potential for adsorbing and immobilizing heavy metals like Cd and As from various environmental samples. Contamination from oily sewage poses a significant challenge to aquatic ecosystems. Iron NPs have shown effectiveness in removing Total Petroleum Hydrocarbons (TPHs) from water, yielding improved results of up to 88.34%. Thus, nanotechnology-based treatments offer the potential to produce high-quality treated water with reduced impurities and toxic substances, along with the removal of heavy metals [30]. Moreover, for the removal of methylene blue dye from water, electrospun polyether sulfone nanofibers embedded with vanadium NPs are employed. Due to their low isoelectric point, these nanofibers exhibit a substantial and highly hydroxylated surface area when in an alkaline medium, facilitating the adsorption of cationic methylene blue molecules [31]. Nanotechnology offers various methods for air purification, including nanofilters, which surpass traditional filters in efficiency. Nanofilters have pores ranging from 1 to 10 nm, effectively removing microorganisms and organic pollutants. Carbon nanotube-based membranes demonstrate superior CO_2 separation capabilities compared to conventional methods. Nanosensors, advanced 3D circular devices, detect and respond to detectable physical changes at the nanometer scale. They function as sensors, alerting to the presence of toxins in a given space [32].

Bioremediation and nanobioremediation

The biological process for the remediation of pollutants is called bioremediation, which helps in the complete degradation of pollutants into basic and non-dangerous waste. Presently, this technique is employed for the management of waste and remediation of polluted soil and groundwater. Mycoremediation (fungi or their compounds to remediate environmental pollutants) has shown to be a cost-efficient, environmentally friendly, and effective method of environmental remediation that includes organic, inorganic, and emerging contaminants, but it has some limitations due to the inability to determine an accurate degradation time for organic pollutants and lack of field studies. Nanobioremediation, an innovative approach at the intersection of nanotechnology and bioremediation, is gaining attention as a promising strategy to address environmental pollution. In general, nanobioremediation is the involvement of NPs or NMs produced by plants, algae, yeast, fungi, and/or bacteria through nanotechnology to eliminate environmental contaminants such as heavy metals and organic and inorganic waste. The scope of nanobiotechnology has a higher potential to support extensively managing major global environmental challenges. One application of nanobioremediation is the enhancement of microbial remediation, where NPs (zero-valent iron) are employed to improve the efficiency of microorganisms in breaking down organic pollutants or immobilizing heavy metals. This synergy between NPs and microbes accelerates the bioremediation process, offering a more effective and targeted approach to pollution mitigation. Thus, the integration of nanobiotechnology along with bioremediation possibly achieves efficient, effective, and sustainable solutions for a clean environment [33].

Another avenue of nanobioremediation is the integration of NMs with phytoremediation, a process that utilizes plants to absorb and sequester contaminants from the environment because exceeding levels of Cr, Cd, Cu, As, Zn, Pb, and Hg influence the living chain and not only causes human damage but also greatly effects animals, plants, and microorganisms. NPs, such as iron oxide or titanium dioxide, are applied to plants to enhance their pollutant uptake and accumulation capabilities. This nano-phytoremediation approach holds promise for addressing soil and water contamination, particularly in areas with high levels of heavy metals or organic pollutants. Additionally, NPs can serve as carriers for essential nutrients, facilitating the growth of pollutant-degrading microorganisms. By acting as delivery vehicles for nutrients, NMs support the proliferation of microbes involved in bioremediation processes, promoting a more sustainable and efficient remediation approach [34].

In the present time, nano-enabled mycoremediation represents another facet of this field, where NPs enhance the remediation capabilities of fungi. By combining fungi with NPs, researchers aim to improve their efficiency in breaking down organic pollutants or immobilizing heavy metals in contaminated environments. Furthermore, nano-bioremediation embraces bioaugmentation on a nanoscale, using engineered NMs as carriers for beneficial microorganisms. This approach facilitates the targeted introduction of microbes with specific pollutant-degrading capabilities into contaminated sites, optimizing their impact and persistence [35]. Additionally, nanobioremediation encompasses nano-enabled enzymatic bioremediation, where NPs mimic or enhance the catalytic activity of enzymes involved in biodegradation processes. These engineered NMs accelerate the breakdown of pollutants, such as pesticides or hydrocarbons, by mimicking or enhancing the action of natural enzymes. Furthermore, nanotechnology-based sensors play a vital role in real-time monitoring of pollutant levels. Nano-scale biosensors offer precise and rapid detection of pollutants, allowing for timely intervention and monitoring of the effectiveness of remediation efforts. It has also contributed significantly to remediating soil, water, and air pollutants into environment-friendly compounds. The emergence of nanobioremediation can nurture the environment through pollution removal and cleanup strategies. As mentioned above, the integration of conventional bioremediation and nanobiotechnological approaches or direct nano-remediation techniques can be the feasible option toremove the contaminants from the environment. However, nanobioremediation holds great potential for addressing pollution challenges, so it is crucial to consider the environmental impact of engineered NPs. Thorough assessments of potential risks are necessary to ensure the safe and responsible use of these technologies. Site-specific optimization and continuous monitoring are key aspects of tailoring nanobioremediation approaches to specific pollutants, environmental conditions, and regulatory requirements, ensuring their effectiveness and minimizing unintended consequences [36, 37].

APPLICATIONS OF NANOBIOREMEDIATION IN REMEDIATION OF POLLUTANTS

Nano-Enhanced Microbial Remediation

Nano-enhanced microbial remediation represents a sophisticated approach where engineered NPs play a pivotal role in amplifying the efficacy and functionality of indigenous microorganisms engaged in bioremediation endeavors. In this innovative process, microorganisms, ranging from bacteria to fungi, undergo a deliberate enhancement through the integration of specifically chosen NPs, such as zero-valent iron NPs. This augmentation strategy is meticulously designed to unleash a spectrum of improvements in the microorganisms' capabilities. By introducing NPs, the microbes become endowed with an augmented proficiency to **degrade** complex organic pollutants more efficiently. Moreover, they exhibit an enhanced ability to immobilize heavy metals, thereby curbing their dispersion and mitigating environmental contamination [38, 39].

Crude oils are one of the major sources of Polycyclic Aromatic Hydrocarbons (PAH), which are also the main pollutants of soil and water bodies. In contrast, nanoremediation technology greatly improved the water and soil quality by reducing pollutants, which results in high-yield products and vegetation and enhances the survival of indigenous microbes. Thus, the merging of nanotechnology with microbial remediation enhances the PAHs remediation potential by several times without producing toxic by-products such as carbon monoxide and metabolites like 9—fluorenone, dibenzene, and anthracene, which inhibits enzymes and cellular activity and prevents the degradation of other PAHs in the system during incomplete bioremediation [38].

Biodegradation of crude oils by strain of *B. licheniformis* has been reported to be enhanced by the amendment of two different types of NPs such as Fe_2O_3 NPs and $Zn_5(OH)_8Cl_2$ NPs [39]. The process for the remediation of benzo(a)pyrene (BaP) from the polluted soil by applying silica NPs coated with lipid 1,2- dimyristoylsn-glycero-3-phosphocholine, bilayers of bacteria (P. aeruginosa), has proven that the biofunctionalized silica NPs effectively adsorbed PAHs [40]. Further, biodegradation of indeno (1,2,3-cd) pyrene, a six-fused benzene ring, has been reported to be enhanced by yeast strain (*Candida tropicalis* NN4) up to 79% in the presence of Fe NPs in 15 days [41]. Similarly, ZnO NPs, along with biosurfactants in the presence of yeast consortium YC01, were applied for the degradation of five fused 50mg/L BaP. It achieved an enhanced degradation rate of 82.67% in 6 days. Additionally, biodegradation of six fused benzene rings benzo(ghi)perylene (BghiP) by yeast consortium YC04 has been enhanced up to 63.8% in the presence of ZnO NPs after 6 days [42, 43].

Strains of yeast consortium YC01 and YC04 with ZnO NPs have been established as a potential candidate in degrading BaP and BghiP of such a high concentration only in 6 days compared to earlier studies reported. Moreover, magnetic NPs are not only reported to enhance the remediation process but also aid as a beneficial tool for the successful isolation of PAH-degrading bacteria like Pseudomonas and Sphingobium sp. from the contaminated sites [44]. Modification of the electrode surface by input of magnetite NPs enhanced the efficiency of bioremediation of PAHs in polluted soil in the electrokinetic system [45].

Nanophytoremediation

Phytoremediation is an environmentally benign technique and is an eventual natural solution to the pollutants. The annexation of NMs with phytoremediation leads to nanophytoremediation. As a green technology, nanophytoremediation, consists of a two-step approach. Firstly, the NMs adsorb the pollutants and degrade them. These degraded materials (which are still pollution-prone) are then accumulated by the plants. The use of NMs in combination with phytoremediation is a remarkable technology that can clean the environment economically in a sustainable way [46].

Nano-Enabled Mycoremediation

Mycoremediation involves utilizing fungi to break down or eliminate environmental toxins, with documented instances of specific fungi playing a crucial role in neutralizing hazardous substances such as weapons and waste. On-going research explores the application of mycoremediation within the realm of national defense, particularly countering chemical and biological warfare, opening up possibilities for its use in restoring environments ravaged by war. Nano-enabled mycoremediation, on the other hand, employs fungi in nanotechnology for NP synthesis, with various fungal strains, including *Aspergillus* spp., *Fusarium* spp., *Penicillium* spp., and *Verticillium* sp., showing promise in fabricating metal NPs through both intracellular and extracellular processes. Notably, the inoculation of diesel oil-contaminated soil with mycelia from oyster mushrooms demonstrated a remarkable 95% conversion of PAHs into non-toxic compounds within four weeks. This suggests that the collaborative action of naturally present microbial communities and fungi leads to the decomposition of pollutants, ultimately resulting in full mineralization of CO_2 and H_2O [47].

Nano-scale Bioaugmentation

Bioaugmentation is a new pollutant control approach that is developed to overcome the inefficiency of biodegradation by microorganisms. One of its primary advantages lies in the ability to tailor treatments to target specific pollutants in the environment. Nano-scale bioaugmentation is a highly effective approach for eliminating both inorganic and organic pollutants, encompassing wastewater. In the context of bioaugmentation, it is important to note that NMs do not inherently confer any advantages as they hinder microbial populations within the contaminated environment. Nevertheless, there is emerging evidence suggesting a significant enhancement in bioaugmentation effectiveness through a novel approach. For instance, a study utilizing carbon nanotubes (CNTs) revealed that the growth inhibition of *Arthrobacter* sp. bacteria depends on the concentration of CNTs, with concentrations below 25 mg/L having no impact on bacterial growth, while concentrations exceeding 100 mg/L exhibited inhibitory effects. *Arthrobacter* sp. is known for biodegrading the organic pollutant atrazine, and the use of CNTs at concentrations > 25 mg/L in a 250 mL-batch reactor resulted in increased atrazine biodegradation compared to the control without CNTs. This enhanced biodegradation rate was linked to the stimulation of bacterial growth. Additionally, at ≤25 mg/L of CNTs, *Arthrobacter* sp. demonstrated the ability to fully utilize atrazine adsorbed in the CNTs. Furthermore, the positive effects of these NMs include the reversible oxidation and reduction of CNTs, providing them with the capability to serve as electron carriers in multiple redox reactions, thereby augmenting the rates of biodegradation reactions [48].

Nano-Enabled Enzymatic Bioremediation

Enzyme-mediated bioremediation harnesses the inherent enzymatic capabilities of microorganisms or plants to degrade and mitigate harmful environmental pollutants, effectively cleansing contaminated areas. Enzymes act as catalysts, expediting the degradation process of substances. Innovations in nanotechnology alongside enzyme technology have introduced single-enzyme NPs, offering heightened productivity, selectivity, and speed compared to traditional enzymatic treatments. This approach is swift, economical, accessible, and precise, outperforming plant and microbial methods. However, challenges like enzyme production costs, stability, and effectiveness across various contaminants limit its widespread use. Despite these constraints, enzymatic bioremediation remains a leading, eco-friendly technique for managing diverse waste materials in soil, water, or air, ensuring a controlled, specific, and manageable degradation process [49]. Moreover, several research findings indicate that NPs exhibit unique characteristics such as enhanced catalytic and adsorptive properties along with heightened reactivity. Presently, microorganisms and their derivatives serve as effective, eco-friendly catalysts for engineered NMs. This integration of technologies, known as nanobioremediation, holds promise for revolutionizing environmental remediation due to its intelligence, safety, environmental friendliness, cost-effectiveness, and sustainability [50].

Nano-scale Biosensors for Pollution Remediation

Biosensors, consisting of a biological element, transducer, and signal detector, utilize recognition elements like enzymes, antibodies, or DNA. The transducer converts signals into current, voltage, heat, or light. Detectors measure signals proportional to analyte-receptor interactions. Widely used in medical, agricultural, environmental, and food industries, biosensors come in different types, including electrochemical, optical, piezoelectric, and thermometric. Offering advantages like rapidity, simplicity, affordability, specificity, and portability, biosensors detect pollutants even in trace amounts from diverse sources such as hospitals, kitchens, agriculture, and sewage water [51]. Table 4 summarizes various applications of nano-bioremediation in the remediation of pollutants.

Table 4. Applications of Nanobioremediation in Remediation of Pollutants

S. N.	Nano-bioremediation	Description	Application
1.	**Nano-Enhanced Microbial Remediation**	Engineered NPs are used to enhance the activity and efficiency of naturally occurring microorganisms in bioremediation processes.	Microbes, such as bacteria or fungi, are augmented with NPs (*e.g.*, zero-valent iron NPs) to increase their capability to degrade organic pollutants or immobilize heavy metals [38, 47].
2	**Nanophytoremediation**	NMs are applied to plants to improve their ability to absorb, translocate, and sequester pollutants from the soil or water.	NPs, such as iron oxide or titanium dioxide, can enhance the phytoremediation potential of plants by improving nutrient uptake and increasing pollutant absorption and accumulation [46].

S. N.	Nano-bioremediation	Description	Application
3.	**Nano-Enabled Mycoremediation**	NPs are utilized to enhance the remediation capabilities of fungi in breaking down or sequestering pollutants.	The application of myconanotechnology is also recommended as a potential future initiative to enhance the effectiveness and rate of mycoremediation. Fungi, such as mycorrhizal species, are combined with NPs to improve their efficiency in degrading organic pollutants or assisting in the immobilization of heavy metals [47].
4.	**Nano-Scale Bioaugmentation**	Engineered NMs are used as carriers for beneficial microorganisms, facilitating their introduction into contaminated sites.	Microorganisms with specific pollutant-degrading capabilities are encapsulated or attached to NPs, enhancing their delivery and persistence in target areas [48].
5.	**Nano-Enabled Enzymatic Bioremediation**	NPs are designed to mimic or enhance the catalytic activity of enzymes involved in bioremediation processes.	NMs with catalytic properties are employed to accelerate the breakdown of pollutants, such as pesticides or hydrocarbons, by mimicking or enhancing the action of natural enzymes [49].
6.	**Nano-Scale Biosensors for Monitoring**	Nanotechnology-based sensors are used for real-time monitoring of pollutant levels in the environment.	Nanoscale biosensors can provide precise and rapid detection of pollutants, allowing for timely intervention and monitoring of the effectiveness of remediation efforts [51].

PHYTONANOTECHNOLOGY: A PROMISING AVENUE FOR ADDRESSING ENVIRONMENTAL POLLUTION

It is an innovative field at the intersection of plant biology and nanotechnology that presents a promising avenue for addressing environmental pollution. This approach utilizes NMs to enhance the phytoremediation capabilities of plants, optimizing their ability to absorb, transport, and accumulate pollutants from the soil or water. One significant application of phyto-nanotechnology is in nano-enhanced phytoremediation, where NPs like iron oxide or zero-valent iron are applied to plants to augment their ability to extract heavy metals from contaminated environments [52].

Phytonanotechnology involves using NMs to influence plant-microbe interactions, particularly in the rhizosphere, where plant roots affect soil. NPs boost microbial activity, aiding in pollutant degradation and nutrient cycling.

They also enhance plant's tolerance against pollution-induced challenges, ensuring their survival for effective remediation. Certain NPs like silica or zinc oxide enable plants to endure adverse conditions caused by pollution. Microbes, such as Bacillus sp., aid in the removal of toxic compounds and heavy metals alongside phytoremediation by secreting enzymes. Bacteria employ both passive (biosorption) and active (bioaccumulation) mechanisms, offering cost-effective remediation with minimal energy consumption and improved air quality. Plants utilize various parts, including leaves, for bioremediation processes like phytovolatilization, converting heavy metals into less toxic forms. Combining bacteria and plants assists in heavy metal remediation, enabling plant growth in contaminated areas. Additionally, bio-adsorbents like sawdust, sugarcane bagasse, and green tea leaves enhance heavy metal removal, constituting a cost-effective remediation approach [53].

However, the application of phytonanotechnology warrants careful consideration of potential environmental impacts. The fate and toxicity of engineered NPs must be thoroughly assessed to ensure their safe and responsible use. Additionally, adherence to regulatory guidelines is crucial for the responsible application of NMs in phytoremediation. As phyto-nanotechnology continues to evolve, ongoing research and a site-specific optimization approach are essential to unlock its full potential and contribute to sustainable pollution remediation practices. The key applications of phyto-nanotechnology as a remedy for pollution are discussed below.

Nano-enhanced Phytoremediation

NMs are utilized to enhance plants' capacity to absorb, transport, and accumulate pollutants from soil or water. Examples include iron oxide or zero-valent iron NPs, which boost the phytoextraction of heavy metals by increasing their availability and uptake in plant tissues. Phytoremediation, a cost-effective method, is most effective when contaminants are present in the plant's root zones. Flax (*Linum usitatissimum*) is a viable candidate for Cu phytoremediation as it can efficiently remove Cu from soils and can be cultivated for seed production. Various phytoremediation techniques, such as phytostabilization, rhizofiltration, phytoextraction, and phytovolatilization, can be employed to remove heavy metal contamination. Alongside plants, rhizospheric bacteria play a crucial role in purifying contaminated areas, utilizing similar phytoremediation principles as plants to reduce organic and inorganic pollutants [54].

Nano-enabled Plant-Microbe Interactions

NMs facilitate interactions between plants and beneficial microbes to enhance remediation processes. Moreover, NPs can improve the microbial activity in the rhizosphere, the soil region influenced by plant roots, promoting the degradation of pollutants and enhancing nutrients. TiO_2 NPs, in association with the rhizobacterium *Pseudomonas fluorescens,* promote phytoremediation in soils contaminated with cadmium (Cd) [55].

NP-Mediated Stress Tolerance

NPs are used to enhance plant stress tolerance, allowing them to thrive in polluted environments. NPs, such as silica or zinc oxide, can help plants tolerate environmental stresses associated with pollution, enabling them to survive and remediate contaminated areas. The TiO_2 NPs can help grow crops in contaminated soils. In addition to improving plant growth and soil quality, TiO_2 is also used to remediate stressful environments of soil [56].

Nano-enabled Plant Signaling

NMs modulate plant signaling pathways to optimize responses to pollution-induced stress. NPs can activate plant defense mechanisms, making them more resilient to pollution and enhancing their ability to cope with environmental stressors [57].

NP-Mediated Soil Improvement

NMs are used to improve soil structure and fertility, creating a more conducive environment for plant growth and pollutant remediation. NPs like biochar or nanoclays improve soil structure, water retention, and nutrient availability, creating a favorable habitat for plants involved in phytoremediation [58].

KEY BENEFITS OF NANO-PHYTOTECHNOLOGY IN THE REMEDIATION OF POLLUTANTS

Nanophytotechnology, the integration of nanotechnology with phytoremediation processes, holds great promise for addressing environmental pollution. This interdisciplinary approach leverages the unique properties of NMs to enhance the efficiency of plants in removing, degrading, or immobilizing pollutants. There are some key benefits of nanophytonanotechnology in the remediation of pollutants, which are described below.

Enhanced Uptake of Pollutants

In the realm of nano-phytotechnology, a key approach involves the deliberate design of NPs to augment the uptake of pollutants by plants, thereby enhancing their phytoremediation efficacy. Nanostructures such as carbon nanotubes, graphene oxide, and iron-based NPs play a pivotal role in improving the absorption and translocation of pollutants, especially heavy metals and organic contaminants. This approach seeks to leverage the unique properties of NMs to facilitate more efficient and targeted pollutant removal by plants [59].

Increased Stress Tolerance in Plants

Nanophytotechnology explores the use of NMs to bolster the stress tolerance of plants, enabling them to thrive in polluted environments. NPs like silica and titanium dioxide have shown promise in enhancing plant resilience against various environmental stressors associated with pollution, including drought, salinity, and exposure to toxic substances. This approach contributes to the development of robust plant species capable of withstanding adverse conditions while actively participating in pollutant remediation [60].

Controlled Release of Nutrients

An innovative approach in nanophytotechnology involves the strategic use of NPs as carriers for essential nutrients, facilitating controlled nutrient release to support plant growth in polluted soils. Nanofertilizers and nutrient-loaded NPs contribute to sustained nutrient availability, promoting the health and vigor of plants in contaminated environments. This approach addresses the nutritional needs of plants, ensuring their optimal growth and enhancing their overall remediation potential [61].

Targeted Delivery of Amendments

Nanophytotechnology employs NPs to enable the targeted delivery of soil amendments, thereby enhancing soil structure and nutrient availability. Engineered NMs, including nanoclays and biochar NPs, play a crucial role in delivering amendments precisely where they are needed. This targeted approach contributes to improving soil quality, supporting both plant growth and the efficiency of pollutant remediation efforts [62].

NP-Plant-Microbe Interactions

In the context of nanophytotechnology, NMs are instrumental in modulating interactions among plants, NPs, and beneficial microbes, fostering enhanced remediation processes. Improved understanding and manipulation of nano-enabled interactions within the rhizosphere—the soil region influenced by plant roots—can lead to heightened microbial activity. This, in turn, promotes pollutant degradation and nutrient cycling, creating a synergistic environment that supports effective phytoremediation [63].

Smart Nanosensors for Monitoring

A cutting-edge application within nanophytotechnology involves the use of nanotechnology-based sensors for real-time monitoring of pollutant levels in the environment. Detecting pollution is crucial for controlling harmful pollutants. The development of rapid and precise sensors capable of measuring pollutant levels plays a vital role in safeguarding both the environment and human health. These advanced sensors have the potential to significantly enhance our ability to monitor and manage pollution, thereby contributing to the sustainability of our ecosystems and the well-being of communities. The development of smart nano-sensors allows for precise and continuous monitoring of pollutant concentrations. This capability enables timely remediation interventions and facilitates the ongoing assessment of phytoremediation efficiency, contributing to adaptive pollution management strategies [64].

Phytoextraction of Heavy Metals

An important aspect of nano-phytotechnology is the application of NPs to enhance the phyto-extraction efficiency of heavy metals from contaminated soils. Engineered NPs, such as zero-valent iron or iron oxide NPs, play a crucial role in mobilizing and increasing the uptake of heavy metals by plant roots, thereby enhancing the overall effectiveness of phytoextraction processes and contributing to the remediation of heavy metal pollution. The United States Environmental Protection Agency (USEPA) has recognized Lead (Pb), Cadmium (Cd), Copper (Cu), Mercury (Hg), Chromium (Cr), Arsenic (As), Zinc (Zn), and Nickel (Ni) as the most commonly found heavy metals in the environment [65]. Various plant-based magnetic adsorbents (Magnetic NPs) prepared by plant-based biomass, which contain phenols, carbonyl, amino acids, glucose, and polysaccharide compounds, have well removal efficacy of adsorbents that is influenced by the effect of time, pH, temperature, concentration of pollutant [66]. These approaches collectively represent the forefront of innovation in pollution remediation, showcasing the potential of nanophytotechnology to revolutionize environmental management strategies. Ongoing collaborative research and interdisciplinary efforts are essential to unlock the full capabilities of these approaches and address the complexities of diverse pollution scenarios.

DIFFERENT DOMAINS OF APPLICATION OF PHYTONANO-TECHNOLOGY

Phytoremediation

Phytonanotechnology is widely used in phytoremediation, where NPs play an important role in enhancing the capacity of plants to remediate contaminated environments [67]. Engineered NMs are designed to improve the absorption and accumulation of pollutants, such as heavy metals and organic contaminants, by plant roots and offer enormous potential in addressing environmental pollution by making phytoremediation more efficient and targeted [68, 69].

Agricultural Nanofertilizers

NMs are utilized to revolutionize traditional fertilizers, giving rise to agricultural nanofertilizers, serving as carriers for nutrients, and facilitating controlled and sustained release of nutrients [70, 71]. This approach not only enhances plant nutrient uptake efficiency but also minimizes environmental runoff, resulting in improved crop yields and more efficient use of fertilizers in agricultural practices [72].

Disease Resistance in Crops

Phyto-nanotechnology contributes to crop health through the development of disease-resistant plants utilizing NPs with antimicrobial properties. These are employed to safeguard plants from pathogenic infections, reducing dependence on chemical pesticides [73, 74]. Nanostructures induce systemic resistance in plants, fortifying them against various diseases and enhancing overall crop resilience [75].

Improved Crop Yield and Quality

The application of nanotechnological approaches leads to advancements in crop yield and quality as NMs enhance stress tolerance in plants, enabling them to thrive in challenging conditions such as drought and salinity [76, 77]. Furthermore, nanostructures aid in biofortification, enhancing the nutritional content of crops and addressing concerns related to food security.

Smart Nanosensors for Plant Monitoring

Nanosensors play a pivotal role in monitoring plant health, nutrient status, and environmental conditions, enabling precision agriculture by offering accurate data for timely decision-making and facilitating environmental monitoring by detecting changes in soil and water conditions that affect plant health and productivity [78 - 80].

Seed Coating for Enhanced Germination

NPs applied in seed coating enhance germination rates and provide early-stage protection against environmental stressors, ensuring higher germination rates, promoting uniform crop establishment, and contributing to the early success of crop cultivation [81, 82].

Nano-Enabled Plant Growth Promoters

NMs act as growth promoters, improving multiple aspects of plant development, including root zone enhancement to improve nutrient and water uptake by fostering robust root growth, as well as photosynthetic efficiency, which leads to increased biomass production and overall plant vigor [83].

Environmental Monitoring and Sensing

Nanoscale sensors are instrumental in environmental monitoring, providing valuable insights into pollutant levels and plant responses [84]. These sensors play a crucial role in pollutant detection, offering real-time data for pollution monitoring and assessment [85]. They also monitor plant stress responses, detecting early signs of stress and enabling proactive interventions for plant health.

FUTURE OF NANOBIOREMEDIATION

The future of nanobioremediation holds promising prospects for addressing environmental pollution challenges more effectively. Nanobioremediation, which combines nanotechnology with bioremediation techniques, offers unique advantages in targeting and degrading pollutants at the molecular level [86]. Furthermore, advancements in understanding the interactions between NPs and biological systems will lead to safer and more environmentally friendly nanobioremediation approaches. Researchers are increasingly focusing on assessing the potential risks associated with the use of NMs in remediation efforts and implementing strategies to mitigate any adverse effects. In summary, the future of nanobioremediation is poised to see notable progress in technology, methods, and practical use, resulting in more eco-friendly and efficient approaches to environmental restoration. Yet, ongoing research, cooperation, and regulation are vital to guarantee the careful and safe implementation of nano-bioremediation methods [87].

Although earlier research has highlighted the potential of nanobioremediation for organic contaminants, recent advancements underscore the importance of precision targeting, responsive NPs, multifunctional NMs, *in situ* nanobioremediation, coupling with traditional methods, environmental fate and safety studies, commercialization, sustainability, collaboration with microbial communities [88], intelligent delivery systems, biodegradability, field-scale implementation, and enhanced monitoring methods. These insights are instrumental in advancing the development of resilient and eco-friendly nanobioremediation strategies tailored for pollutants [89]. The following recent developments highlight the significance of nanobioremediation:

Responsive NPs

They are also called stimuli-sensitive NPs that are designed to react to specific environmental conditions. These NPs can release remediation agents or enhance microbial activity in response to changes in pollutant concentrations, providing a dynamic and adaptive remediation approach [90].

Precision Targeting

Future developments in nanobioremediation may focus on precision targeting of pollutants at the molecular level. Engineered NPs can be designed with greater specificity to target particular contaminants, enhancing remediation efficiency and reducing unintended ecological impacts [91].

Multifunctional NMs

Researchers may explore the development of multifunctional NMs that can address multiple pollutants simultaneously. These advanced NMs can have diverse functionalities, such as capturing heavy metals, degrading organic pollutants, and promoting microbial activity [92].

In Situ Nanobioremediation

Advances in *in situ* nanobioremediation techniques may become more prevalent, allowing for the direct application of engineered NPs in contaminated sites. This approach can streamline remediation efforts by minimizing the need for excavation and transportation of polluted materials [93].

Coupling with Traditional Methods

Subsequent studies might concentrate on merging nanobioremediation with conventional remediation approaches to achieve synergistic outcomes. This fusion of nanotechnology's advantages with well-established methods has the potential to enhance the effectiveness and scope of remediation efforts targeting heavy metal pollutants [94].

Environmental Fate and Safety Studies

Ongoing investigation into the environmental behavior and safety of engineered NPs (ENPs) will be crucial. Subsequent research endeavors may focus on gaining a deeper understanding of the long-term interactions between NPs and ecosystems, thereby ensuring their responsible and sustainable utilization. A growing body of recent research has highlighted the toxic impacts of ENPs on diverse organisms, prompting concerns regarding the behavior and fate of these nanopollutants across different environmental realms. Upon their release into the environment, ENPs interact with various environmental components and undergo dynamic transformation processes, which are interconnected with numerous environmental factors [95].

FUTURE OF PHYTONANOTECHNOLOGY

Phytomediated NPs offer environmentally friendly and cost-effective solutions with long-term safety benefits, primarily focusing on laboratory and greenhouse investigations. However, there are gaps in phytonanotechnology research, particularly in transitioning from lab to field applications. Future studies should address these gaps to enable on-field innovations and broaden applications across various sectors. While some lab experiments show positive effects on plant growth, long-term effects and environmental considerations require thorough assessment before large-scale implementation. Understanding NP uptake, translocation, and fate is essential for addressing ethical and safety concerns and improving field-level applications. Further research is needed to elucidate biomolecular interactions and gene expression regulation, as well as molecular and submolecular functions of NPs [96, 97].

Customized NPs for Plant Species

Subsequent developments may involve tailoring NPs to specific plant species and optimizing their interactions for enhanced pollutant remediation. Customized NMs can ensure compatibility with a diverse range of plants, broadening the applicability of phytonanotechnology [98].

NP-Microbe-Plant Synergies

Research may delve into understanding and harnessing synergies between NPs, beneficial microbes, and plants. This collaborative approach can lead to more robust and resilient ecosystems capable of efficiently remediating pollutants [99].

Phytonanotechnology for Airborne Pollutants

Future applications may expand to address airborne pollutants, utilizing plants and NMs to mitigate air pollution. This can involve the development of plants with enhanced capabilities to absorb and neutralize pollutants present in the atmosphere [100].

Integration with Smart Agriculture Practices

The integration of phytonanotechnology with smart agriculture practices is anticipated. This includes the incorporation of sensors, data analytics, and automation to optimize the deployment of NMs in agricultural settings for both pollutant remediation and crop improvement [101].

Regulatory Guidelines and Standards

The establishment of clear regulatory guidelines and standards for phytonanotechnology applications will become increasingly important. Future efforts may focus on developing internationally recognized frameworks to ensure the safe and responsible use of NMs in agriculture and environmental management [102].

CONCLUSION

The future of nanobioremediation and phytonanotechnology holds exciting possibilities, with ongoing research expected to unlock new insights, overcome challenges, and pave the way for practical, scalable, and sustainable applications in pollution remediation. Interdisciplinary collaboration, ethical considerations, and a commitment to environmental stewardship will be vital in shaping the future landscape of these technologies. Moreover, future initiatives may aim to enhance public awareness and acceptance of phytonanotechnology. Communicating the benefits and potential risks transparently can foster public trust and support for these innovative solutions. In contrast to conventional approaches, phytonanotechnology offers a unique opportunity to understand and modify plants in an unprecedented way. Improved phytonanotechnology research may lead to the advent of "smart plants," offering insight into and interaction with individual plants and their surroundings, thus supporting the sustainability of our planet.

REFERENCES

[1]   Singh Y, Saxena MK. Insights into the recent advances in nano-bioremediation of pesticides from the contaminated soil. Front Microbiol 2022; 13: 982611.
        [http://dx.doi.org/10.3389/fmicb.2022.982611] [PMID: 36338076]

[2]   Roy A, Sharma A, Yadav S. Leta TesfayeJule, RamaswamyKrishnaraj, NMs for Remediation of Environmental Pollutants Bioinorganic Chemistry and Applications. 2021; 2021: p. 16.
        [http://dx.doi.org/10.1155/2021/1764647]

[3]   Tripathi S, Sanjeevi R, Anuradha J, Chauhan DS, Rathoure AK. Nano-bioremediation: nanotechnology and bioremediation. In Research Anthology on Emerging Techniques in Environmental Remediation. IGI Global 2022; pp. 135-49.
        [http://dx.doi.org/10.4018/978-1-6684-3714-8.ch007]

[4]   Gomathi T, Saranya M, Radha E, Vijayalakshmi K, Prasad PS, Sudha PN. Bioremediation: A

Promising Xenobiotics Cleanup Technique Encyclopedia of Marine Biotechnology. DOI 2020; pp. 3139-72.
[http://dx.doi.org/10.1002/9781119143802.ch140]

[5]   Kumar Lakhan, Kamal Pragya, Soni Kaniska, Bharadvaja Navneeta. Phytonanotechnology for Remediation of Heavy Metals and Dyes Book: Phytonanotechnology for Remediation of Heavy Metals and Dyes 2022.
[http://dx.doi.org/10.1002/9783527834143.ch9]

[6]   Gole Apurva, John Diya, Krishnamoorthy Karan, *et al.* Role of Phytonanotechnology in the Removal of Water Contamination Journal of NMs 2022; 202219
[http://dx.doi.org/10.1155/2022/7957007]

[7]   Houshani M, Tarigholizadeh S, Rajput VD, Jatav HS. The degradation of organic and inorganic pollutants Basic concepts in environmental biotechnology. CRC Press. 2021.133-148
[http://dx.doi.org/10.1201/9781003131427-11]

[8]   Kleinteich J, Seidensticker S, Marggrander N, Zarfl C. Refs. Kleinteich J, Seidensticker S, Marggrander N, Zarfl C. Microplastics reduce short-term effects of environmental contaminants. Part II: polyethylene particles decrease the effect of polycyclic aromatic hydrocarbons on microorganisms. Int J Environ Res Public Health 2018; 15(2): 287.
[http://dx.doi.org/10.3390/ijerph15020287] [PMID: 29414906]

[9]   He C, Cheng J, Zhang X, Douthwaite M, Pattisson S, Hao Z. Recent advances in the catalytic oxidation of volatile organic compounds: a review based on pollutant sorts and sources. Chem Rev 2019; 119(7): 4471-568.
[http://dx.doi.org/10.1021/acs.chemrev.8b00408] [PMID: 30811934]

[10]   Tsagkaris AS, Uttl L, Pulkrabova J, Hajslova J. Screening of carbamate and organophosphate pesticides in food matrices using an affordable and simple spectrophotometric acetylcholinesterase assay. Appl Sci (Basel) 2020; 10(2): 565.
[http://dx.doi.org/10.3390/app10020565]

[11]   Herrero J, Puigserver D, Parker BL, Carmona JM. A new method for determining compound specific carbon isotope of chlorinated solvents in porewater. Ground Water Monit Remediat 2021; 41(3): 51-7.
[http://dx.doi.org/10.1111/gwmr.12435]

[12]   Akash MSH, Rehman K, Sabir S, Gul J, Hussain I. Review Potential Risk Assessment of Pharmaceutical Waste: Critical Review and Analysis. Pak J Sci Ind Res Ser A Phys Sci 2020; 63(3): 209-19.
[http://dx.doi.org/10.52763/PJSIR.PHYS.SCI.63.3.2020.209.219]

[13]   Mamontova EA, Mamontov AA. Air Monitoring of Polychlorinated Biphenyls and Organochlorine Pesticides in Eastern Siberia: Levels, Temporal Trends, and Risk Assessment. Atmosphere (Basel) 2022; 13(12): 1971.
[http://dx.doi.org/10.3390/atmos13121971]

[14]   dos Santos CA, de Souza Cruz DR, da Silva WR, *et al.* Heterogeneous electro-Fenton process for degradation of bisphenol A using a new graphene/cobalt ferrite hybrid catalyst. Environ Sci Pollut Res Int 2021; 28(19): 23929-45.
[http://dx.doi.org/10.1007/s11356-020-11913-7] [PMID: 33398742]

[15]   Hoang QA, Trinh HM, Pham DM, *et al.* Analysis and Pollution Assessment of Brominated Flame Retardants (PBDEs, DBDPE, and BTPBE) in Settled Dust from E-waste and Vehicle Processing Areas in Northern Vietnam VNU Journal of Science: Natural Sciences and Technology 2023; 23.

[16]   Feiteiro J, Mariana M, Cairrão E. Health toxicity effects of brominated flame retardants: From environmental to human exposure. Environ Pollut 2021; 285: 117475.
[http://dx.doi.org/10.1016/j.envpol.2021.117475] [PMID: 34087639]

[17]   Dean JR, Wade G, Barnabas IJ. Determination of triazine herbicides in environmental samples. J Chromatogr A 1996; 733(1-2): 295-335.

[http://dx.doi.org/10.1016/0021-9673(95)00691-5]

[18]   Padoley KV, Mudliar SN, Pandey RA. Heterocyclic nitrogenous pollutants in the environment and their treatment options – An overview. Bioresour Technol 2008; 99(10): 4029-43.
[http://dx.doi.org/10.1016/j.biortech.2007.01.047] [PMID: 17418565]

[19]   Vajargah MF. A review on the effects of heavy metals on aquatic animals. J Biomed Res Environ Sci 2021; 2(9): 865-9.
[http://dx.doi.org/10.37871/jbres1324]

[20]   Yalsuyi AM, Hedayati A, Vajargah MF, Mousavi-Sabet H. Examining the toxicity of cadmium chloride in common carp (Cyprinus carpio) and goldfish (Carassius auratus). J Environ Treat Tech 2017; 5(2): 83-6.

[21]   Coffey R, Paul MJ, Stamp J, Hamilton A, Johnson T. A review of water quality responses to air temperature and precipitation changes 2: nutrients, algal blooms, sediment, pathogens. J Am Water Resour Assoc 2019; 55(4): 844-68.
[http://dx.doi.org/10.1111/1752-1688.12711] [PMID: 33867785]

[22]   Melo MJ, Melo E, Pina F. Determination of acid-base equilibria of organic pollutants: the steady state fluorescence emission method. Arch Environ Contam Toxicol 1994; 26(4): 510-20.
[http://dx.doi.org/10.1007/BF00214155]

[23]   Al-Ali IA, Al-Dabbas MA. The effect of variance discharge on the dissolved salts concentration in the Euphrates River upper reach, Iraq. Iraqi J Sci. 2022;63(9):3842-53.
[http://dx.doi.org/10.24996/ijs.2022.63.9.16]

[24]   Mamelkina MA, Herraiz-Carboné M, Cotillas S, *et al.* Treatment of mining wastewater polluted with cyanide by coagulation processes: A mechanistic study. Separ Purif Tech 2020; 237: 116345.
[http://dx.doi.org/10.1016/j.seppur.2019.116345]

[25]   Kuttippurath J, Singh A, Dash SP, *et al.* Record high levels of atmospheric ammonia over India: Spatial and temporal analyses. Sci Total Environ 2020; 740: 139986.
[http://dx.doi.org/10.1016/j.scitotenv.2020.139986] [PMID: 32927535]

[26]   Kumar V, Sharma M, Sondhi S, *et al.* Removal of Inorganic Pollutants from Wastewater: Innovative Technologies and Toxicity Assessment. Sustainability (Basel) 2023; 15(23): 16376.
[http://dx.doi.org/10.3390/su152316376]

[27]   Kabir H, Gupta AK, Tripathy S. Fluoride and human health: Systematic appraisal of sources, exposures, metabolism, and toxicity. Crit Rev Environ Sci Technol 2020; 50(11): 1116-93.
[http://dx.doi.org/10.1080/10643389.2019.1647028]

[28]   Kılıç Ö, Belivermiş M, Sezer N, Kalaycı G, Gözel F. Multi-pollutant Monitoring in a Rehabilitated Estuary: Elements and Radionuclides. Bull Environ Contam Toxicol 2019; 103(2): 354-61.
[http://dx.doi.org/10.1007/s00128-019-02636-8] [PMID: 31119313]

[29]   Roy A, Sharma A, Yadav S, Jule LT, Krishnaraj R. NMs for remediation of environmental pollutants. Bioinorg Chem Appl. 2021 Dec 28;2021:1-18
[http://dx.doi.org/10.1155/2021/5522394]

[30]   Arif Z, Sethy NK, Kumari L, Mishra PK, Upadhyay SN. Recent advances in functionalized polymer-based composite photocatalysts for wastewater treatment. 2020.
[http://dx.doi.org/10.1016/B978-0-12-818598-8.00003-1]

[31]   Homaeigohar S, Zillohu A, Abdelaziz R, Hedayati M, Elbahri M. A novel nanohybrid nanofibrous adsorbent for water purification from dye pollutants. Materials (Basel) 2016; 9(10): 848.
[http://dx.doi.org/10.3390/ma9100848] [PMID: 28773968]

[32]   Skoulidas AI, Sholl DS, Johnson JK. Adsorption and diffusion of carbon dioxide and nitrogen through single-walled carbon nanotube membranes. J Chem Phys 2006; 124(5): 054708.
[http://dx.doi.org/10.1063/1.2151173] [PMID: 16468902]

[33]   Iqbal HM, Bilal M, Nguyen TA, Eds. Nano-Bioremediation: Fundamentals and Applications. Elsevier 2021.

[34]   Aslam F, Mazhar S. Nano-bioremediation of heavy metals from environment using a green synthesis approach. International Journal of Advances in Applied Sciences 2023; 12(1): 7-14.
[http://dx.doi.org/10.11591/ijaas.v12.i1.pp7-14]

[35]   Akpasi SO, Anekwe IMS, Tetteh EK, *et al.* Mycoremediation as a Potentially Promising Technology: Current Status and Prospects—A Review. Appl Sci (Basel) 2023; 13(8): 4978.
[http://dx.doi.org/10.3390/app13084978]

[36]   Tripathi S, Sanjeevi R, Anuradha J, Chauhan DS, Rathoure AK. Nano-bioremediation: nanotechnology and bioremediation. 2022.
[http://dx.doi.org/10.4018/978-1-6684-3714-8.ch007]

[37]   Carata E, Panzarini E, Dini L. Environmental nanoremediation and electron microscopies 2017.
[http://dx.doi.org/10.1007/978-3-319-53162-5_4]

[38]   Rajput VD, Kumari S, Minkina T, Sushkova S, Mandzhieva S. Nano-Enhanced Microbial Remediation of PAHs Contaminated Soil. Air Soil Water Res 2023; 16: 11786221231170099.
[http://dx.doi.org/10.1177/11786221231170099]

[39]   El-Sheshtawy HS, Ahmed W. Bioremediation of crude oil by Bacillus licheniformis in the presence of different concentration nanoparticles and produced biosurfactant. Int J Environ Sci Technol 2017; 14(8): 1603-14.
[http://dx.doi.org/10.1007/s13762-016-1190-1]

[40]   Wang H, Kim B, Wunder SL. Nanoparticle-supported lipid bilayers as an *in situ* remediation strategy for hydrophobic organic contaminants in soils. Environ Sci Technol 2015; 49(1): 529-36.
[http://dx.doi.org/10.1021/es504832n] [PMID: 25454259]

[41]   Ojha N, Mandal S K, Das N. Enhanced degradation of indeno(1,2,3-cd) pyrene using Candida tropicalis NN4 in presence of iron NPs and produced biosurfactant: A statistical approach 3 Biotech 2019; 9(3): 1-13.

[42]   Mandal SK, Ojha N, Das N. Optimization of process parameters for the yeast mediated degradation of benzo[a]pyrene in presence of ZnO nanoparticles and produced biosurfactant using 3-level Box-Behnken design. Ecol Eng 2018; 120: 497-503. a
[http://dx.doi.org/10.1016/j.ecoleng.2018.07.006]

[43]   Mandal SK, Ojha N, Das N. Process optimization of benzo[ghi]perylene biodegradation by yeast consortium in presence of ZnO nanoparticles and produced biosurfactant using Box-Behnken design. Front Biol (Beijing) 2018; 13(6): 418-24. b
[http://dx.doi.org/10.1007/s11515-018-1523-1]

[44]   Li J, Luo C, Zhang G, Zhang D. Coupling magnetic-nanoparticle mediated isolation (MMI) and stable isotope probing (SIP) for identifying and isolating the active microbes involved in phenanthrene degradation in wastewater with higher resolution and accuracy. Water Res 2018; 144: 226-34.
[http://dx.doi.org/10.1016/j.watres.2018.07.036] [PMID: 30032019]

[45]   Pourfadakari S, Ahmadi M, Jaafarzadeh N, *et al.* Remediation of PAHs contaminated soil using a sequence of soil washing with biosurfactant produced by Pseudomonas aeruginosa strain PF2 and electrokinetic oxidation of desorbed solution, effect of electrode modification with Fe_3O_4 nanoparticles. J Hazard Mater 2019; 379: 120839.
[http://dx.doi.org/10.1016/j.jhazmat.2019.120839] [PMID: 31279313]

[46]   Mustafa K, Kanwal J, Farrukh S, Mussaddiq S, Saddiq N, Younas M. Nano-phytoremediation technology in environmental remediation. Phytoremediation technology for the removal of heavy metals and other contaminants from soil and water. Elsevier 2022; pp. 433-59.
[http://dx.doi.org/10.1016/B978-0-323-85763-5.00029-5]

[47]   Gholami-Shabani M, Gholami-Shabani Z, Shams-Ghahfarokhi M, Razzaghi-Abyaneh M. Application

of nanotechnology in mycoremediation: Current status and future prospects Fungal nanobionics: Principles and applications 2018; 89-116.

[48] Nzila A, Razzak S, Zhu J. Bioaugmentation: an emerging strategy of industrial wastewater treatment for reuse and discharge. Int J Environ Res Public Health 2016; 13(9): 846.
[http://dx.doi.org/10.3390/ijerph13090846] [PMID: 27571089]

[49] Kumar L, Bharadvaja N. Enzymatic bioremediation: a smart tool to fight environmental pollutants 2019.
[http://dx.doi.org/10.1016/B978-0-12-818307-6.00006-8]

[50] Hidangmayum A, Debnath A, Guru A, Singh BN, Upadhyay SK, Dwivedi P. Mechanistic and recent updates in nano-bioremediation for developing green technology to alleviate agricultural contaminants. Int J Environ Sci Technol 2023; 20(10): 11693-718.
[http://dx.doi.org/10.1007/s13762-022-04560-7]

[51] Yadav N, Garg VK, Chhillar AK, Rana JS. Detection and remediation of pollutants to maintain ecosustainability employing nanotechnology: A review. Chemosphere 2021; 280: 130792.
[http://dx.doi.org/10.1016/j.chemosphere.2021.130792] [PMID: 34162093]

[52] Srivastav A, Yadav KK, Yadav S, *et al.* Nano-phytoremediation of pollutants from contaminated soil environment: current scenario and future prospects. Phytoremediation. Management of Environmental Contaminants 2018; 6: 383-401.
[http://dx.doi.org/10.1007/978-3-319-99651-6_16]

[53] Linthoingambi Ningombam, Techi Mana, Gemin Apum, Rina Ningthoujam, Yengkhom Disco Singh, Nano-bioremediation: A prospective approach for environmental decontamination in focus to soil, water and heavy metals, Environmental Nanotechnology, Monitoring & Management, Volume 21, 2024,100931,ISSN 2215-1532.
[http://dx.doi.org/10.1016/j.enmm.2024.100931]

[54] Prakash P, S SC. S SC. Nano-phytoremediation of heavy metals from soil: a critical review. Pollutants 2023; 3(3): 360-80.
[http://dx.doi.org/10.3390/pollutants3030025]

[55] De Moraes AC, Ribeiro LD, de Camargo ER, Lacava PT. The potential of NMs associated with plant growth-promoting bacteria in agriculture 3 Biotech 2021; 11(7): 318.

[56] Karnwal A, Dohroo A, Malik T. Unveiling the potential of bioinoculants and NPs in sustainable agriculture for enhanced plant growth and food security. BioMed Res Int 2023; •••: 2023.

[57] Gayathiri E, Prakash P, Pandiaraj S, *et al.* Investigating the ecological implications of nanomaterials: Unveiling plants' notable responses to nano-pollution. Plant Physiol Biochem 2024; 206: 108261.
[http://dx.doi.org/10.1016/j.plaphy.2023.108261] [PMID: 38096734]

[58] Chaudhary P, Ahamad L, Chaudhary A, Kumar G, Chen WJ, Chen S. Nanoparticle-mediated bioremediation as a powerful weapon in the removal of environmental pollutants. J Environ Chem Eng 2023; 11(2): 109591.
[http://dx.doi.org/10.1016/j.jece.2023.109591]

[59] Prakash, P., & S, S. C. (2023). Nano-Phytoremediation of Heavy Metals from Soil: A Critical Review. Pollutants, 3(3), 360-380.
[http://dx.doi.org/10.3390/pollutants3030025]

[60] Munir N, Gulzar W, Abideen Z, Hancock JT, El-Keblawy A, Radicetti E. Nanotechnology improves disease resistance in plants for food security: Applications and challenges. Biocatal Agric Biotechnol 2023; 51: 102781.
[http://dx.doi.org/10.1016/j.bcab.2023.102781]

[61] Neha Agarwal, Vijendra Singh Solanki, Brijesh Pare, Neetu Singh, Sreekantha B. Jonnalagadda, Current trends in nanocatalysis for green chemistry and its applications- a mini-review, Current, Opinion in Green and Sustainable Chemistry, Volume 41, 2023, 100788, ISSN 2452-2236,

[http://dx.doi.org/10.1016/j.cogsc.2023.100788]

[62] Corami A. Nanotechnologies and Phytoremediation: Pros and Cons. InPhytoremediation: Management of Environmental Contaminants 2023; 7: 403-26.

[63] Wahab A, Batool F, Muhammad M, Zaman W, Mikhlef RM, Naeem M. Current knowledge, research progress, and future prospects of phyto-synthesized NPs interactions with food crops under induced drought stress. Sustainability (Basel) 2023; 15(20): 14792.
[http://dx.doi.org/10.3390/su152014792]

[64] Mohamed EF, Awad G. Development of nano-sensor and biosensor as an air pollution detection technique for the foreseeable future 2022.
[http://dx.doi.org/10.1016/bs.coac.2021.11.003]

[65] Kaur S, Roy A. Bioremediation of heavy metals from wastewater using nanomaterials. Environ Dev Sustain 2021; 23(7): 9617-40.
[http://dx.doi.org/10.1007/s10668-020-01078-1]

[66] Kalaiselvi A, Roopan SM, Madhumitha G, Ramalingam C, Elango G. Synthesis and characterization of palladium nanoparticles using Catharanthus roseus leaf extract and its application in the photo-catalytic degradation. Spectrochim Acta A Mol Biomol Spectrosc 2015; 135: 116-9.
[http://dx.doi.org/10.1016/j.saa.2014.07.010] [PMID: 25062057]

[67] Chahar R, Das Mukherji M. Environmental applications of phytonanotechnology: a promise to sustainable future Phytonanotechnology 2022; 141-59.

[68] Wu X, Wang W, Zhu L. Enhanced organic contaminants accumulation in crops: Mechanisms, interactions with engineered nanomaterials in soil. Environ Pollut 2018; 240: 51-9.
[http://dx.doi.org/10.1016/j.envpol.2018.04.072] [PMID: 29729569]

[69] Gavrilescu M. Enhancing phytoremediation of soils polluted with heavy metals. Curr Opin Biotechnol 2022; 74: 21-31.
[http://dx.doi.org/10.1016/j.copbio.2021.10.024] [PMID: 34781102]

[70] Zulfiqar F, Navarro M, Ashraf M, Akram NA, Munné-Bosch S. Nanofertilizer use for sustainable agriculture: Advantages and limitations. Plant Sci 2019; 289: 110270.
[http://dx.doi.org/10.1016/j.plantsci.2019.110270] [PMID: 31623775]

[71] Guo H, White JC, Wang Z, Xing B. Nano-enabled fertilizers to control the release and use efficiency of nutrients. Curr Opin Environ Sci Health 2018; 6: 77-83.
[http://dx.doi.org/10.1016/j.coesh.2018.07.009]

[72] Chien SH, Prochnow LI, Cantarella H. Recent developments of fertilizer production and use to improve nutrient efficiency and minimize environmental impacts. Adv Agron 2009; 102: 267-322.
[http://dx.doi.org/10.1016/S0065-2113(09)01008-6]

[73] Maithani D, Sharma A, Dasila H, Tiwari A, Upadhayay VK. 2023.Nanotechnology for Crop Improvement and Sustainable Agriculture.
[http://dx.doi.org/10.1201/9781003345565-16]

[74] Khan MR, Ahamad F, Rizvi TF. Effect of NPs on plant pathogens. Advances in phytonanotechnology. Academic Press 2019; pp. 215-40.
[http://dx.doi.org/10.1016/B978-0-12-815322-2.00009-2]

[75] Noman M, Ahmed T, Wang J, *et al.* Nano-enabled crop resilience against pathogens: potential, mechanisms and strategies. Crop Health 2023; 1(1): 15.
[http://dx.doi.org/10.1007/s44297-023-00015-8]

[76] Kim DY, Kadam A, Shinde S, Saratale RG, Patra J, Ghodake G. Recent developments in nanotechnology transforming the agricultural sector: a transition replete with opportunities. J Sci Food Agric 2018; 98(3): 849-64.
[http://dx.doi.org/10.1002/jsfa.8749] [PMID: 29065236]

[77] Rehman A, Khan S, Sun F, *et al.* Exploring the nano-wonders: unveiling the role of Nanoparticles in enhancing salinity and drought tolerance in plants. Front Plant Sci 2024; 14: 1324176.
[http://dx.doi.org/10.3389/fpls.2023.1324176] [PMID: 38304455]

[78] John SA, Chattree A, Ramteke PW, Shanthy P, Nguyen TA, Rajendran S. Nanosensors for plant health monitoring. Nanosensors for Smart Agriculture. Elsevier 2022; pp. 449-61.
[http://dx.doi.org/10.1016/B978-0-12-824554-5.00012-4]

[79] Yadav A, Yadav K, Ahmad R, Abd-Elsalam K. Emerging frontiers in nanotechnology for precision agriculture: Advancements, hurdles and prospects. Agrochemicals 2023; 2(2): 220-56.
[http://dx.doi.org/10.3390/agrochemicals2020016]

[80] Kaushik S. 2022.Nanosensor Technology for Smart Intelligent Agriculture.
[http://dx.doi.org/10.1201/9781003268468-14]

[81] Lau ECHT, Carvalho LB, Pereira AES, *et al.* Localization of coated iron oxide (Fe3O4) NPs on tomato seeds and their effects on growth. ACS Appl Bio Mater 2020; 3(7): 4109-17.
[http://dx.doi.org/10.1021/acsabm.0c00216] [PMID: 35025413]

[82] Rai R, Nalini P, Singh Y P. Nanotechnology for sustainable horticulture development: Opportunities and challenges Innovative Approaches for Sustainable Development: Theories and Practices in Agriculture 2022; 191-210.

[83] Fincheira P, Tortella G, Seabra AB, Quiroz A, Diez MC, Rubilar O. Nanotechnology advances for sustainable agriculture: current knowledge and prospects in plant growth modulation and nutrition. Planta 2021; 254(4): 66.
[http://dx.doi.org/10.1007/s00425-021-03714-0] [PMID: 34491441]

[84] Sadik OA, Zhou AL, Kikandi S, Du N, Wang Q, Varner K. Sensors as tools for quantitation, nanotoxicity and nanomonitoring assessment of engineered nanomaterials. J Environ Monit 2009; 11(10): 1782-800.
[http://dx.doi.org/10.1039/b912860c] [PMID: 19809701]

[85] Vaseashta A, Vaclavikova M, Vaseashta S, Gallios G, Roy P, Pummakarnchana O. Nanostructures in environmental pollution detection, monitoring, and remediation. Sci Technol Adv Mater 2007; 8(1-2): 47-59.
[http://dx.doi.org/10.1016/j.stam.2006.11.003]

[86] Singh R, Behera M, Kumar S. Nano-bioremediation: an innovative remediation technology for treatment and management of contaminated sites Bioremediation of Industrial Waste for Environmental Safety. Biological Agents and Methods for Industrial Waste Management 2020; II: pp. 165-82.
[http://dx.doi.org/10.1007/978-981-13-3426-9_7]

[87] Tripathi S, Sanjeevi R, Anuradha J, Chauhan DS, Rathoure AK. Nano-bioremediation: nanotechnology and bioremediation InResearch anthology on emerging techniques in environmental remediation. IGI Global 2022; pp. 135-49.
[http://dx.doi.org/10.4018/978-1-6684-3714-8.ch007]

[88] Ramezani M, Rad FA, Ghahari S, Ghahari S, Ramezani M. Nano-bioremediation application for environment contamination by microorganism. Microorganisms for Sustainability 2021; 26: 349-78.
[http://dx.doi.org/10.1007/978-981-15-7455-9_14]

[89] Dhanapal AR, Thiruvengadam M, Vairavanathan J, Venkidasamy B, Easwaran M, Ghorbanpour M. Nanotechnology Approaches for the Remediation of Agricultural Polluted Soils. ACS Omega 2024; 9(12): 13522-33.
[http://dx.doi.org/10.1021/acsomega.3c09776] [PMID: 38559935]

[90] Parthipan P, Prakash C, Perumal D, Elumalai P, Rajasekar A, Cheng L. Biogenic NPs and strategies of nano-bioremediation to remediate PAHs for a sustainable future. Biotechnology for Sustainable Environment 2021; pp. 317-37.

[91] Chakraborty R, Karmakar S, Ansar W. Advances and Applications of Bioremediation: Network of Omics, System Biology, Gene Editing and Nanotechnology Environmental Informatics: Challenges and Solutions 2022; 167-99.

[92] Liaquat H, Imran M, Latif S, Hussain N, Bilal M. Multifunctional nanomaterials and nanocomposites for sensing and monitoring of environmentally hazardous heavy metal contaminants. Environ Res 2022; 214(Pt 1): 113795.
[http://dx.doi.org/10.1016/j.envres.2022.113795] [PMID: 35803339]

[93] Marcon L, Oliveras J, Puntes VF. *in situ* nanoremediation of soils and groundwaters from the nanoparticle's standpoint: A review. Sci Total Environ 2021; 791: 148324.
[http://dx.doi.org/10.1016/j.scitotenv.2021.148324] [PMID: 34412401]

[94] Kumar SR, Gopinath P. Nano-bioremediation applications of nanotechnology for bioremediation 2017.

[95] Abbas Q, Yousaf B, Amina , *et al.* Transformation pathways and fate of engineered nanoparticles (ENPs) in distinct interactive environmental compartments: A review. Environ Int 2020; 138: 105646.
[http://dx.doi.org/10.1016/j.envint.2020.105646] [PMID: 32179325]

[96] Ali H, Khan E, Sajad MA. Phytoremediation of heavy metals—Concepts and applications. Chemosphere 2013; 91(7): 869-81.
[http://dx.doi.org/10.1016/j.chemosphere.2013.01.075] [PMID: 23466085]

[97] Karupannan SK, Dowlath MJ, Arunachalam KD. Phytonanotechnology: challenges and future perspectives 2020.
[http://dx.doi.org/10.1016/B978-0-12-822348-2.00015-2]

[98] Shang Y, Hasan MK, Ahammed GJ, Li M, Yin H, Zhou J. Applications of nanotechnology in plant growth and crop protection: a review. Molecules 2019; 24(14): 2558.
[http://dx.doi.org/10.3390/molecules24142558] [PMID: 31337070]

[99] Upadhayay VK, Chitara MK, Mishra D, *et al.* Synergistic impact of nanomaterials and plant probiotics in agriculture: A tale of two-way strategy for long-term sustainability. Front Microbiol 2023; 14: 1133968.
[http://dx.doi.org/10.3389/fmicb.2023.1133968] [PMID: 37206335]

[100] Li C, Yan B. Opportunities and challenges of phyto-nanotechnology. Environ Sci Nano 2020; 7(10): 2863-74.
[http://dx.doi.org/10.1039/D0EN00729C]

[101] Jiang M, Song Y, Kanwar MK, Ahammed GJ, Shao S, Zhou J. Phytonanotechnology applications in modern agriculture. J Nanobiotechnology 2021; 19(1): 430.
[http://dx.doi.org/10.1186/s12951-021-01176-w] [PMID: 34930275]

[102] Prakash S, Rajpal VR, Vaishnavi S, Deswal R. Phyto-nanotechnology: Current status, Challenges, and leads for future ScienceOpen Posters 2021; 2.

CHAPTER 9

Bionanomaterials and Environmental Remediation

M.P. Laavanyaa shri[1], R. Margrate Thatcher[1], R. Sakthi Sri[1], Y. Manojkumar[1] and S. Ambika[1,*]

[1] *PG and Research Department of Chemistry, Bishop Heber College, Trichy - 17, Tamil Nadu, India*

Abstract: Environmental pollution is one of the biggest threats to ecosystems and human health around the globe. Over the years, various methods have been implemented for environmental remediation. However, these methods have their limitations and urge the scientific community to find an effective alternate method. The emergence of nanomaterials (NMs) offers tremendous potential for addressing these pollution challenges and promoting sustainable development. Particularly, bioNMs possess unique characteristics such as high surface area, catalytic activity, and selectivity, which make them highly effective in removing contaminants and monitoring environmental conditions. This chapter explores the synthesis of bioNMs from various sources, characterization, their diverse applications in environmental remediation such as water and soil treatment, and air purification. Furthermore, it examines the challenges that need to be addressed and presents prospects for bioNMs in the ongoing battle against environmental pollution.

Keywords: Air pollution, Bionanomaterial, Environmental remediation, Green NMs, Soil pollution, Wastewater management.

INTRODUCTION

Environmental pollution is one of the most serious concerns around the globe. It is increasing gradually and causing a serious impact on living organisms, including humans. It is estimated that by 2030, nearly 3 million tons of waste will be produced globally [1]. According to WHO, every year, 2,70,000 children fall ill due to the toxic effects of environmental pollution [2]. In general, pollution is created by any unwanted change in the physicochemical and biological characteristics of any component of the environment. This includes the major components of water, soil, and air that can cause harmful effects on various forms of life and property [3]. Wastes are generated and discharged into the environment in a wide range of ways. For example, atmospheric pollutants like suspended

* **Corresponding author S. Ambika:** PG and Research Department of Chemistry, Bishop Heber College, Trichy - 17, Tamil Nadu, India; E-mail: chem.ambi@gmail.com

Neha Agarwal, Vijendra Singh Solanki, Neetu Singh & Maulin P. Shah (Eds.)

organic particulates, nitrogen oxides, sulfur oxides, carbon oxides, and hydrocarbons, along with various organic pollutants like pesticides, insecticides, and heavy metals, can be found in soil and water. There are many effective remediation methods to clean it up. Bioremediation is one of the effective approaches to some extent, which involves breaking down the pollutants in the soil by microorganisms [4]. Phytoremediation involves planting species to extract contaminants from the soil, and landfill remediation involves techniques such as capping [5]. Thermal methods such as incineration and thermal desorption involve applying heat to contaminated materials to either volatilize or decompose the contaminants [6]. Chemical remediation involves the oxidation of pollutants into non-toxic forms using strong oxidants such as hydrogen peroxide, ozone gas, potassium permanganate, or persulfates [7]. Physical treatments such as air stripping, filtration, and sedimentation are used to physically separate contaminants from environmental media [7]. Most of the above-mentioned methods involved the removal of contaminants *via* processes like adsorption, absorption, chemical reactions, photocatalysis, and filtration. Here, NMs play a promising role due to their unique physical and chemical properties. Many NMs have been reported for environmental remediation [8]. Compared to classical methods of synthesizing NPs, utilizing biology for the synthesis of NMs, called bioNMs, has become an attractive area of research in recent years [9].

Among the various applications, bio-NMs are considered a promising candidate for environmental remediation due to their low toxicity towards living beings, enhanced biocompatibility, bioavailability, and bioreactivity. It paved the way for a sustainable, clean, and green environment, as they mimic the characteristics of NMs with advancements and remarkable performance, high efficiency, and monitoring of the adverse effects of pollutants. In general, bioNMs are synthesized by biological molecules like proteins, enzymes, nucleic acids, antibodies, secondary metabolites, and microorganisms such as bacteria, fungi, plant extract, *etc* [10]. Their smaller size, high surface area, more reactivity, and specific functionalities enhance environmental remediation ability. BioNMs express unique physical, chemical, structural, biological, and mechanical properties, which make them promising candidates for environmental applications.

SOURCES OF SYNTHESIS OF BIONMS

Synthesis of NMs using physical and chemical methods involves the usage of highly concentrated reductants and stabilizing agents that are highly harmful to the environment [11, 12]. Besides high yield, controlling the uniform size is a challenging task [13]. Over the years, various strategies have been employed to overcome the difficulties faced by traditional methods through the use of eco-friendly solvents, reagents, and methodologies. However, scalability limits their application. Hence, the development of a cost-effective and environmentally friendly yet scalable method is an urgent need. The biological synthesis of NPs is a single-step process that requires less energy, and it is eco-friendly in nature. Therefore, biosynthesis presents a promising alternative approach for the synthesis of nanomaterials. It uses eco-friendly resources such as plants, enzymes, nucleic acids, and microorganisms [13]. Various sources used for synthesizing bioNMs are given in Fig. (**1**).

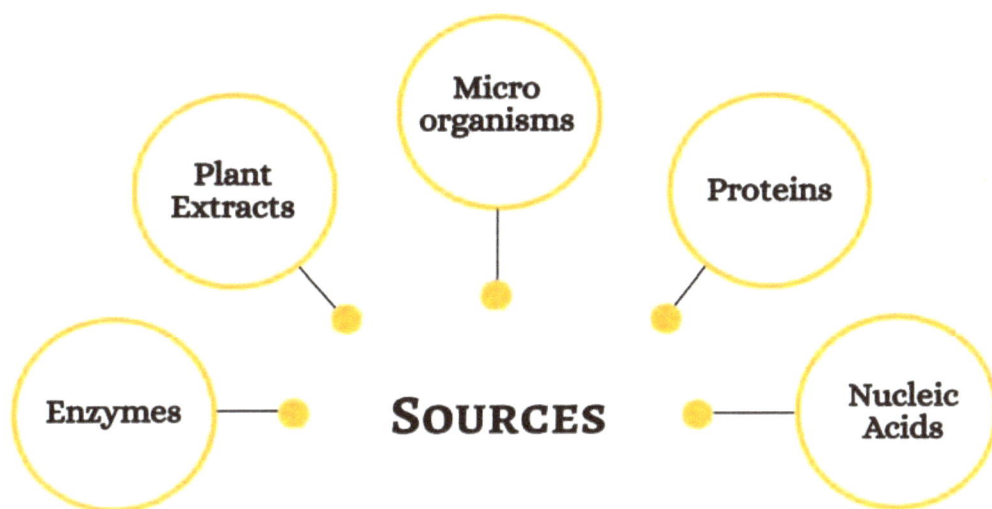

Fig. (1). Pictorial Representation of Source for Synthesis of BioNMs.

Plant Extracts

Plant-derived NPs have garnered huge interest because of their cost-effectiveness, environmental friendliness, and sustainability. In general, choosing the right plant and its components is the most crucial aspect of synthesizing bioNMs using plants. This is because different plants contain different levels of enzyme activity, phytochemicals, and biochemical processes.

BioNPs are prepared by treating the respective metal salts or oxides with plant extracts. After some time, the solution undergoes a visible color change that indicates the formation of NPs, as mentioned in Fig. (**2**). Plant extracts can be obtained using various parts of the plants, like seeds, roots, stems, leaves, flowers, and fruits. Various metal salts and the plant parts used for synthesizing bioNMs are given in Tables **1** and **2**, respectively. Plant extracts obtained from leaves, flowers, roots, *etc.*, contain many bioactive compounds (secondary metabolites) like alkaloids, phenolic compounds, flavonoids, and terpenoids. These secondary metabolites naturally act as stabilizers and reducing agents to produce the NMs [14].

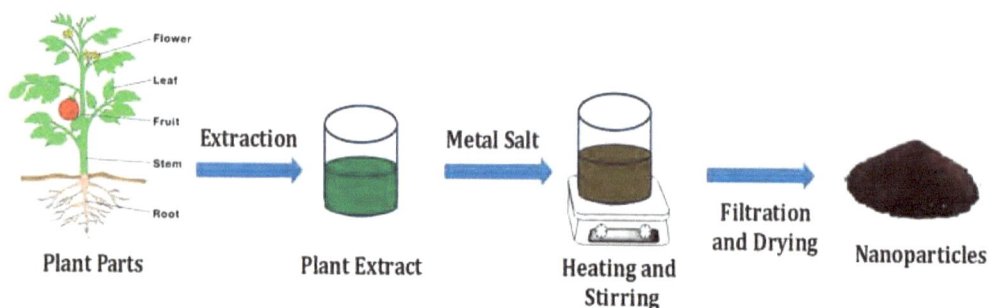

Fig. (2). Schematic representation of the synthesis of plant-derived NPs.

Table 1. Name of the Plants, Starting Materials, and NPs

Plant	Metal Salt (Starting Material)	Nano Particles (Product)	Refs.
Azadirachta indica	$Ag^+NO_3^-$	Ag^o NPs	[15]
Ziziphus ziziphus	$HAu^+Cl_4^-$ $4H_2O$	Au^o NPs	[16]
Atriplex halimus	$H2Pt^+Cl_2^-$ $6H_2O$	Pt^o NPs	[17]
Aloe barbedensis	$CuSO_4$ $5H_2O$	Cu^o NPs	[18]
Cassia fistula	$Zn(NO_3)_2$	ZnO NPs	[19]

Table 2. Name of the Plants, their Parts, and NPs

Plant Name	Parts of the Plant	NP	Size	Refs.
Fenugreek	Seed	Au	12.nm	[20]
Rubus Fairholmianus	Root	ZnO	3.2 nm	[21]
Callicarpa Maingayi	Stem	Ag	5.5 nm	[22]
Mentha Arvensis	Leaves	TiO_2	2.8-9.9 nm	[23]
Melia Azedarach	Flower	Fe_2O_3	23 nm	[24]
Psidium Guajava	Fruit	Cu	18 nm	[25]

Microbes

Microbes like bacteria, fungi, algae, *etc.*, play a major role in the synthesis of NPs [26]. In general, the electronic, magnetic, and physiochemical features mainly focus on the control of the size, shape, and composition of the NMs, which is one of the important aspects in the synthesis of bioNM using microbes. BioNM is prepared using the microbes where they utilize the cell wall that helps to transport metal ions; then, the ions with a positive charge interact with the negative charge wall. These ions are reduced to metal NPs by the enzymes in the cell, which is an intracellular synthesis. Then, the metal ions are aggregated on the surface of the cell and reduce the ions through enzymes, which is called extracellular synthesis. These synthesized NMs undergo various biochemical pathways and enzymatic reactions, which are utilized by the microbes to reduce the pioneer molecules that help the nucleation and growth of NMs [27]. Various compounds like enzymes, proteins, peptides, secondary metabolites, polysaccharides, and cellular components produced by the microbes themselves act as a stabilizer and reductants to produce the NM.

Bacteria

Bacteria are ubiquitous in the environment and, hence, can be cultivated quickly. Bacteria can adapt to different environmental conditions, which facilitates NP synthesis [28]. The synthesis of NPs using bacteria includes common bacterial species like *Escherichia coli, Bacillus subtilis, Pseudomonas aeruginosa, etc.* They are selected and cultivated depending on the desired NM and its application. After a certain period of cultivation, they are introduced to metal precursors, *i.e.*, metal salts, which act as bases for the NM formation. They undergo a reduction process with the use of enzymes like NADH-dependent reductase, nitrate-dependent reductase, and hydrogenases [29, 30]. Once the NMs are formed, they are extracted after filtration. Bioactive compounds like reductases, oxidases, and dehydrogenases produced by the bacteria act as reducing and oxidizing agents to produce bioNMs.

Algae

Algae belong to the kingdom Protista, which can be single or multicellular. Microalgae play an important role in transforming the metal ions into flexible forms, which promotes the green synthesis of NPs [31]. The synthesis of NPs using algae involves the formation of the algal extract, where reagents are used to prepare the algal extract, and the metal precursor is mixed with it. In the beginning, there is a change in the color of the reaction, which shows nucleation, followed by the growth of NPs. Thermodynamically stable NPs of various geometries are formed as the adjacent nucleonic particles join together. Algae contain polysaccharides that are bioactive compounds and they contribute to the reduction and stabilization of the NMs [32].

Fungi

Fungi are microbes that actively participate in the production of NMs [33]. Easy mode of culture, higher biomass, bioaccumulation of metabolites, tolerance to metals, high cell wall binding, and uptake capacity make fungi superior to bacteria for NM synthesis. A great advantage of culturing yeast is that it requires a simple culture medium, making it a promising microorganism for NP synthesis [34]. Additionally, a significant advantage of using yeast compared to other microbes is that a simple encapsulation mechanism is possible using only yeast cells, water, and reagents without any further stabilizers. Like other microbes, NPs are produced intra/extracellularly by fungi, but the NPs that are produced intracellularly are comparatively small to those that are produced extracellularly [35]. Like other microbes, the bioactive compounds like enzymes, proteins, reductase, and oxidase produced from the fungi enable the synthesis of NMs. Various microbes used for the synthesis of bioNMs are given in Table **3**.

Table 3. Different Types of NMs Synthesized from Various Microbes.

Microbe Name	Category	NP	NP size	Refs.
Bacillus subtilis	Bacteria	Au	40	[36]
Lactobacillus Fermentum	Bacteria	Ag	30 nm	[37]
Pseudomonas Aeruginosa	Bacteria	Fe	50-100 nm	[38]
Aspergillus Fumigatus	Fungai	ZnO	1.2- 6.8 nm	[39]
Rhizopus Oryzae	Fungai	Au	10 nm	[40]
Alternaria Alternata.	Fungai	Ag	20 to 60 nm	[41]
Candida glabrata	Yeast	Au	5-20 nm	[42]
Saccharomyces Cerevisiae	Yeast	Zno	5..0 nm	[43]
Schizosaccharomyces Pombe	Yeast	Cds	1.5 nm	[44]
Bifurcaria Bifurcata	Algae	CuO	5-45 nm	[45]
Galaxaura Elongata	Algae	Au	13 nm	[46]
Cyanobacterium Oscillatoria Limnetica	Algae	Ag	3.30 -17.97 nm	[47]

Biomolecules

Proteins

BioNMs synthesized from proteins offer a sustainable and versatile solution for environmental remediation. Soy, silk, collagen, and microorganisms are some of the sources from which the proteins are derived for the synthesis of NMs. These materials possess inherent biocompatibility and can be tailored to exhibit specific properties. For the synthesis of bioNMs, the extracted protein is purified and its solution is prepared by using various techniques. The prepared protein solution is poured into a buffer solution according to their pH level; the synthesis of NPs takes place here by adding specific precursors and metal salts, which are added with the addition of a reducing agent. After synthesizing, the protein solution undergoes precipitation with other precursor solutions to obtain the NM after filtration. Bioactive compounds like amino acids, cysteine residues, motifs, peptides, and glycoproteins present within the proteins act as a template, stabilizer, and reducing agent in the synthesis of NMs.

Enzymes

Biologically synthesized NMs, especially those derived from enzymes, hold tremendous promise for environmental remediation. Enzymes are called natural catalysts capable of facilitating specific chemical reactions, and when harnessed in the synthesis of NMs, they can produce highly efficient and eco-friendly solutions for cleaning up pollutants [48]. They are known for their biocompatibility, scalability, and specificity [49]. Enzyme-synthesized bioNMs are created through some enzymatic reactions where enzymes act as catalysts, facilitating the conversion of precursor metal salts into NPs. Then, the NPs are stabilized and modified by the enzymes. Finally, the resulting bioNMs exhibit tailored properties suitable for environmental remediation, such as biocompatibility and specificity [50]. Bioactive compounds like cysteine, histidine, tyrosine residues, cofactors, ligands, *etc.*, present within the enzymes enable the synthesis of bioNMs from the enzymes.

Nucleic Acids

Nucleic acids are an effective biomacromolecule for synthesizing NMs. NMs synthesized from nucleic acids, particularly iron-based ones, hold great promise for environmental remediation. These NMs can be tailored to efficiently remove pollutants from water, soil, and air due to their high surface area-to-volume ratio and unique properties at the nanoscale [51]. DNA, RNA, and other biological samples act as a source of nucleic acid for the synthesis of NMs, and they are engineered to have specific binding affinities for target pollutants [52]. For synthesizing NMs using the amino acid, the nucleic acid sequences are at first arranged or designed, which will act as the base for the desired NM. These NM strands are then synthesized by certain methods like solid-phase or enzymatic synthesis. It is then mixed under certain conditions, and hybridization between complementary sequences is done to assemble the complex nanostructure, which is processed further and the NMs are obtained. Nucleic acids do not have specific bioactive compounds like other biomacro molecules but their DNA and RNA templates and hybridization act as tools that are essential for the synthesis of NMs. Various types of NMs synthesized using different kinds of biomolecules are shown in Table **4**.

Table 4. Different Types of Nanomaterial Synthesized from Biomacromolecules.

Biomolecules	NP	NP size	Refs.
Protein	Silver	15-20 nm	[53]
Protein	Silver	40 nm	[54]
Protein	Copper	200 nm	[55]
Protein	Gold	20-30 nm	[56]
Protein	Zinc	30 nm	[57]
Enzyme	Silver	13 nm	[58]
Enzyme	Silver	50-100 nm	[59]
Enzyme	Copper	6.5 nm	[60]
Enzyme	Gold	10 nm	[61]
Enzyme	Iron Oxide	7 nm	[62]
Nucleic acid	Silver	8-27 nm	[63]
Nucleic acid	gold	30 nm	[64]

CHARACTERIZATION OF BIONMS

BioNPs can be difficult to detect due to their small size and various shapes and compositions. However, several techniques have been developed to detect and analyze these particles, and characterization of bioNMs was accomplished using UV-Vis spectroscopy, FT-IR, XRD, SEM with EDOX, TEM, and DLS.

UV-visible Spectroscopy

UV-visible spectroscopy serves as a fundamental technique for both quantitative and qualitative analysis of biomolecules, especially NMs. It is used to study the physicochemical properties such as material size, concentration of material, and aggregation. Moreover, it facilitates the exploration of optoelectronic properties and structural conformation in bioNMs. Based on the Beer-Lambert Law, UV-visible spectroscopy establishes linear relationships between absorbance and concentrations. Numerous green-synthesized bioNMs, including Ag, Au, Cu, Pd, *etc.*, have undergone characterization to investigate their structural properties [52]. The UV-visible spectroscopy method was employed for the initial characterization of synthesized NPs. UV-visible absorbance spectrum of noble metal NPs experiences significant changes based on the environment surrounding them, resulting in shifts in the value of λmax [65]. For example, for silver NPs (AgNPs) synthesized from Brassica oleracea (BO) leaf extract, the UV-Vis spectra revealed that the absorption peak at ~ 415 nm indicates the presence of BO-Ag NPs Fig. (**3**) [66].

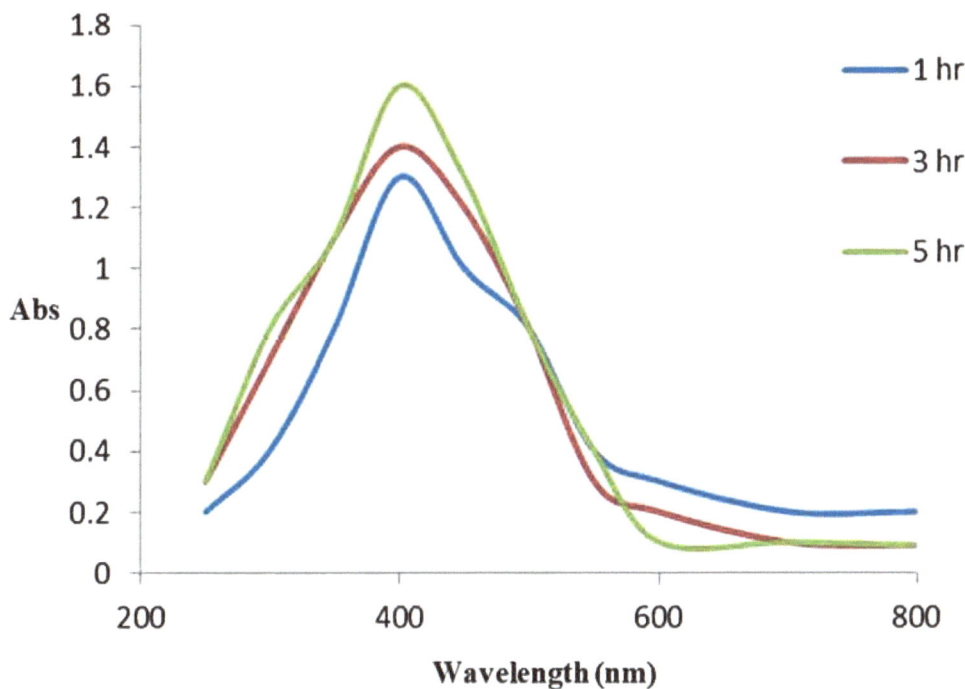

Fig. (3). UV–vis Absorption Spectra of BO-Ag NPs [66].

Fourier-Transformed Infrared Spectroscopy (FT-IR)

Fourier transform infrared spectroscopy (FT-IR) stands out as the most sensitive and efficient analysis method for characterizing bioNMs. It works by analyzing the distinct vibration patterns of individual molecules. It enables the identification of functional groups such as aldehydes, ketones, alcohols, carboxylic acids, and terpenoids and their presence on the surface of the bioNM. Numerous bioNMs, silver, gold, nickel, iron oxide, *etc.*, have been thoroughly analyzed using FT-IR [67]. Using FT-IR, the chemical composition on the NP surface can be determined, and additionally, the active sites responsible for the surface reactivity can be identified [68]. In the case of C. Procumbens extract-mediated silver NPs, there is a significant change in the IR spectra of the NPs compared to the plant extract Fig. (**4**) [69]. The strong absorption peaks observed for the leaf extract at 1500, 1450, and 1100 cm^{-1} indicate the presence of bioactive molecules, including polyphenols, carboxylic acids, and amides. In the case of the NMs, the IR peaks at 1550 and 1450 cm^{-1} are characteristic peaks of primary amide stretches [70]. The peaks at 1550, 1384, and 1100 cm^{-1} indicate the presence of O–H stretching of polyphenols, stretching of carboxylic acids, and amide stretching of proteins [71]. These functional groups from the amides and polyphenols in the leaf extract are responsible for the formation and stabilization of silver NPs.

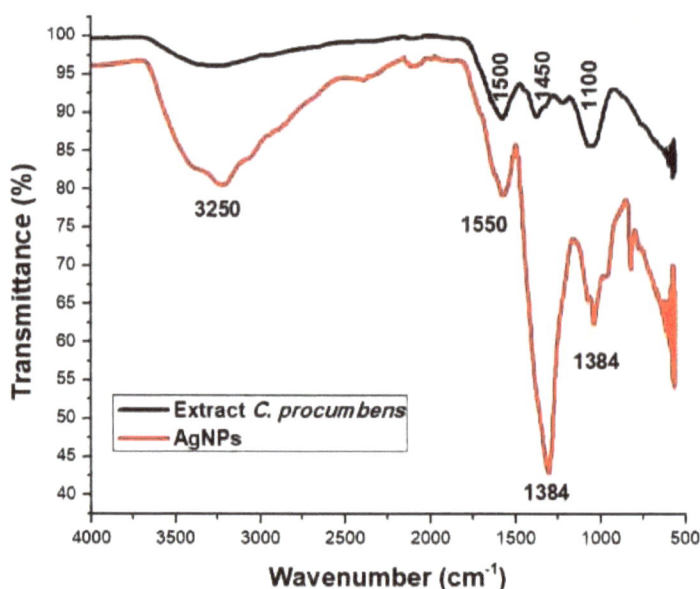

Fig. (4). FT-IR Spectra of C. Procumbens extract and AgNPs [69].

X-ray Diffraction

X-ray diffraction (XRD) is one of the simplest and most useful techniques for the characterization of NPs [72]. It gives important information about the crystallographic planes and average particle size. The X-ray diffraction pattern of Cassia occidentalis L. seed extract-mediated Ag NPs is shown in Fig. (**5**) [73]. The XRD figure shows the composition and crystalline nature of AgNPs. Bragg's reflections for particular 2θ values were recorded at angles of 33.15°, 39.02°, 45.65°, 65.19°, and 78.90°, corresponding to the crystallographic planes 101, 111, 200, 220, and 311. The average particle size of the AgNPs is 19.78 nm [73].

SEM with EDAX Measurements

Scanning electron microscopy (SEM) is used to analyze the surface morphology of materials. When coupled with energy-dispersive X-ray spectroscopy (EDX), it can also examine the chemical composition. SEM offers simple, non-destructive, and rapid measurements for a wide range of materials [74]. There are relatively few differences between SEM and TEM, primarily related to sample thickness and the method of data collection. For EDX analysis, plant extracts containing reduced metal NPs are typically dried and drop-coated onto a carbon film before being analyzed on an SEM instrument equipped with an EDX detector [75]. The results of SEM-EDX analysis for silver NPs synthesized using green tea extract and those coated on a graphene oxide support are presented in Fig. (**6**) [76].

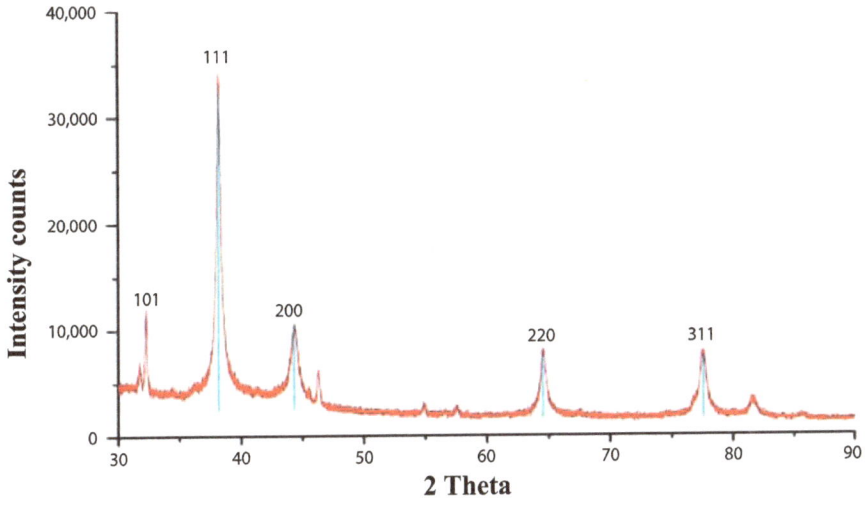

Fig. (5). XRD pattern of Ag NPs [73].

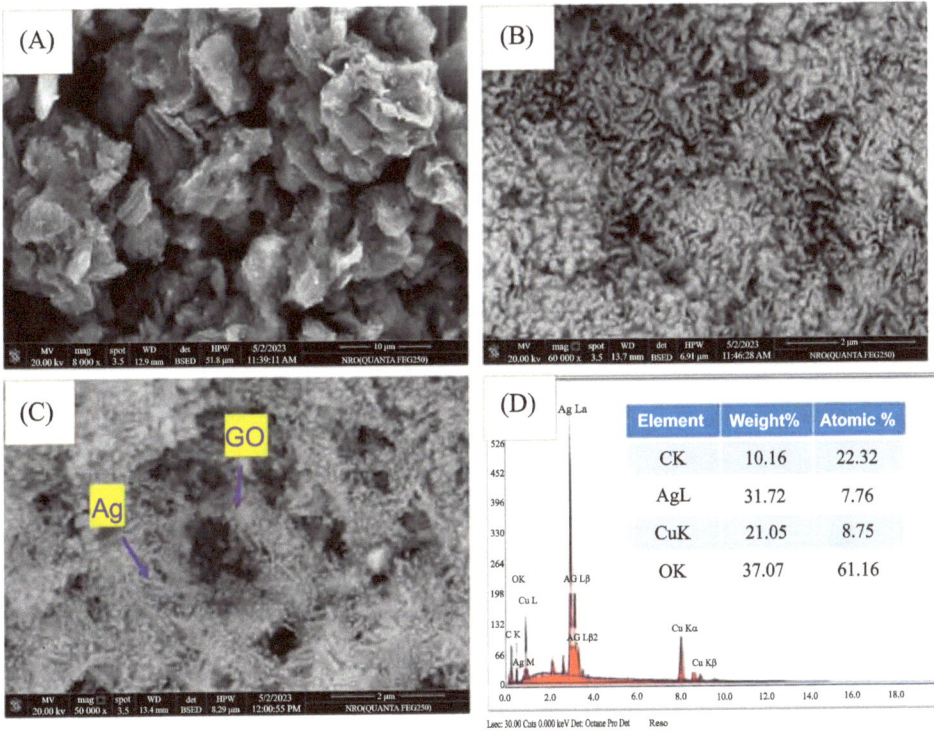

Fig. (6). Representation of SEM and the EDAX of GC/GO/Ag NPs [76].

Transmission Electron Microscope

Transmission electron microscopy (TEM) is one of the most powerful analytical techniques. It produces high-resolution images at the atomic level, allowing visualization of even the smallest particles, a capability beyond the reach of light microscopes and SEM. TEM is used to study the morphological characteristics of materials, such as size and shape. In certain cases, it can also be used to determine the chemical composition of bioNMs. The size and morphology of the synthesized Ag NPs were determined by a high-resolution TEM. The TEM image is given in Fig. 7. It indicated that the synthesized AgNPs using both Tulsi leaf extract and quercetin have a spherical shape with uniform sizes of 14.6 nm and 11.35 nm, respectively [77]. Furthermore, the images confirmed that increasing the pH of the reaction mixture increases the size of the AgNPs.

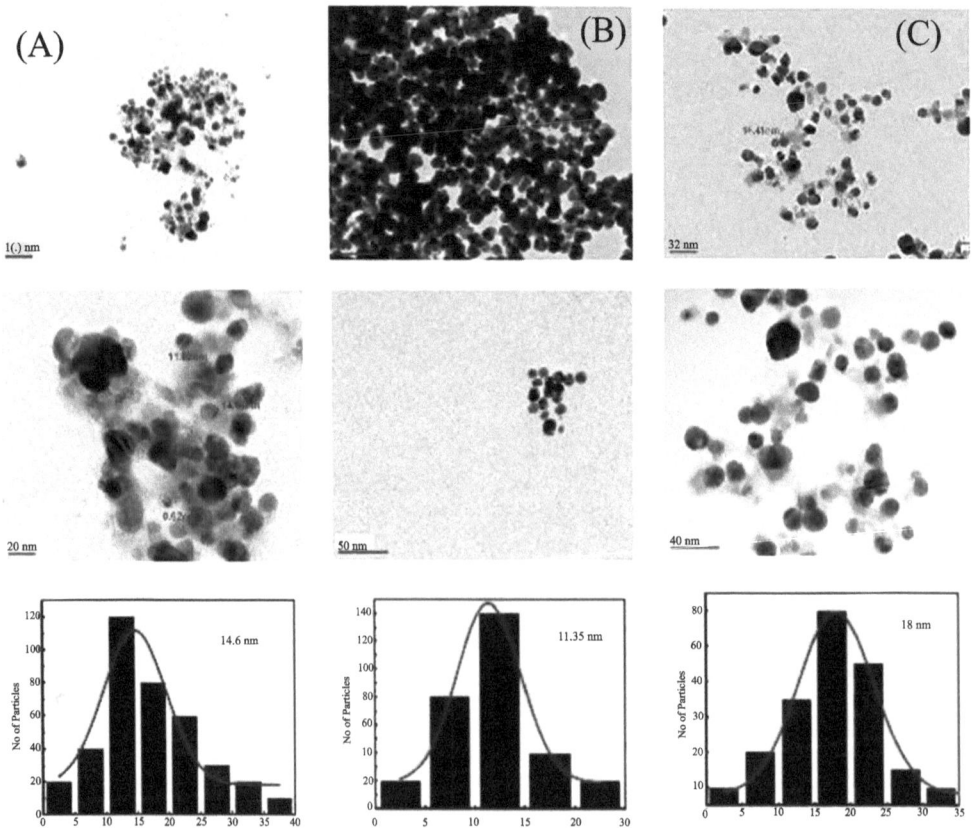

Fig. (7). TEM micrographs and particle size distribution histogram of AgNPs synthesized using Tulsi extract (a), quercetin (b), and quercetin at pH 10 of the reaction mixture (c) [77].

Dynamic Light Scattering (DLS) Analysis

The hydrodynamic size, polydispersity index, and zeta potential of the NM can be determined using DLS analysis. It worked based on the principle of the Brownian movement. It utilizes laser diffraction methods but accounts for multiple scattering from the moment of the different sizes of the NPs. Zeta potential is used to predict the morphology and stability of the NPs. Both DLS and zeta potential measurements are used to evaluate the particle size and potential stability of synthesized NPs in aqueous or physiological solutions. For example, AgNPs were prepared using *Chromolaena odorata* leaf extract, and their average size was 30.5 nm. The zeta potential value was -0.532 mV [78]. The results are shown in Fig. (8). The negative zeta potential of these AgNPs is likely due to the capping of bioactive molecules from the plant extract.

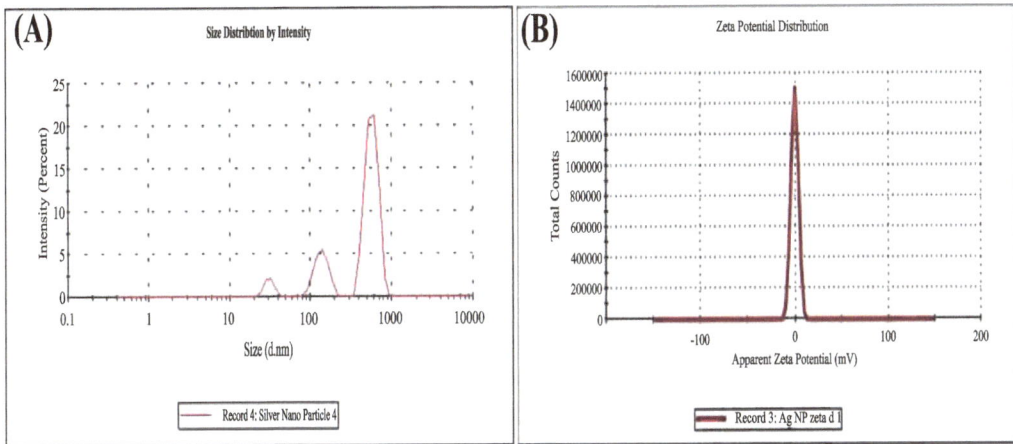

Fig. (8). DLS size distribution by the intensity and (B) zeta potential distribution of AgNP [78].

APPLICATION OF BIONMS FOR ENVIRONMENTAL POLLUTION REMEDIATION

Industrialization and intense agricultural activities are the primary causes of contamination in landfills, water bodies, and the atmosphere. The large surface area binding site and catalytic centers of NMs increase their reactivity towards contaminants. This reactivity leads to the reduction of contaminant concentration in the environment. Reports related to the direct application of biogenic NMs synthesized using microbes and biomolecules for environmental pollution remediation are limited. Most environmental pollution remediation using NPs involves those synthesized using plant extracts. Therefore, this section describes the mechanisms of action of bio-based NMs used for environmental remediation. Some of the bioNM used in the pollution remediation are shown in Table 5.

Table 5. Application of bionanomaterial for pollution remediation.

Source	Metal NP	Pollution Remediation	Refs.
Plant	Ag	Soil	[100]
Plant	Fe_2O_3	Soil	[101]
Microbe	Pd	Soil	[102]
Plant	CuO	Water	[103]
Plant	ZnO	Water	[104]
Microbe	Silver	Water	[105]
Protein	Fe_2O_3	Water	[106]
Protein	Ag	Water	[107]
Plant	Ag	Air	[108]
Plant	CeO_2	Air	[99]
Plant	Fe_2O_3	Air	[109]

Soil Pollution Remediation

The development of bioNMs offers an effective way to restore and remediate polluted soil. These materials utilize various mechanisms, such as absorption, adsorption, redox reactions, and precipitation, to remove pollutants, such as heavy metals, dyes, pesticides, microplastics, *etc.*, from contaminated soil. This pollution detrimentally affects food production, leading to decreased yields and causing serious health issues for humankind. The Fe_3O_4 NPs, when employed alongside soil microorganisms, exhibit the capability to degrade 2,4-dichlorophenoxyacetic acid *via* adsorption from soil [79]. TiO_2NPs (TiO_2 NPs) offer unique capabilities in heavy metal remediation [80]. TiO_2 NP surfaces facilitate the effective adsorption of heavy metal ions through surface complexation/electrostatic interactions. Besides, TiO_2 NPs possess photocatalytic activity when exposed to ultraviolet light, accelerating the breakdown of organic pollutants associated with heavy metal pollution and reducing their mobility and bioavailability in soil [81]. The zero-valent iron NPs effectively immobilize various metals in contaminated soils. Modifications and capping techniques further enhance the effectiveness and environmental compatibility of zero-valent iron NP-based remediation strategies [82]. Biological ZnO and silicon NPs are used to treat the contaminated saline soil with heavy metals, which enhance the growth of plants in saline soil [83, 84]. CuO-NPs were synthesized by leaf extract of *Catharanthus roseus* used for the removal of cadmium and chromium from the soil [85]. Manganese oxide NPs, synthesized using *Brassica Oleraceae,* can be effectively used for the degradation of pharmaceutical chemicals present in the soil [86].

Water Pollution Remediation

The green ZnONPs from *Moringa Oleifera* leaves extract exhibit promising potential for efficiently degrading organic compounds found in synthetic petroleum wastewater *via* photocatalytic activity [87]. Iron NPs are efficiently synthesized from the Mediterranean cypress extract used for the removal of dyes from waste aqueous solutions [88]. The TiO^2NPs are prepared from an aqueous extract of Eichhornia degraded material resulting from the irradiation treatment of reactive organic dye pollutants commonly found in industrial wastewater [89]. Copper NPs are synthesized by ficus carica fruit extract as the reducing agent. These NPs are subsequently applied for the degradation of toxic organic dyes like Alizarin Yellow R, commonly encountered in industrial wastewater [90]. Manganese oxide NPs are synthesized and used for wastewater treatment, which targets reductions of sulfates, phosphates, and other physical parameters in water that significantly influence the potential of cultivating wheat crops [91]. The V_2O_5 NPs synthesized by aqueous extract derived from *Punica granatum* show enhanced adsorption capabilities against methylene blue dye in wastewater [92]. Silver NPs synthesized using plant waste biomass are used for removing dye pollutants from aqueous solutions [93]. Compared to naked ZnO and silver NPs, protein coating enhances the adsorption of dyes in the water [94].

Air Pollution Remediation

In general, lead, nitrogen oxides, sulfur oxides, volatile organic compounds, carbon monoxide, and particulate matter are the major causes of air pollution. Over the years, nanotechnology has been used to provide solutions for air pollution through the use of nanofilters, nanosensors, and nanoadsorbents. $Fe_2O_3$is synthesized from aloe vera and used for the adsorption of arsenic in the air [95]. TiO_2 NPs are synthesized from Psidium Guajava extract and used as a filtrate for the purification of automobile exhaust and waste incinerators [96]. ZnO NPs synthesized *via* Bacillus subtilis are incorporated onto nanofiber membranes, serving as filters to mitigate environmental contamination through air filtration systems [97]. Additionally, these NPs exhibit photocatalytic properties, facilitating the degradation of pollutants. Nanofibrous protein-based materials have been employed in the fabrication of air filters with the capacity to effectively remove particulate matter (PM) and toxic gaseous molecules [98]. *Jatropha curcus* leaves extract mediated CeO_2 NPs used to degrade the indoor air pollutant acetaldehyde [99].

CONCLUSION, OUTLOOK, AND FUTURE DIRECTIONS

The utilization of bioNMs in environmental remediation holds significant importance due to their cost-effectiveness and proficiency in eliminating various pollutants from soil, aqueous streams, and air. These materials possess unique properties such as efficiency and effectiveness, environmental compatibility, versatility, tailor ability, potential for sustainable remediation, and integration with bioremediation and green technologies. The synthesis of biologically derived NMs simplifies the removal of hazardous metals and chemicals from polluted environments, potentially enabling waste elimination at the source through green manufacturing. However, challenges remain in scaling up manufacturing, ensuring mass production, availability, purification, and utilization of NMs without additional matrices. Commercial manufacturing poses challenges in terms of time, economic expenses, and consistent quality. The limited standardized production methods hinder consistent quality and scalability. Additionally, despite established biosynthetic mechanisms, the lack of understanding of molecular mechanisms and interactions presents further challenges in ensuring the reliable production of biosynthetic NMs. Metal and metal oxide NMs show promise in removing pollutants through reduction or oxidation processes, with functionalization enhancing their selectivity. Yet, more research is necessary to improve environmental remediation, focusing on selective removal, pH resistance, stability, and cost optimization.

This chapter comprehensively explored the potential of bioNMs for environmental remediation. It delved into various sources for their synthesis and characterization techniques to determine their properties. Detailed explanations were provided on their applications in treating soil, water, and air pollution. Finally, the chapter addressed the challenges that need to be overcome for real-world implementation. By effectively utilizing bioNMs, we can significantly contribute to environmental remediation efforts and promote sustainable development.

REFERENCES

[1] Kumar A, Agrawal A. Recent trends in solid waste management status, challenges, and potential for the future Indian cities – A review. Current Research in Environmental Sustainability 2020; 2: 100011.
[http://dx.doi.org/10.1016/j.crsust.2020.100011]

[2] Organization WH. Don't pollute my future! The impact of the environment on children's health. World Health Organization 2017.

[3] Chu E, Karr J. Environmental impact: Concept, consequences, measurement. Reference Module in Life Sciences 2017.

[4] Megharaj M, Ramakrishnan B, Venkateswarlu K, Sethunathan N, Naidu R. Bioremediation approaches for organic pollutants: A critical perspective. Environ Int 2011; 37(8): 1362-75.
[http://dx.doi.org/10.1016/j.envint.2011.06.003] [PMID: 21722961]

[5] Kim KR, Owens G. Potential for enhanced phytoremediation of landfills using biosolids – a review. J Environ Manage 2010; 91(4): 791-7.
[http://dx.doi.org/10.1016/j.jenvman.2009.10.017] [PMID: 19939550]

[6] Ding D, Song X, Wei C, LaChance J. A review on the sustainability of thermal treatment for

contaminated soils. Environ Pollut 2019; 253: 449-63.
[http://dx.doi.org/10.1016/j.envpol.2019.06.118] [PMID: 31325890]

[7]   Bennedsen LR. in situ chemical oxidation: the mechanisms and applications of chemical oxidants for remediation purposes Chemistry of Advanced Environmental Purification Processes of Water. Elsevier 2014; pp. 13-74.
[http://dx.doi.org/10.1016/B978-0-444-53178-0.00002-X]

[8]   Das S, Sen B, Debnath N. Recent trends in nanomaterials applications in environmental monitoring and remediation. Environ Sci Pollut Res Int 2015; 22(23): 18333-44.
[http://dx.doi.org/10.1007/s11356-015-5491-6] [PMID: 26490920]

[9]   Saratale RG, Karuppusamy I, Saratale GD, *et al.* A comprehensive review on green nanomaterials using biological systems: Recent perception and their future applications. Colloids Surf B Biointerfaces 2018; 170: 20-35.
[http://dx.doi.org/10.1016/j.colsurfb.2018.05.045] [PMID: 29860217]

[10]   Ozturk M, Roy A, Bhat RA, Sukan FV, Tonelli FMP. Synthesis of BioNMs for Biomedical Applications. Elsevier 2023.

[11]   Ray PC, Yu H, Fu PP. Toxicity and environmental risks of nanomaterials: challenges and future needs. J Environ Sci Health Part C Environ Carcinog Ecotoxicol Rev 2009; 27(1): 1-35.
[http://dx.doi.org/10.1080/10590500802708267] [PMID: 19204862]

[12]   Khoshnamvand M, Hao Z, Fadare OO, Hanachi P, Chen Y, Liu J. Toxicity of biosynthesized silver nanoparticles to aquatic organisms of different trophic levels. Chemosphere 2020; 258: 127346.
[http://dx.doi.org/10.1016/j.chemosphere.2020.127346] [PMID: 32544815]

[13]   Rahimi H-R, Doostmohammadi M. Nanoparticle synthesis, applications, and toxicity. Applications of nanobiotechnology 2019; 10

[14]   Kuppusamy P, Yusoff MM, Maniam GP, Govindan N. Biosynthesis of metallic nanoparticles using plant derivatives and their new avenues in pharmacological applications – An updated report. Saudi Pharm J 2016; 24(4): 473-84.
[http://dx.doi.org/10.1016/j.jsps.2014.11.013] [PMID: 27330378]

[15]   Tripathy A, Raichur AM, Chandrasekaran N, Prathna TC, Mukherjee A. Process variables in biomimetic synthesis of silver nanoparticles by aqueous extract of Azadirachta indica (Neem) leaves. J Nanopart Res 2010; 12(1): 237-46.
[http://dx.doi.org/10.1007/s11051-009-9602-5]

[16]   Aljabali A, Akkam Y, Al Zoubi M, *et al.* Synthesis of gold NPs using leaf extract of Ziziphus zizyphus and their antimicrobial activity. Nanomaterials (Basel) 2018; 8(3): 174.
[http://dx.doi.org/10.3390/nano8030174] [PMID: 29562669]

[17]   Eltaweil AS, Fawzy M, Hosny M, Abd El-Monaem EM, Tamer TM, Omer AM. Green synthesis of platinum nanoparticles using Atriplex halimus leaves for potential antimicrobial, antioxidant, and catalytic applications. Arab J Chem 2022; 15(1): 103517.
[http://dx.doi.org/10.1016/j.arabjc.2021.103517]

[18]   Abbas S, Nasreen S, Haroon A, Ashraf MA. Synthesis of silver and copper NPs from plants and application as adsorbents for naphthalene decontamination. Saudi J Biol Sci 2020; 27(4): 1016-23.
[http://dx.doi.org/10.1016/j.sjbs.2020.02.011] [PMID: 32256162]

[19]   Naseer M, Aslam U, Khalid B, Chen B. Green route to synthesize ZnONPs using leaf extracts of Cassia fistula and Melia azadarach and their antibacterial potential. Sci Rep 2020; 10(1): 9055.
[http://dx.doi.org/10.1038/s41598-020-65949-3] [PMID: 32493935]

[20]   Mittal AK, Chisti Y, Banerjee UC. Synthesis of metallic nanoparticles using plant extracts. Biotechnol Adv 2013; 31(2): 346-56.
[http://dx.doi.org/10.1016/j.biotechadv.2013.01.003] [PMID: 23318667]

[21]   Rajendran NK, George BP, Houreld NN, Abrahamse H. Synthesis of ZnONPs using Rubus

fairholmianus root extract and their activity against pathogenic bacteria. Molecules 2021; 26(10): 3029.
[http://dx.doi.org/10.3390/molecules26103029] [PMID: 34069558]

[22]   Shameli K, Bin Ahmad M, Jaffar Al-Mulla EA, *et al.* Green biosynthesis of silver nanoparticles using Callicarpa maingayi stem bark extraction. Molecules 2012; 17(7): 8506-17.
[http://dx.doi.org/10.3390/molecules17078506] [PMID: 22801364]

[23]   Sunny NE, Mathew SS, Chandel N, Saravanan P, Rajeshkannan R, Rajasimman M, *et al.* Green synthesis of TiO2NPs using plant biomass and their applications-A review. Chemosphere 2022; 300: 134612.
[http://dx.doi.org/10.1016/j.chemosphere.2022.134612] [PMID: 35430203]

[24]   Muzafar W, Kanwal T, Rehman K, *et al.* Green synthesis of iron oxide nanoparticles using Melia azedarach flowers extract and evaluation of their antimicrobial and antioxidant activities. J Mol Struct 2022; 1269: 133824.
[http://dx.doi.org/10.1016/j.molstruc.2022.133824]

[25]   Muhammad A, Umar A, Birnin-Yauri AU, Sanni HA, Elinge CM, Ige AR, *et al.* Green synthesis of copper NPs using Musa acuminata aqueous extract and their antibacterial activity. Asian Journal of Tropical Biotechnology 2023; 20(1)

[26]   Hulkoti NI, Taranath TC. Biosynthesis of nanoparticles using microbes—A review. Colloids Surf B Biointerfaces 2014; 121: 474-83.
[http://dx.doi.org/10.1016/j.colsurfb.2014.05.027] [PMID: 25001188]

[27]   Mughal B, Zaidi SZJ, Zhang X, Hassan SU. Biogenic NPs: Synthesis, characterisation and applications. Appl Sci (Basel) 2021; 11(6): 2598.
[http://dx.doi.org/10.3390/app11062598]

[28]   Marooufpour N, Alizadeh M, Hatami M, Asgari Lajayer B. Biological synthesis of NPs by different groups of bacteria Microbial Nanobionics. State-of-the-art 2019; 1: pp. 63-85.

[29]   Maliszewska I. Microbial synthesis of metal NPs Metal NPs in microbiology. Springer 2011; pp. 153-75.

[30]   Durán N, Marcato PD, Durán M, Yadav A, Gade A, Rai M. Mechanistic aspects in the biogenic synthesis of extracellular metal nanoparticles by peptides, bacteria, fungi, and plants. Appl Microbiol Biotechnol 2011; 90(5): 1609-24.
[http://dx.doi.org/10.1007/s00253-011-3249-8] [PMID: 21484205]

[31]   Kumaresan M, Vijai Anand K, Govindaraju K, Tamilselvan S, Ganesh Kumar V. Seaweed Sargassum wightii mediated preparation of zirconia (ZrO$_2$) nanoparticles and their antibacterial activity against gram positive and gram negative bacteria. Microb Pathog 2018; 124: 311-5.
[http://dx.doi.org/10.1016/j.micpath.2018.08.060] [PMID: 30165114]

[32]   Manivasagan P, Oh J. Marine polysaccharide-based nanomaterials as a novel source of nanobiotechnological applications. Int J Biol Macromol 2016; 82: 315-27.
[http://dx.doi.org/10.1016/j.ijbiomac.2015.10.081] [PMID: 26523336]

[33]   Jha AK, Prasad K. Understanding mechanism of fungus mediated nanosynthesis: a molecular approach. Advances and Applications Through Fungal Nanobiotechnology 2016; pp. 1-23.

[34]   Kumar D, Karthik L, Kumar G, Roa K. Biosynthesis of silver NPs from marine yeast and their antimicrobial activity against multidrug resistant pathogens. Pharmacologyonline 2011; 3: 1100-11.

[35]   Singh S, Vidyarthi AS, Dev A. Microbial synthesis of NPs: an overview. Bio☐NPs: Biosynthesis and Sustainable Biotechnological Implications 2015; 155-86.

[36]   Sun D, Zhang W, Li N, *et al.* Silver nanoparticles-quercetin conjugation to siRNA against drug-resistant Bacillus subtilis for effective gene silencing: *in vitro* and *in vivo*. Mater Sci Eng C 2016; 63: 522-34.
[http://dx.doi.org/10.1016/j.msec.2016.03.024] [PMID: 27040247]

[37]   Garmasheva I, Kovalenko N, Voychuk S, Ostapchuk A, Livins'ka O, Oleschenko L. *Lactobacillus* species mediated synthesis of silver nanoparticles and their antibacterial activity against opportunistic pathogens *in vitro*. Bioimpacts 2016; 6(4): 219-23.
[http://dx.doi.org/10.15171/bi.2016.29] [PMID: 28265538]

[38]   Barsainya M, Singh DP. Green Synthesis of ZnONPs by Pseudomonas aeruginosa and their Broad-Spectrum Antimicrobial Effects. J Pure Appl Microbiol 2018; 12(4)
[http://dx.doi.org/10.22207/JPAM.12.4.50]

[39]   Raliya R, Tarafdar JC. ZnO nanoparticle biosynthesis and its effect on phosphorous-mobilizing enzyme secretion and gum contents in Clusterbean (Cyamopsis tetragonoloba L.). Agric Res 2013; 2(1): 48-57.
[http://dx.doi.org/10.1007/s40003-012-0049-z]

[40]   AbdelRahim K, Mahmoud SY, Ali AM, Almaary KS, Mustafa AEZMA, Husseiny SM. Extracellular biosynthesis of silver nanoparticles using *Rhizopus stolonifer*. Saudi J Biol Sci 2017; 24(1): 208-16.
[http://dx.doi.org/10.1016/j.sjbs.2016.02.025] [PMID: 28053592]

[41]   Gajbhiye M, Kesharwani J, Ingle A, Gade A, Rai M. Fungus-mediated synthesis of silver nanoparticles and their activity against pathogenic fungi in combination with fluconazole. Nanomedicine 2009; 5(4): 382-6.
[http://dx.doi.org/10.1016/j.nano.2009.06.005] [PMID: 19616127]

[42]   Jalal M, Ansari MA, Alzohairy MA, *et al*. Biosynthesis of silver NPs from oropharyngeal Candida glabrata isolates and their antimicrobial activity against clinical strains of bacteria and fungi. Nanomaterials (Basel) 2018; 8(8): 586.
[http://dx.doi.org/10.3390/nano8080586] [PMID: 30071582]

[43]   El-Khawaga AM, Elsayed MA, Gobara M, Suliman AA, Hashem AH, Zaher AA, *et al*. Green synthesized ZnO NPs by Saccharomyces cerevisiae and their antibacterial activity and photocatalytic degradation. Biomass Convers Biorefin. 2023;1-12.
[http://dx.doi.org/10.1007/s13399-023-04599-z]

[44]   Kowshik M, Deshmukh N, Vogel W, Urban J, Kulkarni SK, Paknikar KM. Microbial synthesis of semiconductor CdS nanoparticles, their characterization, and their use in the fabrication of an ideal diode. Biotechnol Bioeng 2002; 78(5): 583-8.
[http://dx.doi.org/10.1002/bit.10233] [PMID: 12115128]

[45]   Abboud Y, Saffaj T, Chagraoui A, *et al*. Biosynthesis, characterization and antimicrobial activity of copper oxide nanoparticles (CONPs) produced using brown alga extract (Bifurcaria bifurcata). Appl Nanosci 2014; 4(5): 571-6.
[http://dx.doi.org/10.1007/s13204-013-0233-x]

[46]   Abdel-Raouf N, Al-Enazi NM, Ibraheem IBM. Green biosynthesis of gold nanoparticles using Galaxaura elongata and characterization of their antibacterial activity. Arab J Chem 2017; 10: S3029-39.
[http://dx.doi.org/10.1016/j.arabjc.2013.11.044]

[47]   Hamouda RA, Hussein MH, Abo-elmagd RA, Bawazir SS. Synthesis and biological characterization of silver nanoparticles derived from the cyanobacterium Oscillatoria limnetica. Sci Rep 2019; 9(1): 13071.
[http://dx.doi.org/10.1038/s41598-019-49444-y] [PMID: 31506473]

[48]   Kulkarni D, Sherkar R, Shirsathe C, *et al*. Biofabrication of nanoparticles: sources, synthesis, and biomedical applications. Front Bioeng Biotechnol 2023; 11: 1159193.
[http://dx.doi.org/10.3389/fbioe.2023.1159193] [PMID: 37200842]

[49]   Salvadori MR, Ando RA, Oller do Nascimento CA, Corrêa B. Intracellular biosynthesis and removal of copper nanoparticles by dead biomass of yeast isolated from the wastewater of a mine in the Brazilian Amazonia. PLoS One 2014; 9(1): e87968.
[http://dx.doi.org/10.1371/journal.pone.0087968] [PMID: 24489975]

[50]   Anboo S, Lau SY, Kansedo J, *et al.* Recent advancements in enzyme□incorporated nanomaterials: Synthesis, mechanistic formation, and applications. Biotechnol Bioeng 2022; 119(10): 2609-38.
[http://dx.doi.org/10.1002/bit.28185] [PMID: 35851660]

[51]   Rather MA, Bhuyan S, Chowdhury R, Sarma R, Roy S, Neog PR. Nanoremediation strategies to address environmental problems. Sci Total Environ 2023; 886: 163998.
[http://dx.doi.org/10.1016/j.scitotenv.2023.163998] [PMID: 37172832]

[52]   Xu W, Yang T, Liu S, *et al.* Insights into the Synthesis, types and application of iron Nanoparticles: The overlooked significance of environmental effects. Environ Int 2022; 158: 106980.
[http://dx.doi.org/10.1016/j.envint.2021.106980]

[53]   Rajora N, Kaushik S, Jyoti A, Kothari SL. Rapid synthesis of silver NPs by Pseudomonas stutzeri isolated from textile soil under optimised conditions and evaluation of their antimicrobial and cytotoxicity properties. IET nanobiotechnology 2016; 10(6): 367-73.

[54]   Kalishwaralal K, Deepak V, Ramkumarpandian S, Nellaiah H, Sangiliyandi G. Extracellular biosynthesis of silver nanoparticles by the culture supernatant of Bacillus licheniformis. Mater Lett 2008; 62(29): 4411-3.
[http://dx.doi.org/10.1016/j.matlet.2008.06.051]

[55]   Upadhyay LSB, Kumar N. Green synthesis of copper nanoparticle using glucose and polyvinylpyrrolidone (PVP). Inorganic and Nano-Metal Chemistry 2017; 47(10): 1436-40.
[http://dx.doi.org/10.1080/24701556.2017.1357576]

[56]   Kaushik V, Lahiri T, Singha S, *et al.* Exploring geometric properties of gold nanoparticles using TEM images to explain their chaperone like activity for citrate synthase. Bioinformation 2011; 7(7): 320-3.
[http://dx.doi.org/10.6026/97320630007320] [PMID: 22355230]

[57]   Polak N, Read DS, Jurkschat K, *et al.* Metalloproteins and phytochelatin synthase may confer protection against zinc oxide nanoparticle induced toxicity in Caenorhabditis elegans. Comp Biochem Physiol C Toxicol Pharmacol 2014; 160: 75-85.
[http://dx.doi.org/10.1016/j.cbpc.2013.12.001] [PMID: 24333255]

[58]   Vaidyanathan R, Gopalram S, Kalishwaralal K, Deepak V, Pandian SRK, Gurunathan S. Enhanced silver nanoparticle synthesis by optimization of nitrate reductase activity. Colloids Surf B Biointerfaces 2010; 75(1): 335-41.
[http://dx.doi.org/10.1016/j.colsurfb.2009.09.006] [PMID: 19796922]

[59]   ADEEYO AO. Green synthesis and antibacterial activities of silver NPs using extracellular laccase of Lentinus edodes. Not Sci Biol 2015; 7(4): 405-11.
[http://dx.doi.org/10.15835/nsb749643]

[60]   Sarkar J, Chakraborty N, Chatterjee A, Bhattacharjee A, Dasgupta D, Acharya K. Green synthesized copper oxide NPs ameliorate defence and antioxidant enzymes in Lens culinaris. Nanomaterials (Basel) 2020; 10(2): 312.
[http://dx.doi.org/10.3390/nano10020312] [PMID: 32059367]

[61]   Gholami-Shabani M, Shams-Ghahfarokhi M, Gholami-Shabani Z, *et al.* Enzymatic synthesis of gold nanoparticles using sulfite reductase purified from Escherichia coli: A green eco-friendly approach. Process Biochem 2015; 50(7): 1076-85.
[http://dx.doi.org/10.1016/j.procbio.2015.04.004]

[62]   Krispin M, Ullrich A, Horn S. Crystal structure of iron-oxide nanoparticles synthesized from ferritin. J Nanopart Res 2012; 14(2): 669.
[http://dx.doi.org/10.1007/s11051-011-0669-4]

[63]   Abbas WS, Atwan ZW, Abdulhussein ZR, Mahdi M. Preparation NPs as antibacterial agents through DNA damage. Mater Technol 2019; 34(14): 867-79.
[http://dx.doi.org/10.1080/10667857.2019.1639005]

[64]   Buchkremer A, Linn MJ, Timper JU, *et al.* Synthesis and internal structure of finite-size DNA–gold

nanoparticle assemblies. J Phys Chem C 2014; 118(13): 7174-84.
[http://dx.doi.org/10.1021/jp412283q]

[65] Ashraf JM, Ansari MA, Khan HM, Alzohairy MA, Choi I. Green synthesis of silver nanoparticles and characterization of their inhibitory effects on AGEs formation using biophysical techniques. Sci Rep 2016; 6(1): 20414.
[http://dx.doi.org/10.1038/srep20414] [PMID: 28442746]

[66] Ansar S, Tabassum H, Aladwan NSM, *et al.* Eco friendly silver nanoparticles synthesis by Brassica oleracea and its antibacterial, anticancer and antioxidant properties. Sci Rep 2020; 10(1): 18564.
[http://dx.doi.org/10.1038/s41598-020-74371-8] [PMID: 33122798]

[67] Dendisová M, Jeništová A, Parchaňská-Kokaislová A, Matějka P, Prokopec V, Švecová M. The use of infrared spectroscopic techniques to characterize nanomaterials and nanostructures: A review. Anal Chim Acta 2018; 1031: 1-14.
[http://dx.doi.org/10.1016/j.aca.2018.05.046] [PMID: 30119727]

[68] Torres-Rivero K, Bastos-Arrieta J, Fiol N, Florido A. Metal and metal oxide nanoparticles: An integrated perspective of the green synthesis methods by natural products and waste valorization: applications and challenges. Compr Anal Chem 2021; 94: 433-69.
[http://dx.doi.org/10.1016/bs.coac.2020.12.001]

[69] González-Pedroza MG, Benítez ART, Navarro-Marchal SA, *et al.* Biogeneration of silver nanoparticles from Cuphea procumbens for biomedical and environmental applications. Sci Rep 2023; 13(1): 790.
[http://dx.doi.org/10.1038/s41598-022-26818-3] [PMID: 36646714]

[70] Fafal T, Taştan P, Tüzün BS, Ozyazici M, Kivcak B. Synthesis, characterization and studies on antioxidant activity of silver nanoparticles using Asphodelus aestivus Brot. aerial part extract. S Afr J Bot 2017; 112: 346-53.
[http://dx.doi.org/10.1016/j.sajb.2017.06.019]

[71] Das B, Dash SK, Mandal D, *et al.* Green synthesized silver nanoparticles destroy multidrug resistant bacteria *via* reactive oxygen species mediated membrane damage. Arab J Chem 2017; 10(6): 862-76.
[http://dx.doi.org/10.1016/j.arabjc.2015.08.008]

[72] Bykkam S, Ahmadipour M, Narisngam S, Kalagadda VR, Chidurala SC. Extensive studies on X-ray diffraction of green synthesized silver NPs. Adv Nanopart 2015; 4(1): 1-10.
[http://dx.doi.org/10.4236/anp.2015.41001]

[73] Arya A, Tyagi PK, Bhatnagar S, Bachheti RK, Bachheti A, Ghorbanpour M. Biosynthesis and assessment of antibacterial and antioxidant activities of silver nanoparticles utilizing Cassia occidentalis L. seed. Sci Rep 2024; 14(1): 7243.
[http://dx.doi.org/10.1038/s41598-024-57823-3] [PMID: 38538702]

[74] Maheshwari R, Kalyane D, Youngren-Ortiz SR, Chougule MB, Tekade RK. Importance of physicochemical characterization of NPs in pharmaceutical product development Basic fundamentals of drug delivery. Elsevier 2019; pp. 369-400.

[75] Mukunthan KS, Elumalai EK, Patel TN, Murty VR. Catharanthus roseus: a natural source for the synthesis of silver nanoparticles. Asian Pac J Trop Biomed 2011; 1(4): 270-4.
[http://dx.doi.org/10.1016/S2221-1691(11)60041-5] [PMID: 23569773]

[76] Ezzat N, Hefnawy MA, Medany SS, El-Sherif RM, Fadlallah SA. Green synthesis of Ag nanoparticle supported on graphene oxide for efficient nitrite sensing in a water sample. Sci Rep 2023; 13(1): 19441.
[http://dx.doi.org/10.1038/s41598-023-46409-0] [PMID: 37945582]

[77] Jain S, Mehata MS. Medicinal plant leaf extract and pure flavonoid mediated green synthesis of silver NPs and their enhanced antibacterial property. Sci Rep 2017; 7(1): 15867.
[http://dx.doi.org/10.1038/s41598-017-15724-8] [PMID: 29158537]

[78]   Bishoyi AK, Sahoo CR, Samal P, *et al.* Unveiling the antibacterial and antifungal potential of biosynthesized silver nanoparticles from Chromolaena odorata leaves. Sci Rep 2024; 14(1): 7513. [http://dx.doi.org/10.1038/s41598-024-57972-5] [PMID: 38553574]

[79]   Ying B, Lin G, Jin L, Zhao Y, Zhang T, Tang J. Adsorption and degradation of 2,4-dichlorophenoxyacetic acid in spiked soil with Fe^0 nanoparticles supported by biochar. Acta Agric Scand B Soil Plant Sci 2015; 65(3): 215-21. [http://dx.doi.org/10.1080/09064710.2014.992939]

[80]   Irshad MA, Nawaz R, Ur Rehman MZ, Adrees M, Rizwan M, Ali S, *et al.* Synthesis, characterization and advanced sustainable applications of TiO2NPs: A review. Ecotoxicol Environ Saf 2021; 212: 111978. [http://dx.doi.org/10.1016/j.ecoenv.2021.111978] [PMID: 33561774]

[81]   Gatou MA, Syrrakou A, Lagopati N, Pavlatou EA. Photocatalytic TiO2-Based Nanostructures as a Promising Material for Diverse Environmental Applications: A Review. Reactions 2024; 5(1): 135-94. [http://dx.doi.org/10.3390/reactions5010007]

[82]   Aliyari Rad S, Nobaharan K, Pashapoor N, *et al.* Nano-microbial remediation of polluted soil: a brief insight. Sustainability (Basel) 2023; 15(1): 876. [http://dx.doi.org/10.3390/su15010876]

[83]   El-Saadony MT, Desoky ESM, Saad AM, Eid RSM, Selem E, Elrys AS. Biological silicon nanoparticles improve Phaseolus vulgaris L. yield and minimize its contaminant contents on a heavy metals-contaminated saline soil. J Environ Sci (China) 2021; 106: 1-14. [http://dx.doi.org/10.1016/j.jes.2021.01.012] [PMID: 34210425]

[84]   Seleiman MF, Alotaibi MA, Alhammad BA, Alharbi BM, Refay Y, Badawy SA. Effects of ZnO NPs and biochar of rice straw and cow manure on characteristics of contaminated soil and sunflower productivity, oil quality, and heavy metals uptake. Agronomy (Basel) 2020; 10(6): 790. [http://dx.doi.org/10.3390/agronomy10060790]

[85]   Verma A, Bharadvaja N. Plant-mediated synthesis and characterization of silver and copper oxide NPs: antibacterial and heavy metal removal activity. J Cluster Sci 2022; 33(4): 1697-712. [http://dx.doi.org/10.1007/s10876-021-02091-8]

[86]   Han B, Zhang M, Zhao D, Feng Y. Degradation of aqueous and soil-sorbed estradiol using a new class of stabilized manganese oxide nanoparticles. Water Res 2015; 70: 288-99. [http://dx.doi.org/10.1016/j.watres.2014.12.017] [PMID: 25543239]

[87]   El Golli A, Contreras S, Dridi C. Bio-synthesized ZnO nanoparticles and sunlight-driven photocatalysis for environmentally-friendly and sustainable route of synthetic petroleum refinery wastewater treatment. Sci Rep 2023; 13(1): 20809. [http://dx.doi.org/10.1038/s41598-023-47554-2] [PMID: 38012203]

[88]   Aragaw TA, Bogale FM, Aragaw BA. Iron-based nanoparticles in wastewater treatment: A review on synthesis methods, applications, and removal mechanisms. J Saudi Chem Soc 2021; 25(8): 101280. [http://dx.doi.org/10.1016/j.jscs.2021.101280]

[89]   El Nemr A, Helmy ET, Gomaa EA, Eldafrawy S, Mousa M. Photocatalytic and biological activities of undoped and doped TiO^2 prepared by Green method for water treatment. J Environ Chem Eng 2019; 7(5): 103385. [http://dx.doi.org/10.1016/j.jece.2019.103385]

[90]   Usman M, Ahmed A, Yu B, Peng Q, Shen Y, Cong H. Photocatalytic potential of bio-engineered copper nanoparticles synthesized from Ficus carica extract for the degradation of toxic organic dye from waste water: Growth mechanism and study of parameter affecting the degradation performance. Mater Res Bull 2019; 120: 110583. [http://dx.doi.org/10.1016/j.materresbull.2019.110583]

[91]   Ishfaq A, Shahid M, Nawaz M, *et al.* Remediation of wastewater by biosynthesized manganese oxide

nanoparticles and its effects on development of wheat seedlings. Front Plant Sci 2023; 14: 1263813.
[http://dx.doi.org/10.3389/fpls.2023.1263813] [PMID: 38126015]

[92] Sabouri Z, Akbari A, Hosseini HA, Hashemzadeh A, Darroudi M. Bio-based synthesized NiO nanoparticles and evaluation of their cellular toxicity and wastewater treatment effects. J Mol Struct 2019; 1191: 101-9.
[http://dx.doi.org/10.1016/j.molstruc.2019.04.075]

[93] Yahyaei B, Azizian S, Mohammadzadeh A, Pajohi-Alamoti M. Preparation of clay/alumina and clay/alumina/Ag nanoparticle composites for chemical and bacterial treatment of waste water. Chem Eng J 2014; 247: 16-24.
[http://dx.doi.org/10.1016/j.cej.2014.02.088]

[94] Jain N, Bhargava A, Panwar J. Enhanced photocatalytic degradation of methylene blue using biologically synthesized "protein-capped" ZnO nanoparticles. Chem Eng J 2014; 243: 549-55.
[http://dx.doi.org/10.1016/j.cej.2013.11.085]

[95] Rodaev VV, Razlivalova SS, Tyurin AI, Vasyukov VM. The nanofibrous CaO sorbent for CO2 capture. Nanomaterials (Basel) 2022; 12(10): 1677.
[http://dx.doi.org/10.3390/nano12101677] [PMID: 35630899]

[96] Akinnawo S. Synthesis, modification, applications and challenges of TiO2NPs. Research Journal of Nanoscience and Engineering 2019; 3(4): 10-22.
[http://dx.doi.org/10.22259/2637-5591.0304003]

[97] Lv D, Wang R, Tang G, *et al.* Ecofriendly electrospun membranes loaded with visible-ligh--responding NPs for multifunctional usages: highly efficient air filtration, dye scavenging, and bactericidal activity. ACS Appl Mater Interfaces 2019; 11(13): 12880-9.
[http://dx.doi.org/10.1021/acsami.9b01508] [PMID: 30869859]

[98] Lin S, Liu W, Fu X, Luo M, Zhong W-H. Protein-based materials for sustainable, multifunctional air filtration. Separ Purif Tech 2023; 126252.

[99] Magudieshwaran R, Ishii J, Raja KCN, *et al.* Green and chemical synthesized CeO2 nanoparticles for photocatalytic indoor air pollutant degradation. Mater Lett 2019; 239: 40-4.
[http://dx.doi.org/10.1016/j.matlet.2018.11.172]

[100] Romeh AAA. Green silver NPs for enhancing the phytoremediation of soil and water contaminated by fipronil and degradation products. Water Air Soil Pollut 2018; 229(5): 147.
[http://dx.doi.org/10.1007/s11270-018-3792-3]

[101] Lin J, Sun M, Su B, Owens G, Chen Z. Immobilization of cadmium in polluted soils by phytogenic iron oxide nanoparticles. Sci Total Environ 2019; 659: 491-8.
[http://dx.doi.org/10.1016/j.scitotenv.2018.12.391] [PMID: 31096378]

[102] Chidambaram D, Hennebel T, Taghavi S, *et al.* Concomitant microbial generation of palladium nanoparticles and hydrogen to immobilize chromate. Environ Sci Technol 2010; 44(19): 7635-40.
[http://dx.doi.org/10.1021/es101559r] [PMID: 20822130]

[103] Nzilu DM, Madivoli ES, Makhanu DS, Wanakai SI, Kiprono GK, Kareru PG. Green synthesis of copper oxide nanoparticles and its efficiency in degradation of rifampicin antibiotic. Sci Rep 2023; 13(1): 14030.
[http://dx.doi.org/10.1038/s41598-023-41119-z] [PMID: 37640783]

[104] Bhattacharjee N, Som I, Saha R, Mondal S. A critical review on novel eco-friendly green approach to synthesize ZnONPs for photocatalytic degradation of water pollutants. Int J Environ Anal Chem 2024; 104(3): 489-516.
[http://dx.doi.org/10.1080/03067319.2021.2022130]

[105] Zhao X, Yan L, Xu X, *et al.* Synthesis of silver nanoparticles and its contribution to the capability of Bacillus subtilis to deal with polluted waters. Appl Microbiol Biotechnol 2019; 103(15): 6319-32.
[http://dx.doi.org/10.1007/s00253-019-09880-2] [PMID: 31115637]

[106] Okoli C, Boutonnet M, Järås S, Rajarao-Kuttuva G, Eds. Protein-functionalized magnetic iron oxide NPs: time efficient potential-water treatment Nanotechnology for Sustainable Development. Springer 2014.

[107] Araújo JN, Tofanello A, Sato JAP, *et al.* Rapid synthesis *via* green route of plasmonic protein-coated silver/silver chloride NPs with controlled contents of metallic silver and application for dye remediation. J Inorg Organomet Polym Mater 2018; 28(6): 2812-8.
[http://dx.doi.org/10.1007/s10904-018-0947-z]

[108] Parmar A, Kaur G, Kapil S, Sharma V, Choudhary MK, Sharma S. Novel biogenic silver nanoparticles as invigorated catalytic and antibacterial tool: A cleaner approach towards environmental remediation and combating bacterial invasion. Mater Chem Phys 2019; 238: 121861.
[http://dx.doi.org/10.1016/j.matchemphys.2019.121861]

[109] Liu WX, Sun JB, Li YN, *et al.* Low-temperature and high-selectivity butanone sensor based on porous Fe_2O_3 nanosheets synthesized by phoenix tree leaf template. Sens Actuators B Chem 2023; 377: 133054.
[http://dx.doi.org/10.1016/j.snb.2022.133054]

CHAPTER 10

Green Nanotechnology and Environmental Remediation: A Critical Review

Saivenkatesh Korlam[1,*] and **Sankara Rao Miditana**[2]

[1] *Department of Botany, S. V. A.Government Degree College (M), Srikalahasti, Tirupati Dt., A.P., India*

[2] *Department of Chemistry, Government Degree College, Puttur, Tirupati Dt., A.P., India*

Abstract: Environmental pollution is a major challenge on a global basis. Traditional methods for cleaning up polluted environments are often associated with certain drawbacks such as high cost, inefficiency, and generation of hazardous by-products. Green nanotechnology has emerged as an innovative approach to synthesizing nanoparticles (NPs) from natural resources like plant extracts, microorganisms, or enzymes. Green-synthesized NPs hold immense potential for the remediation of different environmental matrices due to their unique properties and biodegradable nature.

Green nanotechnology provides sustainable and efficient solutions for environmental remediation and paves the way for a cleaner and healthier planet. In this chapter, the authors have highlighted the principles of green nanotechnology and potential applications of green synthesized NPs in the remediation of air, water, and soil, along with their superiority over other conventional treatment techniques. The authors have also highlighted its limitations and associated challenges so that with continued research and development, green nanotechnology can revolutionize the way we address environmental pollution, ensuring a brighter future for generations to come.

Keywords: Environmental footprint, Environmental remediation, Green nanotechnology, NPs, Natural resources.

INTRODUCTION

Green nanotechnology offers a revolutionary approach to environmental remediation by utilizing nanomaterials (NMs) synthesized through environmentally friendly methods. It can be defined as the design, development, and application of NMs with minimal environmental impact throughout their lifecycle [1]. It prioritizes the use of non-toxic, readily available, or renewable

---

* **Corresponding author Saivenkatesh Korlam:** Department of Botany, S. V. A.Government Degree College (M), Srikalahasti, Tirupati Dt., A.P., India; E-mail: k.saivenkatesh@gmail.com

Neha Agarwal, Vijendra Singh Solanki, Neetu Singh & Maulin P. Shah (Eds.)

resources for the synthesis of NP. This minimizes reliance on harmful chemicals and reduces dependence on depleting resources [2]. Traditional techniques for NP synthesis often involve harsh chemicalsor high energy consumption. Green nanotechnology explores alternative methods like biosynthesis using plant extracts, microbes, or enzymes. These methods are more energy-efficient, produce less waste, and are generally safer for the environment [3, 4]. Ideally, green-synthesized NPs should be biodegradable or easily removable from the environment after performing their remediation function. This prevents long-term environmental contamination concerns associated with some traditional NMs [5].

ENVIRONMENTAL CHALLENGES AND CONVENTIONAL REMEDIATION METHODS

Human activities have significantly impacted the environment, leading to a multitude of challenges. These include:

Water Pollution

Industrial waste, agricultural runoff, and untreated sewage contaminate water bodies with heavy metals, organic pollutants, and pathogens [6].

Soil Contamination

Industrial spills, overuse of pesticides and fertilizers, and improper waste disposal pollute soil with heavy metals, persistent organic pollutants, and radioactive materials [7].

Air Pollution

Emissions from vehicles, industries, and burning fossil fuels release harmful gases like nitrogen oxides, sulfur oxides, and particulate matter, leading to respiratory problems and climate change [8].

Conventional remediation methods have been employed to address these challenges, but they often have several limitations. When the contaminated soil or sediment is removed and transported to landfills, it can be disruptive, expensive, and create a risk of secondary pollution [9]. High energy consumption is required when contaminated groundwater is pumped out, treated above ground, and then reinjected into the ground. This process potentially leaves residual contamination [10]. Chemicals that are used to immobilize or degrade pollutants may be toxic and leave harmful byproducts [11].

GREEN NANOTECHNOLOGY - A PROMISING SOLUTION FOR ENVIRONMENTAL REMEDIATION

Conventional remediation methods for environmental pollution often struggle with high costs, substantial energy consumption, and the generation of secondary waste. Green nanotechnology emerges as a promising alternative, offering a sustainable and potentially more effective approach. Its emergence is driven by several key factors, which are discussed below.

Growing Environmental Concerns

With rising global awareness of environmental issues, there is a strong push for sustainable solutions. Green nanotechnology aligns with this movement by minimizing environmental impact throughout the lifecycle of NMs used in remediation [1].

Limitations of Traditional Methods

Existing remediation techniques often have limitations. Chemical treatments can generate harmful by-products, while excavation and landfilling are disruptive and expensive [9, 10]. Green nanotechnology offers the potential to overcome these limitations by utilizing environmentally friendly synthesis methods and promoting targeted remediation strategies.

Unique Properties of NPs

NPs possess unique properties due to their extremely small size. This includes a significantly larger surface area compared to bulk materials, which allows them to interact more effectively with pollutants [5]. Green nanotechnology leverages these properties to design NPs specifically for environmental remediation applications.

Advancements in Biosynthesis

Research in green nanotechnology has led to significant advancements in eco-friendly synthesis methods. These methods utilize biological resources like plant extracts, microbes, or enzymes to synthesize NPs [2, 4]. It not only reduces reliance on harmful chemicals but also opens doors for potentially scalable and cost-effective production.

GREEN NANOTECHNOLOGY: A SUSTAINABLE APPROACH

Green nanotechnology represents a revolutionary approach to nanotechnology that prioritizes environmental sustainability. It can be defined as the design, development, and application of NMs with minimal environmental impact throughout their lifecycle [1]. A sustainable approach *via* green nanotechnology is shown in Fig. (**1**).

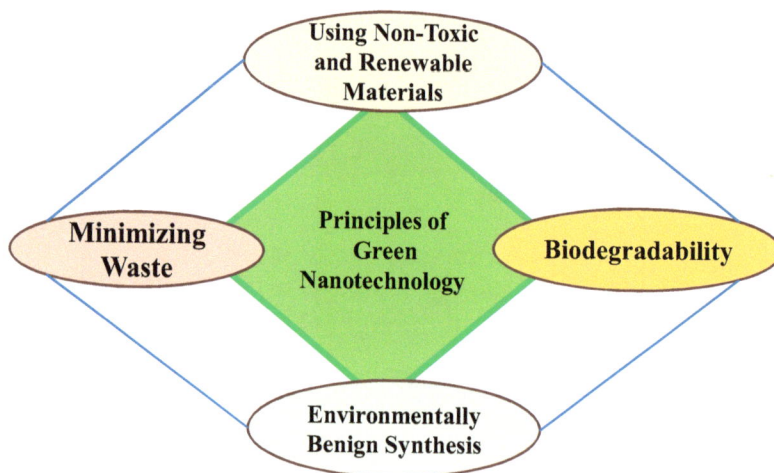

Fig. (1). Principles of green nanotechnology.

Green Nanotechnology Principles

This approach emphasizes several core principles, such as

Minimizing Waste

Green nanotechnology strives to minimize waste generation at all stages of the NM production process. This includes optimizing synthesis methods to reduce the use of solvents and other auxiliary materials, as well as exploring ways to reuse or recycle any by-products [12].

Using Non-Toxic and Renewable Materials

Unlike traditional NM synthesis, which may rely on hazardous chemicals, green nanotechnology prioritizes the use of non-toxic and readily available resources. This can involve utilizing natural materials like plant extracts, biopolymers, or even microorganisms for NP synthesis [2, 4].

Environmentally Benign Synthesis Methods

Green nanotechnology actively seeks out environmentally friendly methods for the synthesis of NPs. This excludes methods relying on harsh chemicals, high energy consumption, or extreme temperatures [5]. Biosynthesis, which utilizes biological processes like those found in plants or microbes, is a prominent example of an environmentally benign synthesis method [6].

Biodegradability or Minimal Environmental Persistence

Ideally, green-synthesized NPs should be biodegradable or easily removable from the environment after fulfilling their purpose in remediation or other applications. This minimizes the risk of long-term environmental contamination associated with some traditional NMs [5].

By adhering to these principles, green nanotechnology offers a more sustainable and environmentally responsible path for the development and application of NMs across various fields, particularly in environmental remediation.

BENEFITS OF GREEN SYNTHESIS OF NPS IN ENVIRONMENTAL APPLICATIONS

Green synthesis offers a compelling approach to NP production for environmental applications compared to traditional methods. Here is a breakdown of its key benefits:

Environmental Sustainability

Reduced Waste Generation

Green synthesis methods often minimize the use of harsh chemicals and solvents, leading to less waste generation compared to conventional techniques [12]. This minimizes environmental pollution associated with waste disposal.

Renewable Resources

Green methods utilize readily available and renewable resources like plant extracts or microorganisms [2]. This reduces reliance on depleting resources commonly used in traditional synthesis.

Biodegradable NPs

Ideally, green-synthesized NPs are biodegradable or easily removable from the environment after use [5]. This minimizes the risk of long-term environmental contamination often associated with some traditional NMs.

Safety and Eco-Friendliness

Non-toxic Precursors

Green synthesis prioritizes non-toxic or biocompatible resources for NP formation [4]. This reduces potential environmental and health risks associated with the use of hazardous chemicals in traditional methods.

Energy Efficient

Green methods often operate under milder conditions like ambient temperature and pressure, requiring less energy compared to high-temperature or high-pressure processes typical in traditional synthesis [5]. This translates to a lower environmental footprint.

Enhanced Functionality

Tailored Properties

Green synthesis offers a double advantage in designing NPs for environmental applications. It allows precise control over the size, shape, and surface properties of the particles. Green synthesis allows for some control over the size, shape, and surface chemistry of the NPs [6]. This enables researchers to tailor the NPs for specific environmental applications, potentially enhancing their effectiveness.

Biocompatibility

NPs synthesized using biological resources may exhibit enhanced biocompatibility with the environment [7]. This is crucial for ensuring their safety in applications like cleaning up polluted sites or monitoring environmental health. This can be advantageous in applications like bioremediation or environmental monitoring.

Scalability and Cost-Effectiveness

Simple Processes

Green synthesis methods can be relatively simple to implement, potentially leading to easier scaling for large-scale production of NPs [13].

Readily Available Resources

Utilizing readily available resources like plant extracts can potentially reduce the overall cost of NP production compared to methods relying on expensive chemicals [14].

DIFFERENT GREEN-SYNTHESIZED NPS FOR ENVIRONMENTAL REMEDIATION

Green nanotechnology offers a promising approach to environmental remediation by providing eco-friendly methods for NP synthesis. Here is a breakdown of some key types of green-synthesized NPs used in remediation applications.

Metal NPs

Metal NPs, like silver or iron, exhibit various remediation capabilities. Metal NPs offer a promising avenue for environmental remediation. Their antimicrobial properties and ability to degrade pollutants or absorb heavy metals make them valuable tools for cleaning up contaminated water. Moreover, the development of green biosynthesis methods paves the way for a more sustainable approach to NP production.

Antimicrobial Properties

Silver NPs are particularly effective against microbes like bacteria and pathogens. They can be used to disinfect water contaminated with harmful organisms, making it safer for consumption [15].

Organic Pollutants' Degradation and Heavy Metal Adsorption

Iron NPs have a dual capability. They can break down organic pollutants present in water through a process called Fenton chemistry. Additionally, they can act like tiny magnets, attracting and absorbing heavy metal contaminants from the water [14].

Green Synthesis Methods

The traditional methods of synthesizing metal NPs can involve harsh chemicals. The passage highlights a greener approach - biosynthesis. Here, biological resources like plant extracts, fungi, or bacteria are used to reduce metal salts into NPs. This method is considered more environmentally friendly [3, 16].

Metal Oxide NPs

Metal oxide NPs, like zinc oxide (ZnO) or titanium dioxide (TiO$_2$), possess high surface area and photocatalytic activity. This allows them to absorb pollutants and degrade them under light irradiation, particularly organic contaminants in water or air [17]. Similar to metal NPs, green synthesis of metal oxides utilizes plant extracts or microorganisms [18, 19].

Super Absorbers

Metal oxide NPs, like zinc oxide (ZnO) and titanium dioxide (TiO$_2$), are incredibly tiny particles with a vast surface area. Imagine a tiny sponge - the more surface area it has, the more water it can soak up. Similarly, these NPs can absorb a lot of pollutants due to their large surface area [17].

Light Activated Cleaners

These NPs also have a special power called photocatalytic activity. When exposed to light, they trigger reactions that break down pollutants into harmless molecules. Think of sunlight activating the NPs to gobble up and destroy pollutants [18].

Cleaning Up with Green Methods

Similar to how we can grow metal NPs using eco-friendly methods, scientists are developing greener ways to create metal oxide NPs. This involves using plant extracts or even microorganisms instead of harsh chemicals [19].

Biochar

Biochar is a porous carbonaceous material produced by the pyrolysis of biomass under limited oxygen conditions. It acts as a powerful adsorbent for various pollutants, including heavy metals, organic contaminants, and pharmaceuticals, due to its high surface area and surface functional groups [20].

GREEN-SYNTHESIZED NPS: UNIQUE PROPERTIES FOR EFFECTIVE REMEDIATION

Green-synthesized NPs offer distinct advantages in environmental remediation due to their inherent properties. Here is a closer look at how these properties contribute to their effectiveness:

High Surface Area

NPs possess a significantly larger surface area compared to their bulk counterparts. This translates to a greater number of active sites available for interacting with pollutants. This massive increase in surface area translates to a greater number of "active sites" available for interaction with pollutants in their environment [21]. The high surface area of green-synthesized NPs allows them to effectively adsorb (capture) pollutants on their surface. This is particularly beneficial for removing heavy metals, organic contaminants, and other toxins from water and soil where NPs act like tiny sponges, effectively adsorbing (capturing) pollutants on their surfaces. This is especially useful for removing harmful heavy metals, organic contaminants, and other toxins from water and soil [22].

Green-synthesized NPs are very promising in the remediation of pollutants. Biochar, for example, is a green material with a highly porous structure and a massive surface area, making it a powerful adsorbent for various pollutants. Similarly, metal oxide NPs like zinc oxide (ZnO) can effectively remove organic dyes from water due to their high surface area. By harnessing the power of tiny surfaces, NPs offer a promising approach to clean our environment [23].

Enhanced Reactivity

Green-synthesized NPs often exhibit higher reactivity compared to their bulk counterparts due to their small size and surface properties [22]. This can involve properties like photocatalytic activity or the ability to generate reactive oxygen species. Enhanced reactivity allows green-synthesized NPs to participate in chemical reactions that degrade pollutants. For example, metal NPs like iron can degrade organic pollutants through Fenton chemistry, while metal oxide NPs like TiO_2 can degrade pollutants under light irradiation (photocatalysis) [24, 25].

Researchers have created iron NPs using green methods that can break down harmful chlorinated organic pollutants found in water [26]. These tiny iron particles are particularly effective due to their high reactivity. In another area of research, scientists are looking at ways to purify air. They have developed titanium dioxide (TiO_2) NPs using bacteria. These NPs have the potential to clean the air by generating reactive oxygen species, which can destroy volatile organic compounds, another type of air pollutant [27]. This research shows promise for developing new green technologies to address water and air pollution.

Biocompatibility and Biodegradability

Biocompatibility and biodegradability are the key features of environment-friendly materials, particularly those used in nano form for cleaning up polluted sites. Certain types of green-synthesized NPs, particularly those derived from biological resources, may exhibit biocompatibility and biodegradability [21]. It minimizes their long-term environmental impact after completing their remediation function. Biocompatible NPs can be particularly advantageous for applications like bioremediation, where they interact with microorganisms to enhance the degradation of pollutants [23]. Biodegradable NPs further minimize potential environmental risks associated with their persistence in the environment. Similarly, biochar is a charcoal-like substance produced by heating organic material in the absence of oxygen. It is often used in soil remediation because it can absorb and trap pollutants. The fact that biochar biodegrades means it will eventually decompose naturally in the soil after it has served its purpose [28].

ENVIRONMENTAL REMEDIATION APPLICATIONS OF GREEN NMS

As already discussed, traditional methods are not apt for degradation of pollutants. Green NMs offer a promising alternative for cleaning up our environment due to their unique properties at the nanoscale. In this section, some specific applications of green NMs are discussed, and the same are presented in Fig. (2).

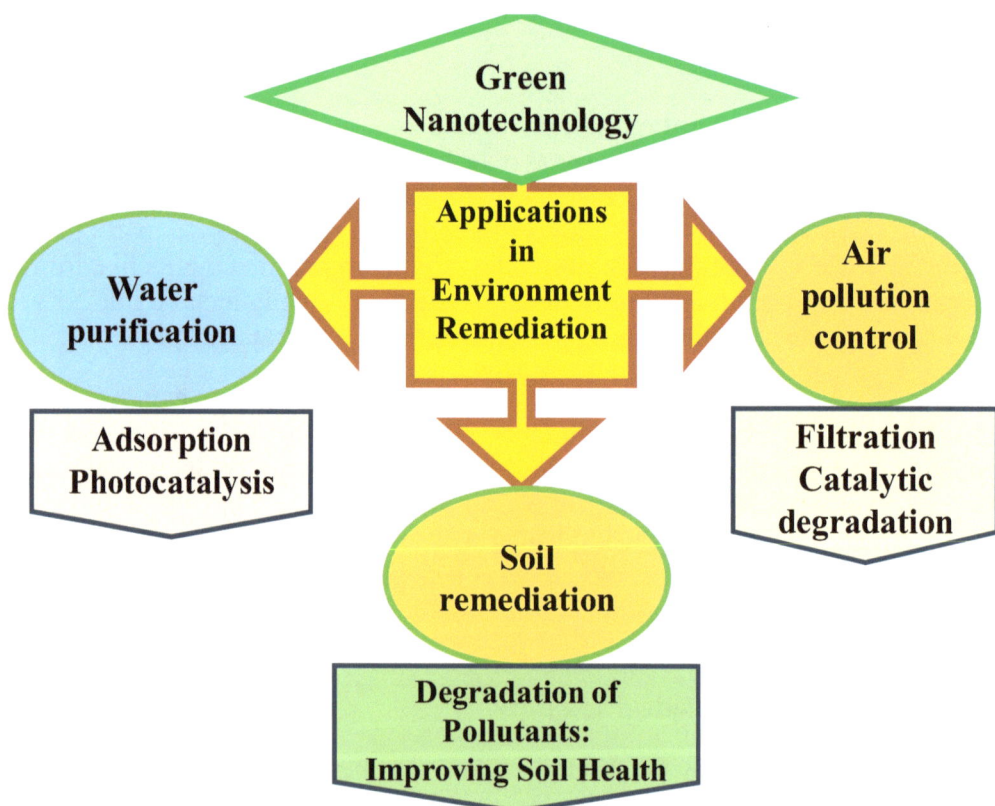

Fig. (2). Applications of green NMs in environmental remediation.

Water Purification

Green NMs provide a new approach to water purification by utilizing their unique properties at the nanoscale to remove pollutants through adsorption or break them down through photocatalysis. This offers a potentially sustainable solution for cleaner water.

Adsorption

Adsorption leverages the high surface area of certain NMs, like cellulose nanofibrils. These can be designed to act like tiny sponges, attracting and clinging onto contaminants like heavy metals and organic pollutants. This offers a way to remove these harmful substances from water [29]. High surface area at the nanoscale allows them to capture large quantities of contaminants.

Photocatalysis

Metal NPs like Titanium dioxide (TiO_2) synthesized using green methods like plant extracts can degrade organic pollutants in water through photocatalysis. Light irradiation activates the NPs, generating reactive oxygen species that break down the pollutants [30].

Air Pollution Control

Both filtration and catalytic degradation represent significant advancements in air pollution control. By harnessing the power of nanotechnology and bio-based materials, these techniques offer promising solutions for a healthier planet.

Filtration

Nanofibrous membranes made from biopolymers can be used as efficient air filters to capture particulate matter (PM) pollutants. Their intricate nanostructure allows them to trap even the finest PM particles [31]. Imagine a filter made from natural materials so fine it can trap even the tiniest dust particles. This is the power of nanofibrous membranes derived from biopolymers. Their complex structure, built on a nanoscale, acts like a sieve, effectively capturing harmful particulate matter (PM) pollutants from the air. This innovation offers a significant advantage as PM exposure is linked to various health problems.

Catalytic Degradation

This method tackles gaseous pollutants like volatile organic compounds (VOCs) often found in paints, solvents, and industrial processes. Green-synthesized metal oxide NPs act as catalysts, accelerating the breakdown of these VOCs into harmless byproducts like carbon dioxide and water vapor. Think of these NPs as tiny green machines that convert harmful pollutants into innocuous elements, leaving the air cleaner. Green-synthesized metal oxide NPs can act as catalysts for the breakdown of gaseous pollutants like volatile organic compounds (VOCs) in the air. These NPs convert VOCs into less harmful molecules like CO_2 and water vapor [32].

These are just a few examples of how green NMs are revolutionizing environmental remediation. As research progresses, we can expect even more innovative applications of these sustainable materials for cleaning up pollutants and creating a cleaner future.

Soil Remediation

Green nanotechnology offers a promising approach to soil remediation by utilizing eco-friendly NPs for degrading toxins and improving soil health. Here are some key applications:

Degradation of Pollutants

Nanoscale Zero-Valent Iron (nZVI) NPs efficiently transform organic contaminants and heavy metals into less harmful forms through reduction reactions [33]. Pollution is a major threat to our planet, contaminating soil and water with harmful organic compounds and heavy metals. Fortunately, innovative technologies like nanoscale zero-valent iron (nZVI) offer promising solutions for environmental remediation. This section will delve into the fascinating world of nZVI and how it helps transform these pollutants into less harmful forms [33]. NPs coated with biosurfactants derived from natural sources like microorganisms enhance contaminant mobilization and degradation by increasing their bioavailability [34].

Improving Soil Health

Biochar, produced by burning biomass like plant residues under limited oxygen, emerges as a champion for soil fertility. These tiny particles, specifically biochar NPs, enhance nutrient retention within the soil. This translates to better access to essential nutrients for plants, leading to improved growth and yield. But biochar's benefits extend beyond just nutrients. It also stimulates microbial activity in the soil. Microbes play a crucial role in decomposing organic matter, releasing nutrients usable by plants, and promoting overall soil health. Biochar's porous structure provides a haven for these beneficial microbes, fostering their growth and activity. Finally, biochar improves soil structure. It helps create a more stable and well-aerated environment for plant roots, promoting better water drainage and retention. Biochar NPs, produced from biomass like plant residues, promote soil fertility by enhancing nutrient retention, microbial activity, and soil structure [35]. These sensors utilize biological elements and NPs to detect soil contaminants and monitor soil health parameters, facilitating targeted remediation strategies [35].

CHALLENGES AND LIMITATIONS OF GREEN NANOTECHNOLOGY

Green nanotechnology offers a beacon of hope in the fight against soil contamination. Harnessing the power of nature to create NPs for remediation presents a sustainable and potentially more effective approach compared to traditional methods. However, this path is not without its challenges. Here, we delve into the limitations that need to be addressed to fully realize the potential of green nanotechnology for soil remediation. Green nanotechnology offers a promising approach to soil remediation, but it faces several limitations that need to be addressed.

Limited Production and Scalability

One of the primary hurdles lies in scaling up production for large-scale remediation projects. Green synthesis methods, while environmentally friendly, often suffer from low yields. Current techniques can be intricate and labor-intensive, making them less cost-competitive compared to established methods. Imagine needing a massive quantity of medicine, yet the current production process only yields a few pills at a time. This is the crux of the scaling issue. Researchers are actively developing more efficient and scalable green synthesis techniques to bridge this gap and make green nanotechnology a viable option for large-scale applications. While green synthesis methods are environmentally friendly, scaling up production for large-scale remediation projects can be challenging. Current techniques often have low yields and require complex processes, making them less cost-effective than traditional methods [36].

Uncertain Environmental Impact

The novelty of green NPs raises concerns about their long-term environmental impact. While they are designed with biodegradability in mind, their behavior within complex ecosystems like soil is not fully understood. Just because something is "green" does not guarantee it is completely harmless. Imagine introducing a new species into an existing ecosystem; the potential consequences for the existing balance need careful consideration. Research is urgently needed to assess the potential effects of green NPs on soil organisms, from microscopic microbes to larger creatures like earthworms. Understanding their impact on overall ecosystem health is crucial for responsible implementation. The novelty of green NPs raises concerns about their long-term environmental impact. While they are often designed to be biodegradable, their behavior in complex ecosystems is not fully understood. Research is needed to assess their potential effects on soil biota and overall ecosystem health [37].

Specificity and Targeting

Another challenge lies in achieving complete selectivity when using green NPs. While some can be designed to target specific pollutants, the ideal scenario where they only interact with the harmful elements remains elusive. There is a possibility of unintended consequences – these green warriors might end up interacting with and potentially disrupting beneficial components of the soil, like the very microbes that help with natural remediation processes. This is akin to accidentally hitting bystanders while trying to take down a villain. Scientists are actively researching ways to enhance the targeting capabilities of green NPs to minimize these non-targeted interactions. While some green NPs can be designed to target specific pollutants, achieving complete selectivity remains a challenge. Non-target interactions with beneficial soil components are possible, potentially disrupting natural remediation processes [37].

Unforeseen Toxicity

Even though green NPs are derived from natural materials, there is a potential for unforeseen toxicity toward soil organisms and even human health. Just because something originates from nature does not mean it is completely safe. Think of a poisonous plant – it is natural but certainly not something you would want in your salad. Thorough testing is a critical step before widespread application. Researchers need to establish rigorous protocols to evaluate the potential toxicity of green NPs on various soil organisms and human health to ensure their safety before large-scale deployment. Even green NPs derived from natural materials may exhibit unforeseen toxicity toward soil organisms or human health. Thorough testing is necessary to ensure their safety before widespread applications [38].

Regulatory Challenges

The regulatory framework for green NMs is still evolving. Currently, there are no well-defined guidelines for assessing their environmental safety and responsible use in remediation projects. This creates a situation where the rules of the game are unclear. Establishing clear and comprehensive regulations is crucial for the sustainable implementation of green nanotechnology. These regulations should encompass environmental safety assessments, responsible use practices, and clear disposal protocols. The regulatory framework for green NMs is still evolving. Establishing clear guidelines for their environmental safety assessment and responsible use is crucial for their sustainable implementation in remediation projects [39].

By addressing these limitations, green nanotechnology can truly fulfill its promise as a powerful tool for sustainable soil remediation. Through continued research, innovative solutions, and robust regulatory frameworks, we can pave the way for a cleaner future, one NP at a time.

CONCLUSION AND FUTURE DIRECTIONS

Green nanotechnology offers exciting possibilities for soil remediation. However, addressing the limitations discussed above is essential for its responsible and sustainable application. Continued research and development are crucial to ensure green NPs become a truly viable and environmentally friendly tool for cleaning up contaminated soils. Green nanotechnology offers a promising path towards a cleaner environment by tackling pollution and promoting sustainability at the nanoscale. This approach has two main thrusts: developing eco-friendly methods to create NMs and using these materials for environmental remediation. On the production side, green nanotechnology focuses on synthesizing NPs using non-toxic materials and minimal energy. This can involve using plant extracts, bacteria, or even sound waves, all of which are less polluting than traditional industrial processes. When it comes to cleaning up the environment, NPs' unique properties shine. Their high surface area makes them excellent adsorbents, capable of capturing pollutants like heavy metals and organic toxins from air and water. Additionally, researchers are developing NPs that can break down pollutants or act as catalysts for cleaner industrial processes.

Green nanotechnology holds immense potential for a greener future. By providing sustainable production methods and powerful environmental remediation tools, it can play a key role in combating pollution and fostering a cleaner planet.

REFERENCES

[1]   Luque R, Casas S, Lucena C. Sustainable Development and Green Technologies in Food Production and Processing. Woodhead Publishing Series in Food Science, Technology and Nutrition 2018.

[2]   Ingle AP, Gade AK, Rai MK. Mycosynthesis of silver NPs from fungus Aspergillus flavus and their antifungal activity against Candida albicans. Curr Nanosci 2009; 5(1): 97-105.

[3]   Iravani S. Green synthesis of metal NPs using various plants and microorganisms. Green Chem 2011; 13(10): 2638-50.
        [http://dx.doi.org/10.1039/c1gc15386b]

[4]   Ahmad N, Sharma S, Singh V, Shamsi S, Mehta S, Faiyaz Ahmad A. Rapid synthesis of silver NPs using dried leaves of Azadirachta indica. J Nanosci Nanotechnol 2010; 10(10): 3101-6.

[5]   Kuppusamy P, Murugesa R, Rajeshkumar S, Thamaraiselvi V. Biosynthesis of silver NPs using Moringa oleifera leaf extract and its application in deactivating multidrug-resistant bacteria. Colloids Surf B Biointerfaces 2016; 141: 235-44.

[6]   Li W, Zhou Q, Luo W, Wang Z. Occurrence and removal of antibiotics in wastewater: A review. Sci Total Environ 2019; 693: 133609.

[7]   McGrath SP, Zhao FJ, Lombi P, Harrison FL. Ground water contamination and remediation: Pb(II)-EDTA. Environ Sci Technol 2001; 35(9): 1800-8.

[8]   Lelieveld J, Klingmüller K, Pozzer A, Burnett RT, Haines A, Ramanathan V, *et al.* Effects of fossil fuel and future scenarios on regional air pollution, ozone and GHGs. Atmos Chem Phys 2015; 15(14): 7759-833.

[9]   Vidali M. Bioremediation of industrial waste sites: A review. Journal of Pollution Control and Management 2001; 2(1): 163-73.

[10]   Aziz H, Zaini MH, Isa MH, Rashid MK. A review on recent developments in pump and treat technology for organic-contaminated groundwater. J Environ Manage 2018; 209: 146-59.

[11]   Lu A, Wang Z, Luo L, Christie P. Chemical remediation technologies for organic contaminated sites. J Hazard Mater 2015; 283: 33-44.

[12]   Apostol YC, Ramirez-Garcia S, Duarte JM, Pereira MFR, Ferrer-Balas D. Waste valorization: Green synthesis of metallic NPs using waste biomass. J Clean Prod 2016; 140: 1741-54.

[13]   Liu C, Majeed S. Green synthesized NPs for efficient heavy metal removal from waste streams. Materials (Basel) 2014; 7(12): 7928-49.
        [PMID: 28788281]

[14]   Khatoon S, Prasad R, Banerjee D, Sastry M. Green synthesis of metallic NPs for environmental applications: a review. J Nanosci Nanotechnol 2019; 19(1): 1-20.
        [PMID: 30326997]

[15]   Ahmad N, Sharma S, Singh A, Shamim B, Mehta A. Rapid synthesis of silver NPs using dried neem leaf extract and their antibacterial activity. J Chem Technol Biotechnol 2010; 85(4): 473-8.

[16]   Ahmad A, Mukherjee P, Mandal D, Senapati S, Khan MI, Rajiv S, *et al.* Fungus-mediated synthesis of silver NPs and their antifungal activity against Trichophyton rubrum. J Appl Microbiol 2003; 94(2): 833-40.

[17]   Hoffmann MR, Stacy AM, Wells DD. Environmental applications of semiconductor photocatalysis. Chem Soc Rev 1993; 22(1): 65-78.

[18]   Dar GN, Ali S, Reddy AV, Sivasankar S. A green approach for the synthesis of ZnO NPs using Aloe vera extract and its application in the degradation of methylene blue dye. Mater Sci Eng C 2014; 42: 512-7.

[19]   Sahu JP, Santra CR. Synthesis of titanium dioxide NPs using fungi and their photocatalytic degradation of methylene blue under UV light irradiation. Green Chem 2014; 16(9): 4313-22.

[20]   Lehmann J, Joseph S. Biochar for environmental management: Science and technology. Abingdon, Oxon; New York, NY: Earthscan from Routledge 2015.
        [http://dx.doi.org/10.4324/9780203762264]

[21]   Debnath S, Das S, Aich S. Synthesis of green NPs for energy, biomedical, environmental, agricultural, and food applications: A review. Environ Sci Pollut Res Int 2020; 27(23): 14049-74.

[22]   Ahmad M, Rajapakshe M, Lim JE, Shahid M, Majeed A, Rizwan M, *et al.* Biochar for environmental remediation: Potential and challenges. Chemosphere 2019; 227: 130-43.

[23]   Jain P, Arora S, Rajwade JM. 'Green' synthesis of metals and their oxide NPs: applications for environmental remediation. J Nanobiotechnology 2021; 19(1): 1-21.
        [PMID: 33397416]

[24]   Studt F, Schaul ZG, Schuster ME. Catalysis by nature. Chem Soc Rev 2007; 36(8): 1173-83.

[25]   Banerjee S, Dionysiou DD, Goswami D. Semiconductor photocatalysis for environmental

applications. J Photochem Photobiol Chem 2015; 307: 67-99.

[26]   Liu J, Zhao G, Sun D, Tang H, Astruc M, Irvine JT. Ultrasmall iron- oxide NPs with tunable near-infrared absorption for photothermal therapy. Angew Chem Int Ed 2016; 55(12): 3982-6.

[27]   Karatepe AA, Bayram A. Biogenic synthesis of TiO2 NPs using Bacillus licheniformis ATCC 14582 and their potential for photocatalytic degradation of volatile organic compounds. Environ Sci Pollut Res Int 2020; 27(16): 20520-32.

[28]   Kalaba M, Mwape L, Chaudhary M. 2021.Green synthesis of NMs: applications in environmental remediation and catalysis

[29]   Lin N, Dufresne A, Thomas L, Dufresne C. Cellulose nanofibril: preparation, properties and applications. Molecules 2018; 23(2): 293.
[PMID: 29385025]

[30]   Chen S, Wang L, Li X, Sun Q, Ren H. Green synthesis of Ag-doped TiO2 NPs for enhanced photocatalytic degradation of organic pollutants. RSC Advances 2017; 7(16): 9870-6.

[31]   Li XM, Yao SC, Fu SQ. Electrospun nanofibrous membranes for filtration. Natl Sci Rev 2007; 4(1): 41-8.

[32]   Karimi G, Zolfaghari A. Green synthesis of ZnO NPs using Robinia pseudoacacia L. flower extract and their application for photocatalytic degradation of gaseous toluene. Mater Res Express 2019; 6(6): 065032.

[33]   Wang J, Chen C. Nanoscale zero-valent iron for removal of heavy metal contaminants from wastewater: A review. Environ Sci Pollut Res Int 2019; 26(8): 6589-602.

[34]   Singh J, Kaur S, Malik R. A perspective review on green nanotechnology in agro-ecosystems: Opportunities for sustainable agricultural practices & environmental remediation. Sustainability (Switzerland) 2017; 9(3): 445.

[35]   Chausali N, Saxena J, Prasad R. Nanobiochar and biochar based nanocomposites: Advances and applications. Journal of Agriculture and Food Research. Elsevier BV 2021; 5: p. 100191.
[http://dx.doi.org/10.1016/j.jafr.2021.100191]

[36]   Khan SH. 2019.Green Nanotechnology for the Environment and Sustainable Development.

[37]   Cao M, Liu M, Xing B. Environmental Risks of Engineered NMs in Water and Soil: Adsorption, Release, and Transformation. Environ Sci Technol 2019; 53(7): 3962-77.
[PMID: 30848581]

[38]   Vance ME, Kuiken TD, Klaine EP. Nanotechnology in the Developing World. Vandenhoeck Ruprecht LLC 2015.

[39]   Ivleva NP, Harmon AA. Regulation of NMs: A Balancing Act. Environ Sci Technol 2014; 48(17): 9982-9.
[PMID: 25084232]

CHAPTER 11

Photocatalytic Activity and Potential of NMs in Environmental Remediation

Sankara Rao Miditana[1,*], A. Ramesh Babu[2], Saivenkatesh Korlam[2], Satheesh Ampolu[3], Neha Agarwal[4] and Vijendra Singh Solanki[2]

[1] *Department of Chemistry, Government Degree College, Puttur, Tirupathi A.P 517583, India*

[2] *Department of Chemistry, SVA Govt. Degree College, Srikalahasti A.P 517644, India*

[3] *Department of Chemistry, Centurion University of Technology and Management, A.P, India*

[4] *Department of Chemistry, Navyug Kanya Mahavidyalaya, University of Lucknow, Lucknow, India*

Abstract: Environmental pollution is a critical global concern that necessitates innovative and sustainable solutions. Among the emerging technologies, photocatalysis using nanomaterials (NMs) has gained significant attention for its potential in environmental remediation. Photoexcitation of wide-bandgap semiconductors like TiO_2, ZnO, SnO_2, CdS, and WO_3 in aqueous media leads to electron-hole pair generation, initiating subsequent photocatalytic processes. The photocatalytic activity is enhanced by the involvement of NMs like Metal oxide NMs, Metal NMs, Graphene-based NMs, and Quantum dots. This chapter explores the photocatalytic activity of various NMs and their applications in addressing environmental challenges. The synergistic effects of NMs in pollutant degradation, wastewater treatment, air purification, and soil remediation are discussed, highlighting the promising prospects for sustainable environmental management. The escalating threats posed by environmental pollution necessitate innovative and sustainable approaches to remediation. The size-dependent properties of NMs result in increased photocatalytic activity, rendering them highly effective in the degradation of diverse environmental pollutants. The interplay between NMs and photocatalysis is elucidated, emphasizing the promising avenues for addressing challenges associated with water, soil, and air quality. As we delve into the applications and mechanisms of NM-based photocatalysis, the chapter also addresses current limitations and future prospects. The insights presented herein contribute to a comprehensive understanding of the photocatalytic activity and potential of NMs, paving the way for sustainable environmental remediation strategies.

Keywords: Contaminants, CNTs, Degradation, Environmental pollution, Graphene oxide, Heavy metals, TiO_2, Photocatalysis, Phytoremediation.

---

* **Corresponding author Sankara Rao Miditana:**Department of Chemistry, Government Degree College, Puttur, Tirupathi A.P 517583, India; E-mail: sraom90@gmail.com

INTRODUCTION

Environmental pollution, a consequence of rapid industrialization, urbanization, and unchecked human activities, has emerged as a pervasive and urgent global challenge [1]. The release of pollutants into air, water, and soil ecosystems poses severe threats to biodiversity, human health, and the overall sustainability of our planet. Common pollutants include hazardous chemicals, heavy metals, greenhouse gases, and various contaminants from industrial and domestic sources. As the scale of pollution continues to escalate, traditional remediation methods struggle to keep pace with the magnitude and complexity of environmental degradation [2]. Conventional approaches, often resource-intensive and limited in efficacy, fall short of addressing the dynamic nature of pollutants and the intricate interplay of environmental systems. The need for innovative and sustainable remediation technologies has never been more pronounced.

In this context, the exploration and development of advanced remediation strategies have become imperative. The advent of innovative technologies capable of efficiently mitigating the impact of pollutants on ecosystems is crucial for safeguarding human health and preserving the delicate balance of nature. One such promising avenue is the application of photocatalysis using NMs, which has gained significant attention for its potential to revolutionize environmental remediation [3]. The urgency to address environmental pollution calls for a paradigm shift in our approach to remediation, emphasizing solutions that are not only effective but also sustainable in the long term [4]. Innovative remediation technologies, such as those incorporating NMs, offer the promise of enhancing pollutant removal while minimizing adverse environmental effects. By harnessing cutting-edge science and engineering, these technologies aim to provide efficient, cost-effective, and environmentally friendly solutions to combat the multifaceted challenges posed by pollution [5].

Photocatalysis, a process reliant on light-induced chemical reactions, stands out as a potent tool in environmental remediation [6]. The photocatalytic mechanism involves the excitation of electrons in semiconductor materials, typically NMs like titanium dioxide (TiO_2) or zinc oxide (ZnO), upon exposure to light [7]. This process generates electron-hole pairs that, in turn, initiate redox reactions capable of breaking down various pollutants. Photocatalysis plays a pivotal role in environmental cleanup through its applications in water purification, air decontamination, and soil remediation. In water treatment, photocatalysis proves effective in degrading organic pollutants and eliminating bacteria [8]. Similarly, in air purification, photocatalytic coatings facilitate the breakdown of airborne contaminants, addressing issues such as volatile organic compounds (VOCs) and nitrogen oxides. Furthermore, photocatalysis exhibits promise in remediating contaminated soil by transforming toxic substances into less harmful forms. Due to its ability to utilize sunlight or artificial light sources for pollutant degradation, photocatalysis emerges as a significant technology for sustainable and energy-efficient environmental remediation.

NMs play a vital role in enhancing photocatalytic activity for environmental remediation due to their unique properties. These materials, characterized by their nanoscale dimensions, offer increased surface area, enabling more active sites for pollutant adsorption and facilitating a higher number of photocatalytic reactions [9]. The quantum size effects of NMs allow for enhanced light absorption across a broader range of wavelengths, including visible light, making them efficient photocatalysts [10]. Additionally, NMs promote improved charge carrier separation, minimizing electron-hole pair recombination and enhancing overall efficiency [11]. The tunable bandgap of semiconducting NMs allows for the optimization of absorption properties, while surface modifications enable tailored chemical and electronic characteristics [12]. The diverse types of NMs, such as metal oxides and carbon-based structures, offer flexibility for specific environmental applications [13, 14]. Compatibility with renewable energy sources, particularly solar energy, aligns NMs with the goal of sustainable and energy-efficient environmental remediation technologies [15]. This collective synergy positions NMs as crucial components in the development of efficient, scalable, and sustainable solutions for mitigating environmental pollution.

NMS FOR PHOTOCATALYSIS

Photocatalysis is a process that utilizes light to initiate and drive chemical reactions. The crucial role of NMs in photocatalysis stems from their distinct physicochemical features, including reduced size, high surface area, quantum confinement effects, and tailorable electronic structures. This diverse toolbox of NMs finds application in various photocatalytic processes, ranging from water purification to air pollution remediation. Some of the NMs are reported in this chapter and are discussed below.

Metal Oxide NMs: Titanium Dioxide (TiO$_2$)

Titanium dioxide stands out as a highly effective material in photocatalysis, largely owing to its distinctive properties that facilitate the degradation of pollutants and the production of clean energy. With a wide bandgap, approximately 3.2 eV, TiO$_2$ exhibits significant photocatalytic activity under ultraviolet (UV) light, generating electron-hole pairs that initiate redox reactions and produce reactive oxygen species (ROS), including superoxide radicals (O$^{2\cdot-}$)

and hydroxyl radicals (˙OH) [16]. This ability makes TiO_2 a key player in environmental remediation, finding applications in the degradation of organic pollutants in air and water [17]. TiO_2-coated surfaces also showcase self-cleaning properties, preventing the accumulation of dirt through the photocatalytic breakdown of organic substances [18]. Furthermore, TiO_2 exhibits potential in photocatalytic water splitting for hydrogen production, presenting a promising avenue for sustainable and renewable energy generation [19].

Zinc Oxide (ZnO)

Zinc oxide emerges as a versatile and efficient material in photocatalysis due to its distinctive characteristics. With a wide bandgap of approximately 3.3 eV, ZnO exhibits effective absorption of UV light, initiating photocatalytic reactions [20]. The material's ability to generate electron-hole pairs and efficiently separate them is crucial for its photocatalytic activity, with ZnO demonstrating notable electron-accepting and donating abilities [21]. This makes ZnO a valuable catalyst for the degradation of various organic pollutants, including dyes and pesticides, in both air and water through the formation of ROS during redox reactions [22]. Beyond environmental applications, ZnO showcases antibacterial properties, making it a candidate for applications in water disinfection and medical settings [23]. Moreover, ZnO is explored in solar energy conversion processes, participating in photocatalytic water splitting for hydrogen production [24]. Its use extends to the field of photovoltaics, contributing to solar cell technologies owing to its high electron mobility [25].

Graphene-based NMs

Graphene-based NMs have emerged as highly effective contributors to photocatalysis, leveraging their exceptional properties for enhanced performance in various applications. Graphene-based materials, encompassing graphene oxide (GO), reduced graphene oxide (rGO), and graphene quantum dots (GQDs), exhibit a confluence of advantageous properties – high surface area, excellent electrical conductivity, and versatile chemical composition. This unique combination significantly impacts their photocatalytic behavior, paving the way for novel and efficient environmental remediation strategies. These materials facilitate efficient charge separation, reduce the recombination of electron-hole pairs, and act as excellent co-catalysts and electron transporters, thereby enhancing overall photocatalytic efficiency [26, 27]. The ability of graphene-based materials to absorb visible light extends the range of photocatalytic activity into the visible spectrum, a crucial advantage for utilizing a broader portion of the solar spectrum [28]. Moreover, the flexibility and versatility of graphene-based

NMs allow for the design of tailored structures, such as composites with semiconductors like TiO_2 or ZnO, resulting in synergistic effects that further boost photocatalytic performance [29 - 31]. The integration of graphene-based NMs extends the realm of photocatalysis to encompass diverse applications, including environmental remediation, solar energy conversion, and the sensitization of other photocatalysts. This multifaceted utilization highlights their significant potential in propelling the development of sustainable technologies.

Metal Nanoparticles (MNPs)

MNPs play a pivotal role in advancing photocatalysis through their distinctive properties, contributing to enhanced efficiency and expanded applications. Gold (Au) and silver (Ag) nanoparticles exhibit localized surface plasmon resonance effects, leading to increased light absorption and the generation of hot electrons, particularly in the visible light range, thereby augmenting photocatalytic activity [32, 33]. Metal nanoparticles on semiconductor photocatalysts improve light absorption, facilitate effective charge separation, and serve as electron sinks, thereby reducing the recombination of electron-hole pairs and prolonging carrier lifetime [34, 35]. Additionally, metal nanoparticles serve as co-catalysts, participating in redox reactions and contributing to the overall degradation of pollutants or the production of valuable products [36]. Combining metal nanoparticles with semiconductors, such as TiO_2 or ZnO, results in synergistic effects, further improving charge transfer and overall photocatalytic performance [37, 38]. These contributions position metal nanoparticles as key components in photocatalytic processes with applications ranging from environmental remediation to renewable energy generation.

Quantum Dots (QDs)

Quantum dots have emerged as versatile and promising materials in photocatalysis, owing to their unique properties that stem from size-dependent electronic and optical characteristics. With a tunable bandgap, QDs can efficiently absorb a broad spectrum of light, including visible light, making them effective in visible-light-driven photocatalytic processes [39, 40]. Their high quantum efficiency enables the conversion of absorbed photons into electron-hole pairs with notable efficiency, facilitating efficient charge separation and utilization in photocatalytic reactions [41]. Quantum confinement effects in QDs result in enhanced charge carrier mobility, contributing to improved charge separation and transport, which enhances overall photocatalytic performance [42]. Quantum dots are known for their versatility in catalytic applications, as they can be engineered with specific surface functionalities, enabling their use in diverse applications such as environmental remediation, water splitting, and organic synthesis [43, 44].

When integrated with semiconductor photocatalysts like TiO_2 or ZnO, QDs contribute to enhanced photocatalytic activity through their size-dependent properties [45]. Furthermore, quantum dots have been explored in photocatalytic water splitting for hydrogen production due to their ability to absorb visible light efficiently [46]. Their reduced recombination rates and ability to generate reactive ROS make quantum dots valuable in the photocatalytic degradation of pollutants, showcasing their potential for sustainable and efficient environmental applications [47, 48].

NMS IN ENVIRONMENTAL REMEDIATION

NMs in environmental remediation have gained significant attention due to their unique properties and potential applications in addressing various environmental challenges. These NMs, which are typically in the range of 1 to 100 nm in size, exhibit distinct chemical, physical, and biological properties that make them suitable for environmental cleanup. The following is a detailed account of the role of NMs in environmental remediation.

NMs for Water Treatment

Heavy Metal Remediation

NMs have emerged as a rapidly advancing field with significant potential for heavy metal remediation in water sources. Their ability to mitigate the adverse effects of metal pollution, particularly from toxic and persistent elements like lead, mercury, cadmium, and arsenic, offers promising solutions for environmental and human health protection. Among the diverse NMs investigated, nanoscale zero-valent iron (nZVI), carbon-based NMs, and metal oxides have demonstrated particular efficacy in heavy metal remediation processes. Nanoscale zero-valent iron is one of the most widely studied NMs for heavy metal remediation due to its high reactivity and ability to reduce a wide range of metal ions to less toxic forms through redox reactions. nZVI nanoparticles can effectively remove heavy metals by adsorption onto their surfaces and subsequent reduction to insoluble metal hydroxides or elemental metals [49]. Additionally, nZVI can also facilitate the precipitation of heavy metal ions, leading to their immobilization in the treatment system.

Carbon-based NMs, including graphene oxide, carbon nanotubes (CNTs), and activated carbon nanoparticles, possess exceptional adsorption capacities due to their high surface area and the presence of abundant functional groups. These features provide numerous adsorption sites and enhance the interaction between the NM and the target pollutant, offering promising potential for water purification applications. These NMs can selectively adsorb heavy metal ions from water through various interactions, including electrostatic attraction, ion exchange, and coordination chemistry [50]. Furthermore, the hierarchical pore structures of carbon-based NMs enhance their accessibility to heavy metal ions, resulting in efficient removal from aqueous solutions.

Metal oxide NMs, including iron oxide (Fe_3O_4), titanium dioxide (TiO_2), and manganese dioxide (MnO_2), possess unique physicochemical properties that enable them to adsorb, oxidize, and/or reduce heavy metal contaminants in water [51]. These metal oxide nanoparticles can chemically bind to heavy metal ions through surface complexation or ion exchange processes, leading to their immobilization and subsequent removal from the aqueous phase. Table **1** summarizes the applications of NMs in the environmental remediation of water contaminants.

Table 1. NMs and their applications in environmental remediation of water contaminants.

Type of NMs	Material Used	Applications	Refs.
Metal-based NMs	Ag NPs	Water disinfectant	[52 - 56]
-	TiO_2 NPs	Aromatic hydrocarbons, biological nitrogen, Water disinfectant, soil-MS-2 phage, E. coli, hepatitis B virus.	[57 - 60]
-	Fe NPs	Chlorinated organic solvents and removal of heavy metals from water.	[61 - 64]
Graphene-based NMs	Pristine graphene	Water: Fluoride	[65]
-	Graphene Oxide	Heavy metals, pesticides, Water/Gaseous-SOx, H_2, NH_3	[66 - 68]
-	ZnO-graphene	Water-heavy metals.	[69]
Polymer-based NMs	PAMAM dendrimers	Wastewater-heavy metals	[70]
-	Polymer nanocomposites	Water-Dyes, metal ions, and microorganisms.	[71 - 73]

Organic Pollutant Degradation

NMs have emerged as promising candidates for the degradation of organic pollutants in water treatment processes due to their unique properties and high reactivity. Organic pollutants, including dyes, pesticides, pharmaceuticals, and industrial chemicals, pose significant environmental and health risks, necessitating effective remediation strategies. NMs offer several advantages for the degradation of organic pollutants in water, including large surface area-t--volume ratios, high catalytic activity, and tunable surface properties. Among the various types of NMs investigated for this purpose, photocatalytic NMs, such as Titanium dioxide, Zinc oxide, and Graphene-based materials, have garnered significant attention due to their ability to harness solar or UV light energy for pollutant degradation through photocatalytic reactions [74, 75].

Photocatalytic NMs utilize light energy to generate electron-hole pairs, which can initiate redox reactions on their surfaces, leading to the degradation of organic pollutants into harmless byproducts, such as carbon dioxide and water. TiO_2 nanoparticles, in particular, have been extensively studied for their photocatalytic activity and stability under various environmental conditions [76]. Under UV or visible light irradiation, TiO_2 nanoparticles undergo photocatalysis, leading to the generation of reactive oxygen species (ROS), including hydroxyl radicals ($^{\bullet}OH$) and superoxide radicals ($O_2^{\bullet-}$). These ROS effectively oxidize organic pollutants, converting them into simpler, non-toxic molecules. (Detailed mechanism is reported in **Section 3.1.3**.

Graphene-based NMs, including graphene oxide and reduced graphene oxide (rGO), have also shown promise for organic pollutant degradation due to their large surface areas, excellent electron transport properties, and chemical stability [77]. These NMs can serve as photocatalytic supports or co-catalysts to enhance the photocatalytic activity of semiconductor nanoparticles, such as TiO_2 or ZnO, through synergistic effects. Additionally, graphene-based NMs can adsorb organic pollutants onto their surfaces, facilitating their degradation *via* photocatalytic or non-photocatalytic pathways.

NMs have the potential for organic pollutant degradation in water treatment but face challenges like limited efficiency under sunlight and potential agglomeration. Research is focused on developing novel nanocomposites, hetero-structured materials, and surface modifications to improve efficiency and stability. Integrating NMs with other water treatment technologies can also reduce energy consumption.

Dye Degradation and Its Mechanism

Fine particles acting as catalysts are crucial for generating hydroxyl radicals ($HO^{\bullet}$), which degrade pollutants [78, 79]. Fig. (**1**) illustrates the proposed mechanism involved, where metal nanoparticles (MNPs) play a key role.

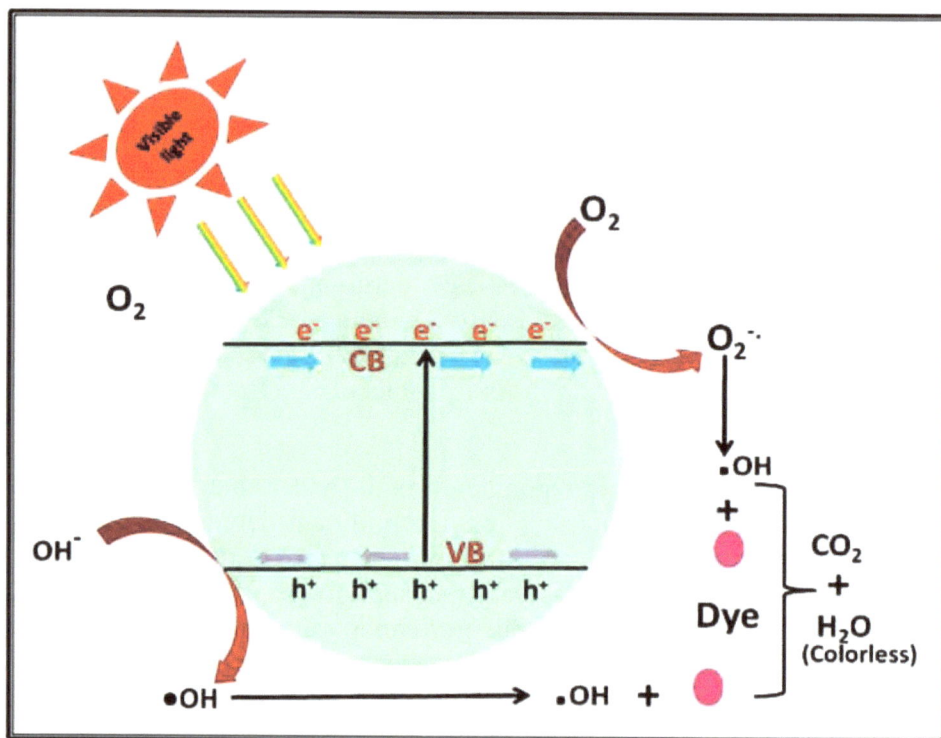

Fig. (1). Scheme for the photocatalytic degradation of Dyes.

The formation of hydroxyl radicals (HO$^{\bullet}$) is pivotal in photocatalytic pollutant degradation. MNPs irradiated with visible light undergo electron excitation from the valence band to the conduction band, creating vacant holes in the valence band. The effectiveness of the catalyst relies on hindering the recombination of these generated electrons and holes [80]. Therefore, dopant ions attract the excited electrons, preventing their recombination with the holes.

$$\text{MNP} + h\nu \longrightarrow \text{MNP}\,(h^+) + \text{MNP}\,(e^-)$$

The photogenerated holes within the metal nanoparticles participate in reactions with either surface-bound hydroxyl groups or water molecules adsorbed on the MNP surface. These reactions yield hydroxyl radicals ($^{\bullet}$OH) and hydrogen ions (H$^+$).

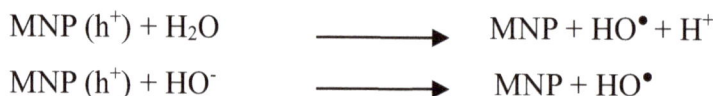

$$\text{MNP}\,(h^+) + H_2O \longrightarrow \text{MNP} + HO^{\bullet} + H^+$$

$$\text{MNP}\,(h^+) + HO^- \longrightarrow \text{MNP} + HO^{\bullet}$$

Electron transfer to adsorbed oxygen triggers the formation of a superoxide ion.

$$\text{MNP (e}^-) + O_2 \longrightarrow \text{MNP} + O_2^{\bullet-}$$

The interaction of superoxide anions with adsorbed water molecules leads to the formation of peroxide radicals (HOO$^{\bullet}$) and hydroxyl ions (OH$^-$) through a series of reactions.

$$\text{MNP (e}^-) + O_2^{\bullet-} + H_2O \longrightarrow \text{MNP} + HO_2^{\bullet} + HO^-$$

Peroxide radicals undergo a protonation reaction with hydrogen ions (H$^+$), leading to the formation of additional hydroxyl radicals (OH$^{\bullet}$) and hydroxyl ions (OH$^-$). Hydrogen peroxide (H$_2$O$_2$) is also generated as a transient intermediate in this process.

$$\text{MNP (e}^-) + HO_2^{\bullet} + H^+ \longrightarrow \text{MNP} + H_2O_2$$

$$\text{MNP (e}^-) + H_2O_2 \longrightarrow \text{MNP} + HO^{\bullet} + HO^-$$

Photogenerated holes oxidize hydroxyl ions (OH$^-$) to hydroxyl radicals (HO$^{\bullet}$), ultimately leading to the degradation of pollutants. All previously mentioned species contribute to the generation of these highly reactive HO$^{\bullet}$ radicals.

$$\text{Dye} + HO^{\bullet} \longrightarrow \text{Products}$$

The formation and subsequent utilization of hydroxyl radicals (HO$^{\bullet}$) play a vital role in the degradation mechanism. These highly reactive species act as potent oxidants, directly attacking and breaking down pollutants.

NMs for Air Quality Improvement

NMs hold significant promise for improving air quality by facilitating the removal of airborne nanoparticles, which can have adverse effects on human health and the environment. Airborne nanoparticles, often generated from combustion processes, industrial activities, and vehicle emissions, pose considerable risks due to their small size, high surface area, and potential for carrying toxic compounds. NMs offer several advantages for airborne nanoparticle removal, including their large surface area, high reactivity, and tunable surface properties. Among the various types of NMs investigated for this purpose, porous materials, metal-organic frameworks (MOFs), and nanofibers have shown particular promise due to their ability to efficiently capture and adsorb nanoparticles from the air [81].

Porous materials, such as activated carbon, zeolites, and silica aerogels, possess interconnected pore structures and high surface areas, which enable them to effectively adsorb airborne nanoparticles through physical and chemical interactions. These materials can be engineered to have specific pore sizes and surface chemistries tailored for capturing nanoparticles of different sizes and compositions. Additionally, the large surface area-to-volume ratios of porous materials enhance their adsorption capacities, making them suitable for air purification applications. MOFs represent a class of nanoporous materials composed of metal ions or clusters coordinated with organic ligands, offering unique properties such as high surface area, tunable pore sizes, and selective adsorption capabilities. MOFs can be functionalized with various functional groups to enhance their affinity towards specific airborne nanoparticles, enabling efficient removal from the air [82]. Furthermore, the modularity and versatility of MOF synthesis allow for the design of custom-tailored materials optimized for nanoparticle capture and removal.

Nanofibers, including electrospun polymers, carbon nanotubes, and nanofiber membranes, offer another approach for airborne nanoparticle removal through filtration mechanisms. These NM-based filters can effectively capture nanoparticles from the air by physical mechanisms such as interception, diffusion, and electrostatic attraction. Additionally, the functionalization of nanofiber surfaces with specific coatings or nanoparticles can enhance their filtration efficiency and selectivity towards target nanoparticles [83]. Despite their potential, the practical application of NMs for airborne nanoparticle removal faces several challenges. One major concern is the scalability and cost-effectiveness of NM-based air purification technologies, which can limit their widespread adoption. Additionally, the long-term stability and recyclability of NMs in air purification systems require further optimization to ensure continuous and efficient nanoparticle removal. To address these challenges and advance NM-based approaches for air quality improvement, ongoing research efforts are focused on developing scalable synthesis methods, engineering multifunctional nanocomposites, and optimizing air purification system designs. Furthermore, interdisciplinary collaborations between materials scientists, engineers, and environmental researchers are essential for translating fundamental research findings into practical solutions for mitigating airborne nanoparticle pollution. Some recent applications of NMs used in air quality improvement are listed in Table **2**.

Table 2. Applications of NMs used in air quality enhancement.

Type of Adsorptive NM	Target Contaminant	Adsorption Results	Refs.
Carbon Nanotubes (CNTs)	NO and NO_2	NO and O_2 travel through CNTs, where NO oxidizes to NO_2 and gets adsorbed onto nitrate surfaces.	[84]
	CO_2	Amine-rich CNT surfaces offer numerous CO_2 adsorption sites, boosting CO_2 capture at low temperatures (20–100°C).	[85]
	CO and CH_3OH	The enhanced electronic properties of SWCNTs lead to improved gas adsorption, regardless of the binding mechanism.	[86]
	CO_2	Amine-functionalized CNTs reign supreme in CO_2 adsorption, setting a new benchmark for capturing this climate-warming gas.	[87]
	HC, NOx, CO_2, and CO	For HC, NOx, CO_2, and CO capture, efficiencies varied from 5.0% to 60.0% with similar CNT amounts.	[88]
Fullerene	CO_2	Strong chemisorption enables high CO_2 uptake by the material.	[89]
Graphene	CO_2, NH_3, SO_2	GO's functional groups enable gas adsorption and synergize with metal centers, enhancing gas capture on its surface.	[90]
Metal Oxide NPs (TiO_2)	CO_2	Both TiO_2 disks and rods exhibit strong interaction with CO_2.	[91]
	SO_2	Smaller NMs inherently adsorbed strongly interacting species more efficiently.	[92]
Metal Oxide NPs (ZnO)	SO_2	ZnO NM shape and size significantly impact their SO_2 adsorption capacity.	[93]
	H_2S	Compared to material synthesized without ultrasonic treatment (3.60 mg/g), UZnO treated with ultrasound exhibited a significantly enhanced hydrogen sulfide adsorption capacity of 29.50 mg/g.	[94]
Metal NPs (Ag and Zn)	Airborne microbes	The study demonstrated the potential for low-cost, silver/zinc oxide-coated air filters to effectively remove germ contaminants from indoor air.	[95]

NMs for Soil Remediation

Nanoparticle-assisted phytoremediation represents a promising approach for soil remediation, leveraging the unique properties of NMs to enhance the efficiency of plant-based remediation strategies. Soil contamination with heavy metals, organic pollutants, and other toxic substances poses significant environmental and human health risks, necessitating effective remediation technologies. Phytoremediation, the use of plants to remove, degrade, or stabilize contaminants in soil and water, offers a sustainable and cost-effective approach to soil remediation [96]. However, the effectiveness of phytoremediation can be limited by factors such as low plant uptake efficiency, slow contaminant degradation rates, and restricted plant growth in contaminated soils. Nanoparticle-assisted phytoremediation aims to overcome these limitations by enhancing the availability and uptake of contaminants by plants, promoting plant growth, and facilitating contaminant degradation processes.

NMs, such as nanoparticles of nZVI, TiO_2, CNTs, and metal oxides, can be applied to soil either alone or in combination with plants to improve remediation outcomes. These NMs possess unique physicochemical properties, including high surface area-to-volume ratios, reactivity, and surface functionalities, which make them suitable for interacting with contaminants and enhancing phytoremediation processes [97].

One of the primary mechanisms by which NMs enhance phytoremediation is by improving the availability and uptake of contaminants by plant roots. Nanoparticles can adsorb or immobilize contaminants in the soil, reducing their mobility and bioavailability to plants. Additionally, NMs can modify soil properties, such as pH, organic matter content, and microbial activity, creating favorable conditions for plant growth and contaminant uptake [98]. Furthermore, NMs can serve as carriers for delivering nutrients, water, and beneficial microorganisms to plants, promoting their growth and metabolic activities. Functionalized nanoparticles can also facilitate the degradation or transformation of contaminants in the rhizosphere through catalytic or redox reactions, enhancing the overall remediation efficiency of phytoremediation systems. Some of the NMs, which are used in soil remediation, are reported in Table **3**.

Table 3. NMs used in soil remediation.

NM used	Targeted pollutant	Result	Refs.
CNTs	Polycyclic aromatic hydrocarbons (PAHs)	66.1% removal within 24 hours.	[99]
TiO_2	Diphenylarsinic acid (DPAA)	82% removal within 3 hours	[100]
TiO_2	PAHs	Half-life reducing from 45.9 to 31.36 hours	[101]
Fe_3O_4	Co	31% removal within 60 days	[102]
Fe_3O_4	PAHs	85.2% of removal at pH 3.5	[103]
Fe_3O_4	As	Reducing 93% of water-leachable As (V) within 150 hours.	[104]
Fe_3O_4	PAHs	Degrading almost 90% within 7 days.	[105]
nZVI	PAHs	Removing 82.21% of PAHs within 104 days.	[106]
nZVI	DDT	Removing 45% of DDT within 5 days.	[107]

Despite their potential, the practical application of NMs for nanoparticle-assisted phytoremediation faces several challenges. One major concern is the potential toxicity and environmental risks associated with the release of nanoparticles into soil ecosystems. Studies have shown that certain nanoparticles may exhibit phytotoxicity or bioaccumulation in plants, raising questions about their long-term impacts on soil health and ecosystem functioning. Additionally, the scalability and cost-effectiveness of nanoparticle-assisted phytoremediation technologies require further optimization to enable their widespread adoption in contaminated sites. Furthermore, the fate and transport of nanoparticles in soil environments, including their interactions with soil components and potential leaching into groundwater, need to be thoroughly investigated to ensure the safety and sustainability of nanoparticle-based remediation approaches.

FUTURE STRATEGIES FOR ENVIRONMENT-SAFE NANO REMEDIATION

Sustainable nano-remediation strategies necessitate a holistic approach, leveraging the power of nanotechnology while ensuring environmental sustainability and human safety. Despite the extensive application of NMs in environmental remediation, concerns regarding their potential adverse effects on human health and ecosystems arise, particularly when released in large quantities and integrated into the food chain. Therefore, developing sustainable technologies that effectively remediate pollutants while minimizing risks to human health and the environment is crucial. Employing NMs for toxic pollutant remediation demands a thorough analysis of their advantages, disadvantages, dispersion characteristics, and retention properties. Different NM types possess unique pros and cons, posing a significant challenge in selecting the optimal choice for environmental remediation. Chemically synthesized NMs have exhibited risks to human health and ecosystems, emphasizing the need for sustainable and highly efficacious alternatives. Polymer-based and green-synthesized NMs are currently under active research and development. Researchers exploring the use of supporting materials like plant waste and polymers to improve the efficacy of NMs by manipulating their structure and composition. Readily available and cost-effective materials, such as bone char, charcoal ash, fly ash, rice hull ash, and pomegranate peels, can be utilized as adsorbents and supporting materials to enhance efficiency and mitigate the downsides of chemically synthesized NMs [108, 109]. However, comparing the adsorption capacity of different NMs remains challenging due to variations in experimental parameters and tested adsorbates. Nevertheless, numerous NMs demonstrate promising capabilities for robust metal ion sorption.

Future research in NMs remediation focuses on surfactant-coated variants, enhancing pollutant access and self-destructing post-task. However, responsible and eco-friendly practices necessitate prioritizing human and environmental impacts during NMs selection and deployment. Robust regulatory frameworks and stakeholder engagement are crucial for transparency, accountability, and ethical considerations throughout the nano-remediation process. Ultimately, these strategies integrate innovative technologies, risk assessment methodologies, and stakeholder engagement to achieve responsible and effective solutions for environmental challenges, minimizing ecosystem and human health impacts.

CONCLUSION

NMs hold immense significance in advancing sustainable environmental management due to their unique properties and versatile applications. These materials offer unprecedented opportunities for addressing complex environmental challenges more efficiently and effectively. One of the key advantages of NMs lies in their high surface area-to-volume ratio, which enables enhanced reactivity and catalytic activity. This property can be harnessed for various environmental remediation processes, such as pollutant degradation, wastewater treatment, and soil remediation. Additionally, NMs exhibit exceptional adsorption capabilities, allowing them to selectively capture and remove contaminants from air, water, and soil. Moreover, the tunable properties of NMs enable customization to target specific pollutants or environmental conditions, thereby optimizing remediation processes. This chapter explores water purification, air quality enhancement, and soil remediation using NMs through the process of photocatalysis and also reports their applications in the aforementioned fields. Moving forward, continued research and innovation in NM synthesis, characterization, and application are essential for realizing the full potential of photocatalytic NMs in environmental remediation and contributing to a cleaner and healthier planet. Overall, the integration of NMs in sustainable environmental management holds great promise for achieving cleaner air, water, and soil while minimizing the ecological footprint and safeguarding human health for future generations.

REFERENCES

[1]   United Nations Environment Programme. Global Environment Outlook - GEO-6: Healthy Planet. Healthy People 2019.

[2]   National Research Council. Environmental Engineering for the 21st Century: Addressing Grand Challenges 2019.

[3]   Wang WN, Serpone N. Visible light photocatalysis of organic dyes in the presence of inorganic anions over highly dispersed polycrystalline titania catalysts. Appl Catal B 2006; 63: 221-31.

[4]   Steffen W, Richardson K, Rockström J, *et al*. Planetary boundaries: Guiding human development on a changing planet. Science 2015; 347(6223): 1259855.

[http://dx.doi.org/10.1126/science.1259855]

[5]   Chen F, Yang L, Dong Y, Zhang Y. Recent advances in the development of noble-metal-free NMs for photocatalytic applications. Catal Sci Technol 2021; 11(4): 1032-49.

[6]   Chen X, Mao SS, Wang L. Titanium dioxide NMs: synthesis, properties, modifications, and applications. Chem Rev 2010; 110(11): 6503-70.
        [http://dx.doi.org/10.1021/cr1001645] [PMID: 21062099]

[7]   Schneider J, Matsuoka M, Takeuchi M, *et al.* Understanding TiO2 photocatalysis: mechanisms and materials. Chem Rev 2014; 114(19): 9919-86.
        [http://dx.doi.org/10.1021/cr5001892] [PMID: 25234429]

[8]   Wang Y, Cao R, Yang W, Zhang J. Enhanced photocatalytic activity of TiO2-graphene nanocomposites for organic degradation. ACS Appl Mater Interfaces 2017; 9(34): 29536-46.

[9]   Zeebaree AYS, Zeebaree SYS, Rashid RF, Zebari OIH, Albarwry AJS, Ali AF, et al. Sustainable engineering of plant-synthesized TiO☐ nanocatalysts: Diagnosis, properties and their photocatalytic performance in removing methylene blue dye from effluent. A review. Curr Res Green Sustain Chem. 2022;5:100312.
        [http://dx.doi.org/10.1016/j.crgsc.2022.100312]

[10]   Tahoon, Mohamed A., Saifeldin M. Siddeeg, Norah Salem Alsaiari, Wissem Mnif, and Faouzi Ben Rebah. 2020. "Effective Heavy Metals Removal from Water Using Nanomaterials: A Review" Processes 8, no. 6: 645.
        [http://dx.doi.org/10.3390/pr8060645]

[11]   Fujishima A, Zhang X. Titanium dioxide photocatalysis: present situation and future approaches. C R Chim 2005; 9(5-6): 750-60.
        [http://dx.doi.org/10.1016/j.crci.2005.02.055]

[12]   Hoffmann MR, Martin ST, Choi W, Bahnemann DW. Environmental applications of semiconductor photocatalysis. Chem Rev 1995; 95(1): 69-96.
        [http://dx.doi.org/10.1021/cr00033a004]

[13]   Pare B, Barde VS, Solanki VS, Agarwal N, Yadav VK, Alam MM, et al. Green synthesis and characterization of LED-irradiation-responsive nano ZnO catalyst and photocatalytic mineralization of malachite green dye. Water. 2022;14(20):3221.
        [http://dx.doi.org/10.3390/w14203221]

[14]   Dostanić, Jasmina, Davor Lončarević, Milica Hadnađev-Kostić, and Tatjana Vulić. 2024. "Recent Advances in the Strategies for Developing and Modifying Photocatalytic Materials for Wastewater Treatment" Processes 12, no. 9: 1914.
        [http://dx.doi.org/10.3390/pr12091914]

[15]   Zhang Y, Tang ZR, Fu X, Xu YJ. TiO2-graphene nanocomposites for gas-phase photocatalytic degradation of volatile aromatic pollutant: is TiO2-graphene truly different from other TiO2-carbon composite materials? ACS Nano 2013; 7(12): 9323-38.
        [PMID: 21117654]

[16]   Fujishima A, Honda K. Electrochemical photolysis of water at a semiconductor electrode. Nature 1972; 238(5358): 37-8.
        [http://dx.doi.org/10.1038/238037a0] [PMID: 12635268]

[17]   Chen X, Mao SS. Titanium dioxide nanomaterials: synthesis, properties, modifications, and applications. Chem Rev 2007; 107(7): 2891-959.
        [http://dx.doi.org/10.1021/cr0500535] [PMID: 17590053]

[18]   Fujishima A, Rao TN, Tryk DA. Titanium dioxide photocatalysis. J Photochem Photobiol Photochem Rev 2000; 1(1): 1-21.
        [http://dx.doi.org/10.1016/S1389-5567(00)00002-2]

[19]   Ahmadiasl R, Moussavi G, Shekoohiyan S, Razavian F. Synthesis of Cu-doped TiO☐ nanocatalyst for

the enhanced photocatalytic degradation and mineralization of gabapentin under UVA/LED irradiation: Characterization and photocatalytic activity. Catalysts. 2022;12(11):1310.
[http://dx.doi.org/10.3390/catal12111310]

[20]   Özgür Ü, Alivov YI, Liu C, *et al.* A comprehensive review of ZnO materials and devices. J Appl Phys 2005; 98(4): 041301.
[http://dx.doi.org/10.1063/1.1992666]

[21]   Su PG, Lee YH. Photocatalytic degradation of phenol on ZnO particles with different size. Appl Catal B 2004; 54(1): 19-25.

[22]   Zuo, Fanjiao, Yameng Zhu, Tiantian Wu, Caixia Li, Yang Liu, Xiwei Wu, Jinyue Ma, Kaili Zhang, Huizi Ouyang, Xilong Qiu, and et al. 2024. "Titanium Dioxide Nanomaterials: Progress in Synthesis and Application in Drug Delivery" Pharmaceutics 16, no. 9: 1214.
[http://dx.doi.org/10.3390/pharmaceutics16091214]

[23]   Brayner R, Ferrari-Iliou R, Brivois N, Djediat S, Benedetti MF, Fiévet F. Toxicological impact studies based on Escherichia coli bacteria in ultrafine ZnO nanoparticles colloidal medium. Nano Lett 2006; 6(4): 866-70.
[http://dx.doi.org/10.1021/nl052326h] [PMID: 16608300]

[24]   Nabi I, Bacha AUR, Li K, *et al.* Complete Photocatalytic Mineralization of Microplastic on TiO_2 Nanoparticle Film. iScience 2020; 23(7): 101326.
[http://dx.doi.org/10.1016/j.isci.2020.101326] [PMID: 32659724]

[25]   Law M, Greene LE, Johnson JC, Saykally R, Yang P. Nanowire dye-sensitized solar cells. Nat Mater 2005; 4(6): 455-9.
[http://dx.doi.org/10.1038/nmat1387] [PMID: 15895100]

[26]   Liu Y, Zhang L, Wei Y. Graphene and graphene oxide: two ideal choices for the preparation of nanocomposites with conducting polymers for applications in fuel cells. Chem Soc Rev 2015; 44(11): 3056-75.
[http://dx.doi.org/10.1039/C4CS00478G] [PMID: 25793455]

[27]   Zhang Y, Tang ZR, Fu X, Xu YJ, Zhu YF. TiO2-graphene nanocomposites for gas-phase photocatalytic degradation of volatile aromatic pollutant: is TiO2-graphene truly different from other TiO2-carbon composite materials. ACS Nano 2012; 6(1): 977-85.
[PMID: 21117654]

[28]   Wang Y, Shi R, Lin J, Huang Y, Wang Z, Fu X. Recent progress in covalent organic framework thin films: fabrications, applications and perspectives. J Mater Chem A Mater Energy Sustain 2014; 2(5): 16811-31.

[29]   Xu Y, Shi G. Modified graphene oxide for photocatalytic degradation of organic pollutants. Phys Chem Chem Phys 2011; 13(16): 7056-64.

[30]   Zhang W, He F, Li J. Graphene oxide: preparation and photocatalytic activity. J Nanopart Res 2010; 12(7): 2313-8.
[PMID: 21170131]

[31]   Dong Y, Shao J, Chen C, *et al.* Blue luminescent graphene quantum dots and graphene oxide prepared by tuning the carbonization degree of citric acid. Carbon 2012; 50(12): 4738-43.
[http://dx.doi.org/10.1016/j.carbon.2012.06.002]

[32]   Kelly KL, Coronado E, Zhao LL, Schatz GC. The Optical Properties of Metal Nanoparticles: The Influence of Size, Shape, and Dielectric Environment. J Phys Chem B 2003; 107(3): 668-77.
[http://dx.doi.org/10.1021/jp026731y]

[33]   Christopher P, Xin H, Linic S. Visible-light-enhanced catalytic oxidation reactions on plasmonic silver nanostructures. Nat Chem 2011; 3(6): 467-72.
[http://dx.doi.org/10.1038/nchem.1032] [PMID: 21602862]

[34]   Zhang L, Zhou M, Wang A, Zhang T. Selective Hydrogenation over Supported Metal Catalysts: From

Nanoparticles to Single Atoms. Chem Rev 2020; 120(2): 683-733.
[http://dx.doi.org/10.1021/acs.chemrev.9b00230] [PMID: 31549814]

[35]   Chen X, Liu L, Huang F, Blackwood DJ. Metal nanoparticle-enhanced photocatalysis for environmental remediation. J Hazard Mater 2020; 141(3): 581-90.

[36]   Kansal S, Singh M, Sud D. Studies on photodegradation of two commercial dyes in aqueous phase using different photocatalysts. J Hazard Mater 2007; 141(3): 581-90.
[http://dx.doi.org/10.1016/j.jhazmat.2006.07.035] [PMID: 16919871]

[37]   Sun Y, Xia Y. Shape-controlled synthesis of gold and silver nanoparticles. Science 2002; 298(5601): 2176-9.
[http://dx.doi.org/10.1126/science.1077229] [PMID: 12481134]

[38]   Bai S, Shen X, Lu Z, Li D. Metal nanoparticle-embedded 2D TiO2 nanosheets for enhanced photocatalysis and field emission. J Mater Chem A Mater Energy Sustain 2017; 5(5): 2272-80.

[39]   Alivisatos AP. Semiconductor clusters, nanocrystals, and quantum dots. Science 1996; 271(5251): 933-7.
[http://dx.doi.org/10.1126/science.271.5251.933]

[40]   Cotta MA. Quantum Dots and Their Applications: What Lies Ahead? ACS Appl Nano Mater 2020; 3(6): 4920-4.
[http://dx.doi.org/10.1021/acsanm.0c01386]

[41]   Kamat PV. Quantum dot solar cells. Semicond Sci Technol 2008; 23(3): 034004.

[42]   Hines MA, Guyot-Sionnest P. Synthesis and Characterization of Strongly Luminescing ZnS-Capped CdSe Nanocrystals. J Phys Chem 1996; 100(2): 468-71.
[http://dx.doi.org/10.1021/jp9530562]

[43]   Rogach AL, Klar TA, Lupton JM, Meijerink A, Feldmann J. Energy transfer with semiconductor nanocrystals. J Mater Chem 2009; 19(9): 1208-21.
[http://dx.doi.org/10.1039/B812884G]

[44]   Wang X, Qu K, Xu B, Ren J, Qu X. Multicolor luminescent carbon dots for simultaneous ion sensing and cellular imaging. J Am Chem Soc 2014; 136(8): 2715-8.
[PMID: 24467448]

[45]   Zhang Q, Han B, Li D, Shi JA. Quantum-Dot-Based Antimony-Doped Tin Oxide Photoanode for Efficient Photoelectrochemical Water Splitting. Angew Chem Int Ed 2017; 56(28): 8155-9.

[46]   Robel I, Subramanian V, Kuno M, Kamat PV. Quantum dot solar cells. harvesting light energy with CdSe nanocrystals molecularly linked to mesoscopic TiO2 films. J Am Chem Soc 2006; 128(7): 2385-93.
[http://dx.doi.org/10.1021/ja056494n] [PMID: 16478194]

[47]   Luther JM, Law M, Beard MC, *et al.* Schottky solar cells based on colloidal nanocrystal films. Nano Lett 2008; 8(10): 3488-92.
[http://dx.doi.org/10.1021/nl802476m] [PMID: 18729414]

[48]   Liu J, Cao Z, Lu Y, Zhang Y. Quantum dots: synthesis, functionalization, and sensing applications in analytical chemistry. Analyst (Lond) 2013; 138(8): 2506-15.

[49]   Di L, Chen X, Lu J, Zhou Y, Zhou Y. Removal of heavy metals in water using nano zero-valent iron composites: A review. J Water Process Eng 2023; 53: 103913.
[http://dx.doi.org/10.1016/j.jwpe.2023.103913]

[50]   Saleh NB, Gupta VK. Synthesis and characterization of alumina-coated carbon nanotubes and their application for lead removal. J Hazard Mater 2012; 211-212: 567-77.

[51]   Tran HN, You SJ, Nguyen TV, Chao HP. Insight into the adsorption mechanism of cationic dye onto biosorbents derived from agricultural wastes. Chem Eng Commun 2017; 204(9): 1020-36.
[http://dx.doi.org/10.1080/00986445.2017.1336090]

[52]   Chou KS, Lu YC, Lee HH. Effect of alkaline ion on the mechanism and kinetics of chemical reduction of silver. Mater Chem Phys 2005; 94(2-3): 429-33.
[http://dx.doi.org/10.1016/j.matchemphys.2005.05.029]

[53]   Gupta A, Silver S. Molecular Genetics: Silver as a biocide: Will resistance become a problem? Nat Biotechnol 1998; 16(10): 888.
[http://dx.doi.org/10.1038/nbt1098-888] [PMID: 9788326]

[54]   Bosetti M, Massè A, Tobin E, Cannas M. Silver coated materials for external fixation devices: *in vitro* biocompatibility and genotoxicity. Biomaterials 2002; 23(3): 887-92.
[http://dx.doi.org/10.1016/S0142-9612(01)00198-3] [PMID: 11771707]

[55]   Xiu Z, Zhang Q, Puppala HL, Colvin VL, Alvarez PJJ. Negligible particle-specific antibacterial activity of silver nanoparticles. Nano Lett 2012; 12(8): 4271-5.
[http://dx.doi.org/10.1021/nl301934w] [PMID: 22765771]

[56]   Pal S, Tak YK, Song JM. Does the antibacterial activity of silver nanoparticles depend on the shape of the nanoparticle? A study of the Gram-negative bacterium Escherichia coli. Appl Environ Microbiol 2007; 73(6): 1712-20.
[http://dx.doi.org/10.1128/AEM.02218-06] [PMID: 17261510]

[57]   Cho M, Chung H, Choi W, Yoon J. Different inactivation behaviors of MS-2 phage and Escherichia coli in TiO2 photocatalytic disinfection. Appl Environ Microbiol 2005; 71(1): 270-5.
[http://dx.doi.org/10.1128/AEM.71.1.270-275.2005] [PMID: 15640197]

[58]   Alizadeh Fard M, Aminzadeh B, Vahidi H. Degradation of petroleum aromatic hydrocarbons using TiO_2 nanopowder film. Environ Technol 2013; 34(9): 1183-90.
[http://dx.doi.org/10.1080/09593330.2012.743592] [PMID: 24191451]

[59]   Bessa da Silva M, Abrantes N, Nogueira V, Gonçalves F, Pereira R. TiO 2 nanoparticles for the remediation of eutrophic shallow freshwater systems: Efficiency and impacts on aquatic biota under a microcosm experiment. Aquat Toxicol 2016; 178: 58-71.
[http://dx.doi.org/10.1016/j.aquatox.2016.07.004] [PMID: 27471045]

[60]   Gu J, Dong D, Kong L, Zheng Y, Li X. Photocatalytic degradation of phenanthrene on soil surfaces in the presence of nanometer anatase TiO2 under UV-light. J Environ Sci (China) 2012; 24(12): 2122-6.
[http://dx.doi.org/10.1016/S1001-0742(11)61063-2] [PMID: 23534208]

[61]   Hooshyar Z, Rezanejade Bardajee G, Ghayeb Y. Sonication enhanced removal of nickel and cobalt ions from polluted water using an iron-based sorbent. J Chem 2013; 2013(1): 786954.
[http://dx.doi.org/10.1155/2013/786954]

[62]   Ebrahim SE, Sulaymon AH, Saad Alhares H. Competitive removal of Cu2+, Cd2+, Zn2+, and Ni2+ ions onto iron oxide nanoparticles from wastewater. Desalination Water Treat 2016; 57(44): 20915-29.
[http://dx.doi.org/10.1080/19443994.2015.1112310]

[63]   Poguberović SS, Krčmar DM, Maletić SP, *et al.* Removal of As(III) and Cr(VI) from aqueous solutions using "green" zero-valent iron nanoparticles produced by oak, mulberry and cherry leaf extracts. Ecol Eng 2016; 90: 42-9.
[http://dx.doi.org/10.1016/j.ecoleng.2016.01.083]

[64]   Guo M, Weng X, Wang T, Chen Z. Biosynthesized iron-based nanoparticles used as a heterogeneous catalyst for the removal of 2,4-dichlorophenol. Separ Purif Tech 2017; 175: 222-8.
[http://dx.doi.org/10.1016/j.seppur.2016.11.042]

[65]   Li Y, Zhang P, Du Q, *et al.* Adsorption of fluoride from aqueous solution by graphene. J Colloid Interface Sci 2011; 363(1): 348-54.
[http://dx.doi.org/10.1016/j.jcis.2011.07.032] [PMID: 21821258]

[66]   Song B, Zhang C, Zeng G, Gong J, Chang Y, Jiang Y. Antibacterial properties and mechanism of graphene oxide-silver nanocomposites as bactericidal agents for water disinfection. Arch Biochem Biophys 2016; 604: 167-76.

[http://dx.doi.org/10.1016/j.abb.2016.04.018] [PMID: 27170600]

[67]   Gao P, Ng K, Sun DD. Sulfonated graphene oxide–ZnO–Ag photocatalyst for fast photodegradation and disinfection under visible light. J Hazard Mater 2013; 262: 826-35.
[http://dx.doi.org/10.1016/j.jhazmat.2013.09.055] [PMID: 24140534]

[68]   Deng CH, Gong JL, Zhang P, Zeng GM, Song B, Liu HY. Preparation of melamine sponge decorated with silver nanoparticles-modified graphene for water disinfection. J Colloid Interface Sci 2017; 488: 26-38.
[http://dx.doi.org/10.1016/j.jcis.2016.10.078] [PMID: 27821337]

[69]   Zhang N, Yang MQ, Tang ZR, Xu YJ. CdS–graphene nanocomposites as visible light photocatalyst for redox reactions in water: A green route for selective transformation and environmental remediation. J Catal 2013; 303: 60-9.
[http://dx.doi.org/10.1016/j.jcat.2013.02.026]

[70]   Diallo MS, Christie S, Swaminathan P, Johnson JH Jr, Goddard WA III. Dendrimer enhanced ultrafiltration. 1. Recovery of Cu(II) from aqueous solutions using PAMAM dendrimers with ethylene diamine core and terminal NH2 groups. Environ Sci Technol 2005; 39(5): 1366-77.
[http://dx.doi.org/10.1021/es048961r] [PMID: 15787379]

[71]   Khare P, Yadav A, Ramkumar J, Verma N. Microchannel-embedded metal–carbon–polymer nanocomposite as a novel support for chitosan for efficient removal of hexavalent chromium from water under dynamic conditions. Chem Eng J 2016; 293: 44-54.
[http://dx.doi.org/10.1016/j.cej.2016.02.049]

[72]   Mittal H, Maity A, Ray SS. Synthesis of co-polymer-grafted gum karaya and silica hybrid organic–inorganic hydrogel nanocomposite for the highly effective removal of methylene blue. Chem Eng J 2015; 279: 166-79.
[http://dx.doi.org/10.1016/j.cej.2015.05.002]

[73]   Sun XF, Liu B, Jing Z, Wang H. Preparation and adsorption property of xylan/poly(acrylic acid) magnetic nanocomposite hydrogel adsorbent. Carbohydr Polym 2015; 118: 16-23.
[http://dx.doi.org/10.1016/j.carbpol.2014.11.013] [PMID: 25542101]

[74]   Ma J, Zhang H, Wang Z. Photocatalytic degradation of organic pollutants in water with carbon-based NMs. Environ Sci Pollu Res 2018; 25(25): 24515-30.

[75]   Zhang L, Li J, Tang Y. Recent progress on graphene-based photocatalysts for degradation of organic pollutants in water. Adv Sustain Syst 2018; 2(9): 1800088.

[76]   Miditana SR, Tirukkovalluri SR, Raju IM, Alim SA. Photocatalytic degradation of Orange-II by surfactant assisted Mn/Mg co-doped TiO2 nanoparticles under visible light irradiation. Current Chemistry Letters 2024; 13(1): 265-76.
[http://dx.doi.org/10.5267/j.ccl.2023.6.003]

[77]   Rajamanickam D, Kumar P, Chen SM, Gopinath SCB. Graphene-based NMs for the removal of organic pollutants from wastewater. Nanomater 2020; 10(8): 1604.

[78]   Aroob S, Carabineiro SAC, Taj MB, *et al.* Green synthesis and photocatalytic dye degradation activity of CuO nanoparticles. Catalysts 2023; 13(3): 502.
[http://dx.doi.org/10.3390/catal13030502]

[79]   Miditana SR, Tirukkovalluri SR, Raju IM. Synthesis and antibacterial activity of transition metal (Ni/Mn) co-doped TiO2 nanophotocatalyst on different pathogens under visible light irradiation. Nanosystems: Physics, Chemistry, Mathematics 2022; 13(1): 104-14.
[http://dx.doi.org/10.17586/2220-8054-2022-13-1-104-114]

[80]   Miditana S, Tirukkovalluri S, Imandi M. A, B., A, R. Review on the synthesis of doped TiO2 NMs by Sol-gel method and description of experimental techniques. Journal of Water and Environmental Nanotechnology 2022; 7(2): 218-29.

[81]   Li X, Yang S, Chen Y, Zhang Z. Metal–Organic Frameworks for Air Purification. Cryst 2018; 8(11):

421.

[82]   Yu H, Wang Z, Sun X, *et al.* Recent advances in metal–organic frameworks for air pollution control. Environ Sci Nano 2018; 5(1): 38-51.

[83]   Zhang Y, Cao B, Zhao S, Zhang Y. A review on the development of electrospun nanofibers for air pollution control. J Mater Sci 2016; 51(24): 12569-86.

[84]   Zhang XX, Bing Y, Dai ZQ, Luo CC. The gas response of hydroxyl modified SWCNTs and carboxyl modified SWCNTs to H2S and SO2. Prz 2012; 88: 311-4.

[85]   Su F, Lu C, Cnen W, Bai H, Hwang JF. Capture of CO2 from flue gas *via* multiwalled carbon nanotubes. Sci Total Environ 2009; 407(8): 3017-23.
[http://dx.doi.org/10.1016/j.scitotenv.2009.01.007] [PMID: 19201012]

[86]   Azam MA, Alias FM, Tack LW, Seman RNAR, Taib MFM, Taibb FM. Electronic properties and gas adsorption behaviour of pristine, silicon-, and boron-doped (8, 0) single-walled carbon nanotube: A first principles study. J Mol Graph Model 2017; 75: 85-93.
[http://dx.doi.org/10.1016/j.jmgm.2017.05.003] [PMID: 28531817]

[87]   Gui MM, Yap YX, Chai SP, Mohamed AR. Multi-walled carbon nanotubes modified with (3-aminopropyl)triethoxysilane for effective carbon dioxide adsorption. Int J Greenh Gas Control 2013; 14: 65-73.
[http://dx.doi.org/10.1016/j.ijggc.2013.01.004]

[88]   Romero-Guzmán L, Reyes-Gutiérrez LR, Romero-Guzmán ET, Savedra-Labastida E. Carbon nanotube filters for removal of air pollutants from mobile sources. J Miner Mater Charact Eng 2018; 6(1): 105-18.
[http://dx.doi.org/10.4236/jmmce.2018.61009]

[89]   Dong H, Lin B, Gilmore K, Hou T, Lee ST, Li Y. B40 fullerene: An efficient material for CO2 capture, storage and separation. Curr Appl Phys 2015; 15(9): 1084-9.
[http://dx.doi.org/10.1016/j.cap.2015.06.008]

[90]   Wu S, He Q, Tan C, Wang Y, Zhang H. Graphene-based electrochemical sensors. Small 2013; 9(8): 1160-72.
[http://dx.doi.org/10.1002/smll.201202896] [PMID: 23494883]

[91]   Tumuluri U, Howe JD, Mounfield WP III, *et al.* Effect of surface structure of TiO_2 nanoparticles on CO_2 adsorption and SO_2 resistance. ACS Sustain Chem& Eng 2017; 5(10): 9295-306.
[http://dx.doi.org/10.1021/acssuschemeng.7b02295]

[92]   Baltrusaitis J, Jayaweera PM, Grassian VH. Sulfur dioxide adsorption on TiO2 nanoparticles: influence of particle size, coadsorbates, sample pretreatment, and light on surface speciation and surface coverage. J Phys Chem C 2011; 115(2): 492-500.
[http://dx.doi.org/10.1021/jp108759b]

[93]   Wu CM, Baltrusaitis J, Gillan eg, Grassian VH. Sulfur dioxide adsorption on ZnO nanoparticles and nanorods. J Phys Chem C 2011; 115(20): 10164-72.
[http://dx.doi.org/10.1021/jp201986j]

[94]   Huy NN, Thanh Thuy VT, Thang NH, *et al.* Facile one-step synthesis of zinc oxide nanoparticles by ultrasonic-assisted precipitation method and its application for H2S adsorption in air. J Phys Chem Solids 2019; 132: 99-103.
[http://dx.doi.org/10.1016/j.jpcs.2019.04.018]

[95]   Pokhum C, Intasanta V, Yaipimai W, *et al.* A facile and cost-effective method for removal of indoor airborne psychrotrophic bacterial and fungal flora based on silver and zinc oxide nanoparticles decorated on fibrous air filter. Atmos Pollut Res 2018; 9(1): 172-7.
[http://dx.doi.org/10.1016/j.apr.2017.08.005]

[96]   Khosravi M, Kiamahalleh MV, Shojaosadati SA, Ghazi-Khansari M. A review of phytoremediation of heavy metals and utilization of nanoparticles to enhance efficiency. J Environ Manage 2018; 220: 345-

54.

[97] Ma X, Gurung A, Deng Y, Phan T, Li J. Phytoremediation of heavy metal–contaminated soil with the remediation potential of metal-accumulating plants and metal nanoparticles. Environ Rev 2019; 27(3): 285-300.

[98] Mirzajani A, Shariatmadari H. The role of nanoparticles in enhancing plant growth and yield: A review. J Plant Nutr 2020; 43(12): 1907-23.

[99] Zhang T, Lowry GV, Capiro NL, *et al. in situ* remediation of subsurface contamination: opportunities and challenges for nanotechnology and advanced materials. Environ Sci Nano 2019; 6(5): 1283-302.
[http://dx.doi.org/10.1039/C9EN00143C]

[100] Wang A, Teng Y, Hu X, *et al.* Diphenylarsinic acid contaminated soil remediation by titanium dioxide (P25) photocatalysis: Degradation pathway, optimization of operating parameters and effects of soil properties. Sci Total Environ 2016; 541: 348-55.
[http://dx.doi.org/10.1016/j.scitotenv.2015.09.023] [PMID: 26410709]

[101] Bidast S, Golchin A, Baybordi A, Zamani A, Naidu R. The effects of non-stabilised and Na-carboxymethylcellulose-stabilised iron oxide nanoparticles on remediation of Co-contaminated soils. Chemosphere 2020; 261: 128123.
[http://dx.doi.org/10.1016/j.chemosphere.2020.128123] [PMID: 33113646]

[102] Pare B, Nagraj G, Solanki VS, Albakri GS, Alreshidi MA, Abbas M, et al. Eco-friendly LEDs radiation-assisted photocatalytic mineralization of toxic azure B dye using Bi☐MoZnO☐ nanocomposite. Inorg Chim Acta. 2024;568:122109.
[http://dx.doi.org/10.1016/j.ica.2024.122109]

[103] Barzegar G, Jorfi S, Soltani RDC, *et al.* Enhanced Sono-Fenton-Like Oxidation of PAH-Contaminated Soil Using Nano-Sized Magnetite as Catalyst: Optimization with Response Surface Methodology. Soil Sediment Contam 2017; 26(5): 538-57.
[http://dx.doi.org/10.1080/15320383.2017.1363157]

[104] Liang Q, Zhao D. Immobilization of arsenate in a sandy loam soil using starch-stabilized magnetite nanoparticles. J Hazard Mater 2014; 271: 16-23.
[http://dx.doi.org/10.1016/j.jhazmat.2014.01.055] [PMID: 24584068]

[105] Usman M, Faure P, Ruby C, Hanna K. Remediation of PAH-contaminated soils by magnetite catalyzed Fenton-like oxidation. Appl Catal B 2012; 117-118: 10-7.
[http://dx.doi.org/10.1016/j.apcatb.2012.01.007]

[106] Song Y, Fang G, Zhu C, *et al.* Zero-valent iron activated persulfate remediation of polycyclic aromatic hydrocarbon-contaminated soils: An *in situ* pilot-scale study. Chem Eng J 2019; 355: 65-75.
[http://dx.doi.org/10.1016/j.cej.2018.08.126]

[107] El-Temsah YS, Oughton DH, Joner EJ. Effects of nano-sized zero-valent iron on DDT degradation and residual toxicity in soil: a column experiment. Plant Soil 2013; 368(1-2): 189-200.
[http://dx.doi.org/10.1007/s11104-012-1509-8]

[108] Asghar N, Hussain A, Nguyen DA, *et al.* Advancement in nanomaterials for environmental pollutants remediation: a systematic review on bibliometrics analysis, material types, synthesis pathways, and related mechanisms. J Nanobiotechnology 2024; 22(1): 26.
[http://dx.doi.org/10.1186/s12951-023-02151-3] [PMID: 38200605]

[109] Visa M. Synthesis and characterization of new zeolite materials obtained from fly ash for heavy metals removal in advanced wastewater treatment. Powder Technol 2016; 294: 338-47.
[http://dx.doi.org/10.1016/j.powtec.2016.02.019]

CHAPTER 12

Insights into the Impact of Nanocomposite TiO$_2$ Photocatalyst in Wastewater Effluents

Ajay Kumar Tiwari[1,*] and **Sheerin Masroor**[2]

[1] *Department of Chemistry, School of Applied Sciences, Uttaranchal University, Dehradun, Uttarakhand 248007, India*

[2] *Department of Chemistry, A.N. College, Patliputra University, Patna 800013, Bihar, India*

Abstract: Impurities of hazardous organic components are of growing concern for water, which is considered the primary operating parameter used in photodegradation investigations. Even in low quantities, the presence of hazardous chemicals in the water system can pose threats to living organisms' health and the environment. Traditional remediation methods are inefficient in eliminating the toxicity of hazardous chemicals containing wastewater effluents from the dye industry, the chemical industry, the pharma industry, and the cosmetic industry. Nanocomposites (NCs) of titanium dioxide (TiO$_2$) act as promising environmentally friendly photocatalysts for reducing water pollution. This chapter is focused on the discussion of the intermediate products that are produced during the photodegradation process using TiO$_2$ NCs and determining the impact of adding new elements on the TiO$_2$ energy gap. The pace at which photogenerated electron-hole pairs recombine, along with the suppression of the anatase-to-rutile phase transition is also discussed. The benefit of conducting comprehensive comparisons with a variety of photocatalytic reactions involving many substrates; utilizing a solar simulator to clarify the effectiveness of doped materials is also included in this chapter. The authors have tried to prove the idea of modulating the photocatalytic process and anticipated the potential for using this process to accomplish the utilization of wastewater effluent resources.

Keywords: Nanocomposite, Photocatalysts, Hazardous, Eliminating, Photodegradation, Utilization, Wastewater effluent.

INTRODUCTION

Titanium dioxide (TiO$_2$) is a semiconducting and inert metallic oxide that exhibits photocatalytic activity under solar radiation. Due to its unique characteristics, TiO$_2$ is commonly known as *'titania'* and has gained considerable attention in a vast range of environmental applications. Additionally, its relatively low cost and

* **Corresponding author Ajay Kumar Tiwari:** Department of Chemistry, School of Applied Sciences, Uttaranchal University, Dehradun, Uttarakhand 248007, India; E-mail: ajay.tiwari1591@gmail.com

Neha Agarwal, Vijendra Singh Solanki, Neetu Singh & Maulin P. Shah (Eds.)

easy processing have made it a popular choice over the years. TiO_2 has been classified as chemically and biologically nonreactive material in animals and human beings [1]. TiO_2 has been formed from various minerals. There has been an exponential increase in the direct production of nano-sized TiO_2 during the last few years, with the annual production of TiO_2 powder estimated to be around 5 million tons worldwide in 2005, a 2.5% increase in 2009, and a 10% increase in 2015 [2]. TiO_2 in powdered form has characteristic properties such as opacity in many commercial products; such as paint texture, pharmaceuticals, paper texture, and cosmetics. It offers an advanced technological potential due to its high refractive index and light-scattering qualities. Nano-sized TiO_2 has been the basis for technological advancements that have made it possible to employ it in a variety of applications, including the plastics sector, confectionary, glass antifogging, and self-cleaning coatings. Furthermore, TiO_2 NPs and their NCs play a vital role in food and bakery products and pharmaceutical drugs [3]. Concerns have been raised regarding the increasing possible effects of TiO_2 powder (nano-sized) on humans and the environment due to its manufacturing process.

PROPERTIES AND POTENTIAL OF TIO_2 NPS

Physical and Chemical Characteristics of TiO_2 NPs

The nano range of TiO_2 particles is smaller than 100 nm with various shapes under modification, such as nano-powder, nano-film, nanowire, nano-tubes, nanorods, nano-sheet, *etc.* TiO_2 particles in micron and nano sizes have the same properties because of their larger specific surface area, but nano-powders have different physical characteristics. A recent study examined the size-related characteristics of metallic oxide NPs and discovered special features when their diameters were less than 30 nm, which improves their reactivity across interfaces [4]. NPs bind together to create soft aggregates with strong bonds. The zeta potential, which varies greatly for TiO_2 particles across a broad pH range (3.5-8.8), determines how soft agglomerates disperse [5]. In the range of physiological pH, TiO_2 is claimed to significantly impact its bioavailability due to its isoelectric nature. Regretfully, the majority of research on the interaction between TiO_2 NPs and biological systems has virtually ignored the size of the particles and their properties checked by zeta potential [6].

Naturally occurring crystalline TiO_2 comes in 3 forms: anatase, brookite, and rutile. Rutile is the most stable form out of the three crystalline forms. The average particle size of TiO_2 powder under visible light is 230 nm while 60 nm under UV light. The anatase form of TiO_2 exhibits*-*- greater photocatalytic activity than that of rutile and brookite forms. The TiO_2 NPs energy gap creates a band where an electron jumps from the valence level to the conduction level [7]. Photo-activation of nano-TiO_2 can occur when it is exposed to UV-A to C, visible, fluorescent, and X-ray radiation. Highly reactive photo radicals are produced as a result of photocatalytic activity, and these free radicals can react frequently with organic materials.

The cellular function of NPs possesses various characteristics that are important from a toxicological standpoint. Among these are their dimensions, surface area, chemistry and charge on the surface, crystallinity, shape, solubility, and state of agglomeration or aggregation. The surface of NPs can make them hydrophilic/lipophobic or hydrophobic/lipophilic as well as active or passive in catalytic activity. Active absorption by endocytosis and passive uptake by free diffusion are the two primary mechanisms of uptake of NPs in the cells. The "professional" phagocytes, such as macrophages, are known for their action-dependent endocytic function, known as phagocytosis [8]. Phagocysts are capable of phagocytosing particles that are smaller than 500 nm, which causes a persistent strain on other cells and tissues. According to studies, when rats are exposed to TiO_2 powders through inhalation, it efficiently removes 3-6 μm particles but 20 nm-sized particles are difficult to remove. Human keratinocytes have been shown to take up TiO_2 NPs endocytotically through *in vitro* investigations [9]. Reported experimental research has evaluated that the TiO_2 NPs of less than 200 nm size easily enter the human blood cells. However, particles of size greater than 200 nm couldn't enter; rather they accumulate on the surface of the human blood cells [10]. These findings demonstrated that the size and aggregation state of NPs affects their cellular uptake and subcellular localization, which ultimately determines their toxicity.

Increasing human population, urbanization, expansion of agricultural lands, climate change, and pollutants from industrial effluents create wide problems concerning harm to humans as well as the environment [11]. As a result, clean water resources are becoming scarce, making the water crisis an urgent issue that needs to be resolved. Furthermore, it is confirmed that there is widespread water pollution as a result of organic and inorganic contaminants and hazardous toxic elements (HTEs) traveling through different channels. A vast array of organic active moieties is included in organic contaminants, all of which are poorly and ineffectively removed by conventional wastewater treatment facilities through activated sludge [12]. As a result, they are released into the environment, where they have been found to be carcinogenic to the endocrine system, and mutagenic.

The Energy Gap of the Valence Band (VB) and Conduction Band (CB)

In the electronic band structure of the semiconducting material, the valence band

and the conduction band are the energy levels of TiO_2 (Fig. **1**). The valence band has electrons that are compactly bonded to their corresponding atomic species, while the conduction band has electrons that are free to transfer and participate in conduction processes.

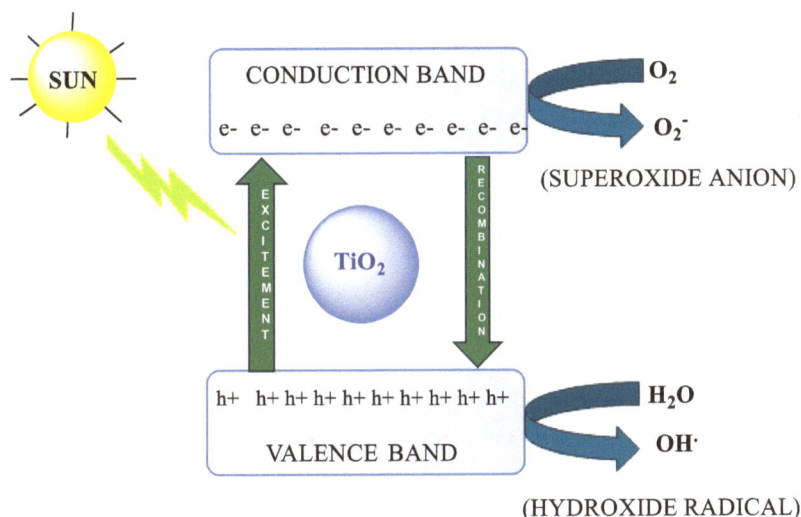

Fig. (1). Photocatalytic Activity of TiO_2 NPs.

Pure Tio₂ Photocatalysis for Water Purification

TiO_2 NPs have proven to be a promising photocatalytic agent in the degradation of numerous toxic pollutants and contaminants available in the water system. When exposed to UV light, these NPs generate hydroxyl radicals ($\cdot OH$) that can cleave these organic complex compounds into non-hazardous simpler compounds [13]. TiO_2 is used to purify wastewater under advanced water purification technologies due to its significant availability in the nano range. In addition, TiO_2 NPs also act as a disinfection agent when exposed to UV light. This application is particularly relevant in treating water for potable consumption where microbial contamination is a great concern. The water treatment chemical industry incorporates TiO_2 NPs for the removal of toxic pollutants such as endocrine-disrupting chemicals, organic contaminants, heavy metals, and dyes [14]. This is especially valuable for industries that generate wastewater containing challenging-to-treat substances. TiO_2 acts as a catalyst in Oxidation Processes (OPs), which are designed to oxidize and remove recalcitrant organic compounds from water [15]. OPs can be effective in treating water sources contaminated with persistent organic pollutants (POPs) or emerging contaminants that are not easily removed by conventional treatment methods. The increasing requirement for clean fresh water, coupled with stricter environmental regulations, has driven the adoption of advanced water treatment technologies [16]. As a result, the use of TiO_2 NPs in wastewater purification is gaining momentum.

Nano-TiO₂ (Anatase/Rutile) Photocatalysis for Environmental Remediation

TiO_2 absorbs the violet color of visible light that is near UV light. Rutile TiO_2 absorbs visible light (violet light) at 415 nm, and anatase TiO_2 absorbs near-UV light (UV-A) at 385 nm. On earth, TiO_2 is available in mixed form (20% rutile and 80% anatase) under UV illumination [17]. As compared to individual rutile TiO_2, a mixture of both rutile and anatase TiO_2 performs better photocatalytic work under visible light. Photocatalytic degradation is a promising technique to remove organic pollutants from the environment. Photocatalysts absorb solar radiation to form electron-hole systems, which generate hydroxyl radicals ($\cdot OH$), that are very reactive and can cleavage the organic pollutants [18]. Photocatalysis is a process in which a photocatalyst material is used to accelerate a chemical reaction by absorbing light energy. This phenomenon is commonly employed for the sustainable development of environmental and environmental energy-based applications, such as water purification, air detoxification, hydrogen generation, and solar energy conversion. A photocatalyst NM refers to a nanoscale-sized material that possesses photocatalytic properties. Mixtures of semiconductors consist of crucial photocatalytic processes in various fields of research. Semiconductor NMs are selected due to their unique electronic structure, which allows them to harness light energy and drive chemical reactions [19].

POPS IN WASTEWATER, THEIR SOURCES AND REMEDIATION METHODS

Sources of Waste Water Effluents (POPs)

Essentially, a type of hazardous substance that hurts human health is POPs due to their characteristic properties. These are synthetic organic chemicals that are hazardous to the environment and living organisms. POPs are unique in their potential to remain in the environment for long periods, resisting destruction by natural processes such as photolysis, microbial breakdown, or hydrolysis. Instead, if not degraded, they persist in the environment for years. POPs can migrate to remote areas like the polar regions, thousands of miles from their release sources, and cover great distances through the atmosphere and water bodies (Fig. **2**). This phenomenon is also referred to as "long-range transport," which is caused by evaporation, volatilization, and re-condensation. POPs are therefore a global concern since they might have an impact in areas that are far from their sources [20].

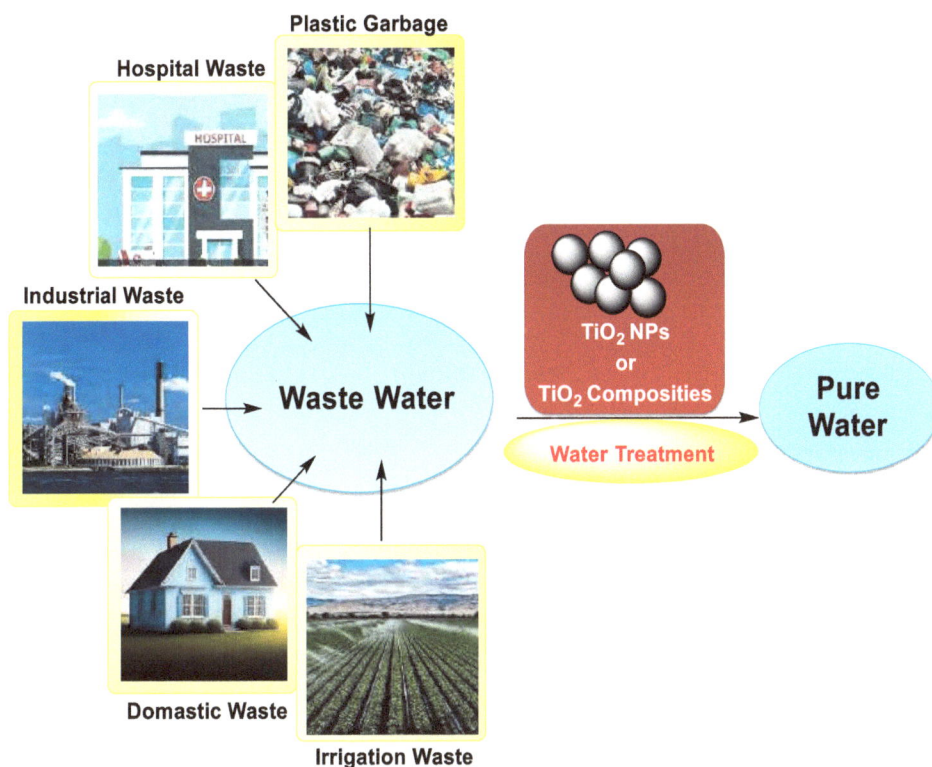

Fig. (2). Various sources of wastewater effluents.

Various Methods of Removal of POPs from Water Systems

The elimination of (POPs) present in the water is a challenging task due to their persistent nature and potential toxicity. Various methods have been developed to treat water contaminated with POPs. Some of the commonly used methods include activated carbon, which is highly effective in adsorbing organic compounds and POPs from water. It works by attracting and binding the pollutants to its porous surface. Chemical precipitation is also used, which involves adding chemical reagents to the water to form insoluble complexes with the POPs, causing them to precipitate and settle out of the water. In some cases, certain microorganisms can naturally degrade POPs. Thus, bioremediation techniques can be employed to encourage the growth of these microorganisms and enhance the degradation process [21]. Ozone treatment involves exposing the contaminated water to ozone gas. Ozone reacts with POPs, breaking them down into less harmful by-products.

The removal of contaminants from water can also be accomplished by Advanced Oxidation Processes (AOPs), which combine powerful oxidizing agents, ozone, and hydrogen peroxide under solar irradiation to generate highly reactive radicals that can break down POPs, as mentioned in Fig. (**3**). Membrane filtration methods, like osmosis and nano-scale filtration, are capable of effectively removing POPs by physically separating them from water; based on size and molecular weight. Solid phase extraction (SPE) uses specific resins or sorbents to selectively bind and remove POPs from water samples [22].

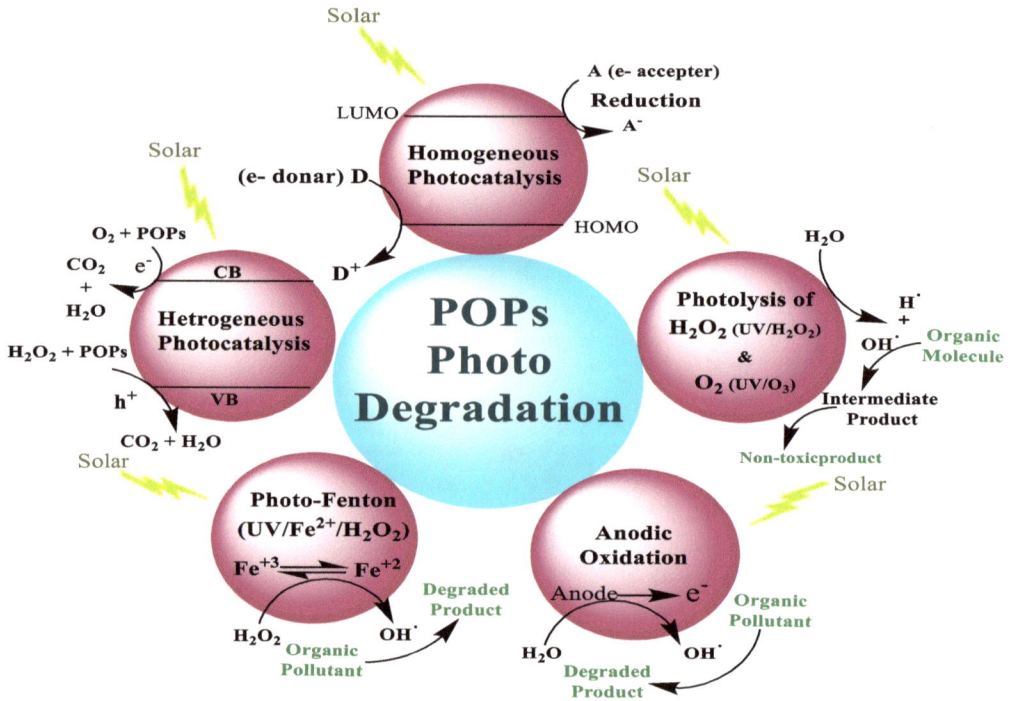

Fig. (3). Various photodegradation methods for the removal of POPs from the water system.

In some cases, contaminated water can be treated by passing it through soil or constructed wetlands, where specific POPs can be absorbed or degraded. This process is called soil and water remediation. Electrochemical methods, like electrocoagulation and electrochemical oxidation, can remove POPs by causing chemical changes at the electrodes. Specific chemical treatments can be used to degrade POPs into less harmful substances, which is also known as chemical degradation [23]. It's necessary to keep in mind that the effect of these methods varies depending on the specific types of POPs present, their concentrations, and the water quality. Often, a combination of several treatment methods may be required to achieve adequate removal and remediation. Additionally, the selection of the most suitable method will depend on the scale of contamination, available resources, and regulatory requirements.

Nanocomposites for Wastewater Treatment

The nanocomposites (NCs) were designed to address the challenge of treating wastewater from various industrial effluents and an experimental study was performed on the photocatalytic degradation of p-nitro phenol (p-NP) as mentioned in Fig. (**4**). The NCs synthesis includes incorporating the TiO_2 nanowires onto a carbon-based matrix as commercial cellulose acetate (C-Ac) and waste carbon sourced materials derived from coconut husk. For the wastewater treatment study, the nanomaterial TiO_2 was blended with biopolymer C-Ac and layered double hydroxide (LDH) to form a membrane of the composite. The performance of NCs was evaluated through photocatalytic degradation experiments using model organic pollutants from nitrophenol using UV light [24].

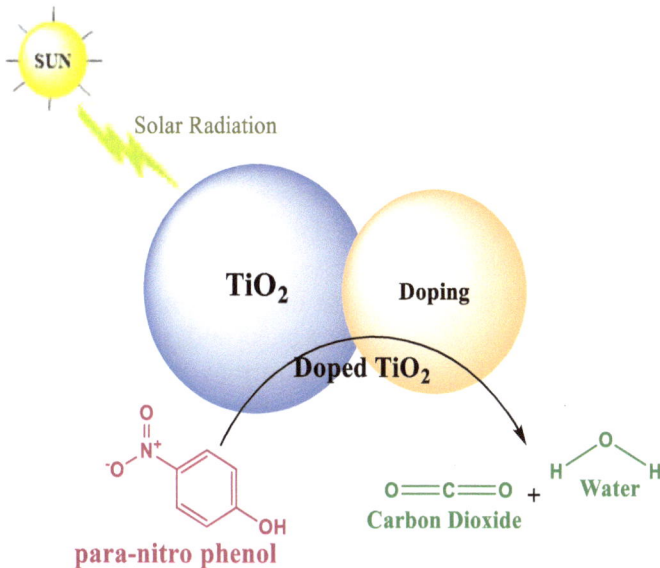

Fig. (4). Photodegradation of Para-nitrophenol.

The NCs exhibit excellent photocatalytic activity, attributed to the synergistic effects between TiO_2 and carbon materials, enabling effective degradation and mineralization of pollutants. Furthermore, the NCs were tested for water treatment application also using a membrane for pure water flux determination. The effect of the TiO_2 nanowire and an LDH and hydrotalcite $Mg_6Al_2CO_3(OH)_{16} \cdot 4H_2O$ was also studied in the membrane of the composite material [25]. Overall, this study highlights that the TiO_2 and carbon-based NCs (C-Ac, coconut husk fibers) efficiently worked. Its ability to efficiently remove pollutants under UV light/visible light irradiation makes it a sustainable and effective solution for treating wastewater from diverse industrial effluents.

Photocatalytic Degradation of Toxic Textile Waste/Dyes

Biphenyl was widely used in electrical equipment, transformers, and industrial applications. They are highly persistent and can contaminate soil and water for long periods. Dioxins are unintentional by-products of various industrial processes and specific chemical reactions [26]. Some other toxic pollutants from textile industries are aldrin, dieldrin, hexachlorobenzene (HCB), and chlordane, which are formed during the combustion of organic matter in the presence of chlorine. Aldrin and dieldrin are the pesticides that were commonly used in the past for agricultural purposes. Though their use has been restricted, they persist in the environment. HCB was used as a fungicide and an industrial chemical. It is highly toxic and can be released into the environment through various processes. Chlordane is a pesticide that controls termites and other pests. It is persistent and can contaminate soil and water. Mirex was used as a pesticide for fire ant control and in specific industrial applications [27, 28].

A variety of treatment techniques are available to eliminate harmful components of human origins from industrial waste effluents [29 - 31]. The primary concern with industrial wastewater is textile wastewater, which is produced in enormous quantities and contains a wide range of chemical changes. Inefficacies in dyeing, pigments, and printing materials in the textile and pulp industrial process cause approximately 10-50% of the damage to the environment. Unfortunately, textile dyes are often washed away in river water, and they are highly resistant to temperature and light [32]. Low concentrations of certain dyes can be very noticeable, and the degradation products of azo dyes have the potential to be poisonous, allergenic, or carcinogenic, which can have a catastrophic effect on the ecosystem [33]. The challenges associated with treating wastewater containing synthetic textile dyes are due to their complex aromatic molecular structure, stability, and resistance to conventional treatment methods [34]. This discussion leads towards various approaches such as the adsorption process through membrane and electrochemical treatment [35], advanced oxidation processes, coagulation and flocculation, and biological treatment. However, these methods have limitations in fully oxidizing resistant organic dyes and may produce harmful by products while consuming significant energy. Additionally, this challenging problem can be solved by the use of heterogeneous photocatalysis, which has emerged as a promising technique for environmental remediation due to its low cost, high decomposition performance, non-toxicity, chemical stability, and easy availability [36]. Improving solar light irradiation has enabled quick progress in nano-photocatalytic degradation of toxic pollutants.

Photocatalytic Degradation of Endocrine-Disrupting Chemicals (EDCs) in Water

Photodegradation of several estrogen derivatives that are available in wastewater effluents is harmful to human health (Table 1). Multicomponent estrogen contamination degrades through TiO_2 NCs photocatalysts under UV and visible illumination Fig. (5).

Table 1. Source of EDCs and their side effects.

S.No.	Endocrine-disrupting chemicals (EDCs)	Source	Side Effect	Refs.
1.	Nonylphenol (NP)	Coastline wastewater, municipal wastewater, industrial effluents	Irritation to the eye, human respiratory, reduce egg formation of female zebrafish	[40]
2.	Estrogen EDCs	Urban wastewater and Freshwater	Breast cancer, Infertility, and Animal hermaphroditism	[41]
3.	17α-estradiol and 17β-estradiol	Wastewater effluents		[42, 43]
4.	Bisphenol-A (BPA)	Urban wastewater, River water, Sewage water, Municipal wastewater,	Acute, sub-chronic toxicity, effects on the liver, kidney and body weight; induce mutation	[44, 45]
5.	Phthalate (benzyl butyl phthalate, diethylhexyl phthalate)	Industrial wastewater effluent	Inhalation, ingestion and dermal exposure	[46, 47]

Multicomponent estrogen contamination degrades through Au/TiO_2 photocatalyst under UV and visible illumination. Au/TiO_2 photocatalyst composites are kinetically and efficiently photodegraded through the attacking of OH radical water effluent estrogen derivatives [37]. Various EDCs are widely present in wastewater bodies in the form of pharmaceutical waste, plastic products, and industrial waste [38]. The physical and chemical characteristics of the EDCs directly affect the environment, especially water systems. EDC_S can be removed from water by various techniques such as biodegradation and photodegradation. Photocatalytic degradation is far better than biodegradation and is highly efficient with low energy utilization. If a photocatalyst is combined with bacteria, fungi, and algae, then its degradation efficiency increases [39].

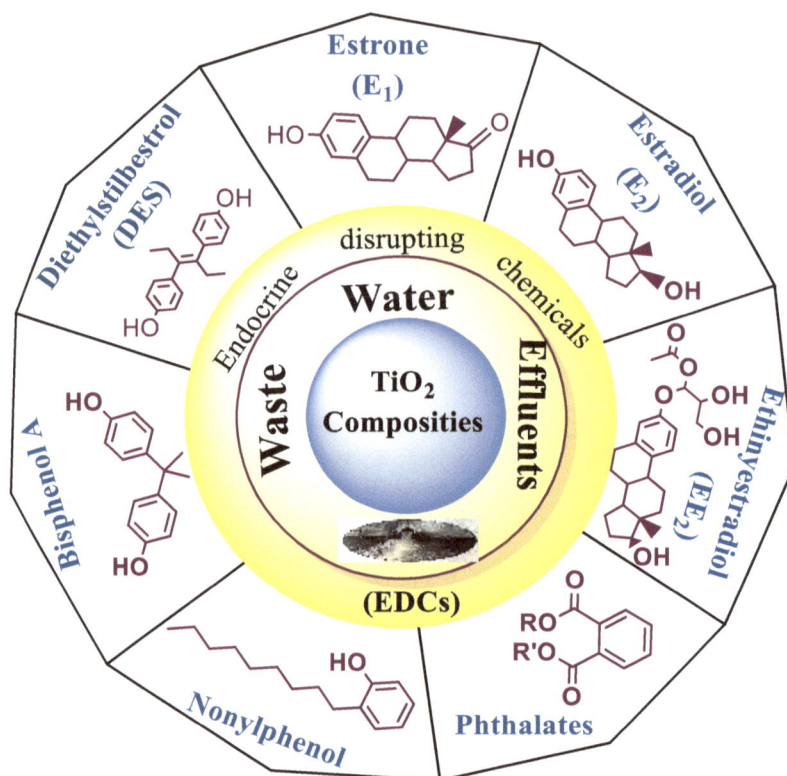

Fig. (5). Waste Water Effluents containing EDC_s and its Derivatives.

Some inorganic ions like HPO_4^{2-}, NH_4^+, and HCO_3^- in water systems inhibit the efficiency of photocatalysts. The polar compounds increase EDC activity. So, the photocatalytic activity of individual photo-NPs doped with rare earth metals is used for the degradation of ions containing EDCs [48]. EDCs are classified into several groups such as phthalates (from personal care products), plasticizers, resins, bisphenols (BP) (from food packaging), clothing, furniture, polychlorinated biphenyls (PCBs) from fungicides, pesticides, herbicides, and solvents, non-steroidal estrogens (NSEs) from synthetic hormones and pharmaceutical waste [49, 50].

Several studies on the deterioration of dye pollutants have been carried out recently, each using a unique combination of semiconductors and magnetic materials. A study using a hetero-coagulation technique examined the functional colloidal microbeads of $TiO_2/Fe_3O_4/SiO_2$ [51]. TiO_2 NPs have been investigated as the most efficient photocatalyst for the degradation of organic dye. However, TiO_2 NPs offer a variety of challenges for the adsorption, recovery, and separation due to their optical gap and interfacial area [52]. To improve the photocatalytic applications in wastewater treatment, numerous non-magnetic semiconductor NPs are acting as photocatalysts [53]. TiO_2 NPs (band gap-3.0-3.2 eV) absorb UV light of the solar spectrum, which is a drawback in photocatalysis because only 5% of solar light is composed of UV illumination. Therefore, the latest research focuses on doping TiO_2 with TiO_2 NCs to modify the band gap to less than 3eV under visible light [54] (Fig. **6**) Modifying the band gap can achieve modification in the texture of the particle surface, and it becomes sensitive to the degradation of organic dyes [55]. The modified nano-catalysts are photo-sensitive in aqueous medium for the treatment of wastewater effluents in water systems [56].

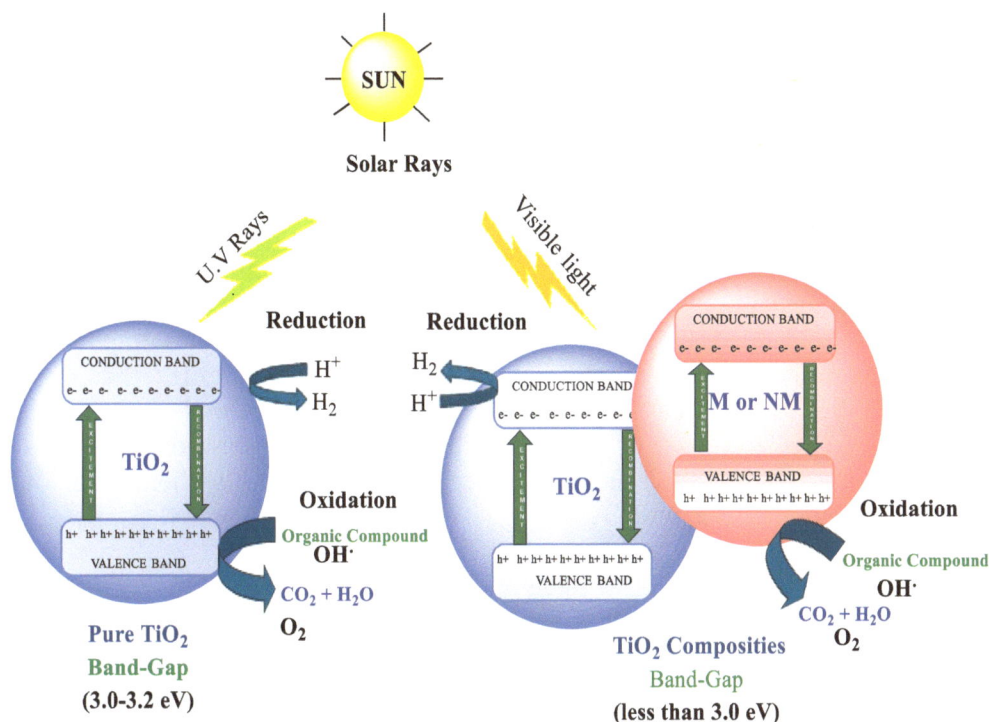

Fig. (6). Photocatalytic properties of pure TiO_2 and its composites.

NANO COMPOSITES (DOPED TIO_2 NPS) PHOTOCATALYSIS

Doping with periodic elements *i.e.* metals (M) and non-metals (N-M) with TiO_2 creates a new energy gap between TiO_2 particles. In TiO_2 particles doped with elements/compounds, a significant band gap is observed as compared to the pure TiO_2 particles. Some metallic elements like vanadium (V), iron (Fe), rhodium (Rh), palladium (Pd), and silver (Ag) as well as some non-metals like carbon (C), fluorine (F), iodine (I), sulfur (S) and nitrogen (N_2) have been doped successfully with bulk TiO_2 particles under visible rays. Such composites enhance TiO_2 NPs' photocatalytic activity. Under visible rays, the above-mentioned composites achieve various chemical degradation techniques in an aqueous medium. Moreover, Ag doping of photocatalysts under visible rays is attributed to a shift towards longer wavelength (red) and it directly affects the band gap of the composites [57]. Rh doping followed by carbon-doped TiO_2 photocatalyst shows microcystin-LR degradation under visible light [58]. In contrast, the Pt-doped nano-TiO_2 photocatalyst exhibited inefficient degradation of platinum chloride bond (homolytic bond cleavage) from chloride radicals under visible light. The synthetic route of hydrothermally V-doped TiO_2 composites catalyst formed from the sol-gel method acquires the potential for the degradation of n-butanol in an aqueous medium [59]. The properties and activity of the N_2-doped TiO_2 catalyst are significantly affected by heat treatment. Decomposition of ammonia occurs at 650°C and the dark green color of Ti^{4+} (anatase) changes to Ti^{3+} (color change) due to recombination nature [60]. Similarly, sulfur (S)-doped TiO_2 was used in visible light photocatalytic degradation of water-toxic pollutants [61, 62]. Some factors that influence the activity of the doped nano-TiO_2 catalyst include charge transfer, crystal phase, dopant concentration, and the absorption of light intensity. Nano-TiO_2 photocatalysts doped in UV light have been reported for their photocatalytic activity for the reduction of nitrate pollutants and bacterial disinfection [63]. Currently, the primary technological obstacles impeding its complete commercialization are related to the catalyst particles that can be recovered after water treatment, especially pollutant photooxidation [64]. In keeping with the "zero" waste strategy in the water and wastewater industries, it has a huge capacity, reasonably priced, environmentally beneficial, and sustainable treatment technology.

The kinetics, mechanism, simulation, and modeling associated with photocatalytic and photo-disinfection wastewater effluent were studied by Olivo and coworkers [65]. For example, ammonia-based rectors absorb textile dyes and bind with TiO_2 NPs/ TiO_2 composites under solar radiation. The photo-reactor performs a cyclic photo-oxidation reaction of the hydrogel by degrading textile dyes to carbon dioxide and water molecules. After completion of the photo-oxidation reaction, the water quality was found to be improved and the photo reactor could be reused for water treatment (Fig. 7).

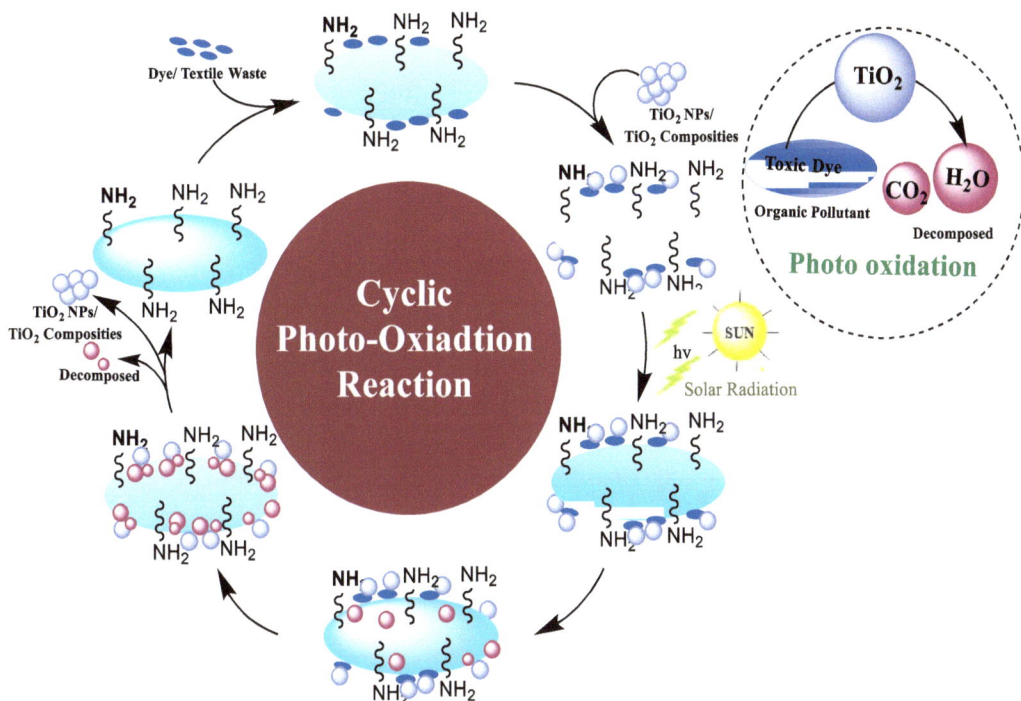

Fig. (7). Illuminating cyclic photo-oxidation reaction of hydrogel used in water treatment plant.

PHOTOCATALYTIC WATER PURIFICATION BY TIO$_2$-METAL COMPOSITES

Metal and Metal Oxide TiO$_2$ Composites

On the nanoscale, many metals exhibit high chemical reactivity, making it difficult to harness their properties effectively. Additionally, some metals may not possess properties that are useful for making composites with TiO$_2$. As a result, it is often not necessary to incorporate such metals into these composites.

Group I and II Metals

Elements of groups I and II in the periodic table and their oxides are unstable in nature and have high oxidative potential. In a mixed K$_2$O/TiO$_2$-ZrO$_2$ system, reactivity (60.59%) and selectivity (97.04%) could be enhanced by the addition of K$_2$O for the dehydrogenation of ethylbenzene to styrene [66]. Furthermore, the adsorption process on MgO/TiO$_2$/Ag composites was characterized through the Freundlich adsorption isotherm at approximately 7pH. The degradation of methyl blue dye by the above-mentioned composites exhibited good photocatalytic activity. Coal fly ash of industrial waste effluents was also used for bio-diesel production with active photocatalysts [67]. Such composites possess important properties (acidity/basicity) for many photo-emerging fields (catalysis and electrochemical cells). This recombination technique studied and monitored the degradation of various organic molecules under UV irradiation with different weight percentages of MgO loaded on TiO_2. MgO clusters of 30% weight are the optimal weight percentage for significant activity [68, 69]. The combination of these findings has inspired new research into these MgO/TiO_2 composite structures as affordable materials with improved catalytic performance.

Group II metals have not been used much, which is likely due to the abundance of research on the more useful and interesting systems with strontium (Sr) and barium (Ba) doped TiO_2. Some researchers have been using CaO with TiO_2 mixed heterogeneous catalysts in ceramic research due to basic oxide recombination. TiO_2@SrOcore@shell nanowires were prepared and their photocatalytic properties were tested for dye degradation [70]. Calcination of TiO_2 nanowires with $Sr(NO_3)_2$ was done at 450°C to prepare TiO_2 dip-coated SrO nanowires. It is worth noting that compared to pure TiO_2 nanowires, the dye degradation rate was improved upon optimizing the shell coating. NCs of Al_2O_3/ ZrO_2- TiO_2 were synthesized for the reduction of NOx [71]. Photo-degradation of herbicides was done through $BaTiO_3/TiO_2$ NCs for water remediation [72]. Adsorption of NO_x (NO^+, NO_2) on BaO (5%)/TiO_2 photocatalyst exhibited good specific photo-adsorption capacity (SPC) under UV illumination [73].

Early Transition Metals

Although oxides of scandium have not been commonly used in composites with TiO_2, other early transition metal oxides have seen considerable use. Some composite of Lanthanide-oxide (Y_2O_3): Yb^{3+}/TiO_2 photocatalysts form calcination at 500°C in ethanol-containing dye [74]. When exposed to sunlight; the optimal mol% of Yb^{3+} is 2 mol% with TiO_2. This is the rate at which methyl orange (MO) completely degrades [75]. Interaction between ZrO_2 with TiO_2 brings photocatalytic degradation of organic compounds through various mechanisms *via* photo-oxidation, dehydrogenation, and NOx reduction. ZrO_2/TiO_2 photocatalyst combination increases the surface area, TiO_2 porosity, volume, and long-range stabilization. The photocatalytic activity was further tested *via* the visible light degradation of Rhodamine B (Rh-B) [76]. Mesoporous visible light $ZrO_2/CeO_2/TiO_2$ composites are thermally stable, having the highest photocatalytic activity as compared to other combinations in the light degradation of Rh-B [77]. Hafnium oxide (HfO_2)–TiO_2 composite has recently been extensively researched as a potential dielectric material due to the high relative permittivity of HfO_2 with TiO_2 [78]. Various metal/metal oxide composites under various light sources with certain band gaps Table **2**.

Table 2. TiO$_2$ with metal/metal oxide composites under solar irradiation.

Sr. No.	TiO$_2$ with Metal/Metal oxide composites	Band Gap (eV)	Light Source Used	Refs.
1.	TiO$_2$ + F-Pd	3.01-2.07eV	HAL-320	[79]
2.	TiO$_2$ + Graphene/Pd	-	400 W Osram lamp	[80]
3.	TiO$_2$ + BiVO$_4$	-	50 W halogen lamp	[81]
4.	TiO$_2$ + Nb	3.15 eV	400 W Fe-doped Metallic halide UV bulb	[82]
5.	TiO$_2$ + ZrO$_2$	3.30-3.20eV	-	[83]
6.	TiO$_2$ + Al$_2$O$_3$	-	8 W mercury lamps	[84]
7.	TiO$_2$ + ZnFe$_2$O$_4$	-	8 W-visible-light lamps	[85]
`8.	TiO$_2$ + SnO$_2$	2.92–2.96eV	400 W lamp	[86]

Middle Transition Metals

Vanadium metal oxides (V$_2$O$_5$) have seen frequent usage when combined with TiO$_2$ as a supportive material for the photo-catalytic reduction of NOx and oxidation of organic pollutants. Niobium metal oxide (Nb$_2$O$_5$)/ TiO$_2$ photocatalyst is the selective photo-oxidative agent used to degrade toxic pollutant dyes [87]. Unfortunately, this incorporation also yields an increase in recombination centers, which is not ideal and diminishes the photocatalytic properties. However, chromium (Cr/CrO$_2$) dopants in TiO$_2$ NPs enhance visible light photoactivity for photocatalytic reduction of numerous NOx compounds. CrO$_2$ is a poor thermodynamic stable metal oxide but with TiO$_2$ its stability is significantly enhanced. Oxides of both molybdenum (Mo) and tungsten (W) doped on TiO$_2$ form metal oxides composite materials. The promise in photocatalysis has drawn a considerable amount of attention to these systems, especially in the case of tungsten (VI) oxide. On the other hand, molybdenum oxides (MoO$_3$)/TiO$_2$ composites' photocatalytic efficiency is comparatively less than WO$_3$/TiO$_2$ composites. Therefore, WO$_3$/TiO$_2$ photo composites are frequently used for water treatment [88].

TiO$_2$ composites containing manganese (Mn)/ manganese oxide (MnO$_2$) and rhenium (Re)/rhenium oxide (ReO$_2$) are frequently used in supporting the photo-catalytic reduction of NOx in wastewater. In some cases, ReO$_2$ performs a water–gas shift reaction (WGSR). MnO$_2$/TiO$_2$ composites are used as photocatalysts as well as electrochemical and supercapacitors. Some pure metals like Pt and Re have been used as supporting co-catalysts on TiO$_2$ and have shown significant enlacement in WGSR. Re/TiO$_2$ facilitates the redox reaction and oxidation of CO to form CO$_2$ and the reduction of H$_2$O to form H$_2$ [89]. The use of

Re/ReO_2 on TiO_2 is limited; however, due to its catalytic enhancement, this is a medium in which some improvement can still be seen with the rational design of the catalyst [90].

Late 3d Transition Metals

The transition metals like Fe, Co, and Ni show the highest magnetic properties with TiO_2 and its oxides [91]. A variety of applications are made possible by the quick separation that can be accomplished when magnetic NPs are employed in conjunction with the external magnetic fields [92]. Iron oxide is a metallic substrate that is used frequently and the most familiar iron oxides are magnetite (Fe_3O_4), hematite (α-Fe_2O_3), and maghemite (γ-Fe_2O_3) [93]. TiO_2 is also one of the best photocatalysts because it is efficient, stable, non-toxic, and abundant. It is widely used in self-cleaning surfaces, water treatment, and air purification because of its different energy gap (3.2 Anatase) and absorption of light in the UV region (3.2-3.8 eV) [94]. Several studies on the deterioration of dye pollutants have been carried out recently, each using a unique combination of semiconductors and magnetic materials. A study using $TiO_2/Fe_3O_4/SiO_2$ as photocatalysts was used to degrade methylene blue (MB) dye [95]. The instability of Fe oxide/TiO_2 composites has been extensively implemented to enable the magnetic recoverability and cyclability of TiO_2. The modified $Fe_3O_4@TiO_2$ composites were then used for photocatalytic degradation of organic pollutants under UV irradiation [96].

The addition of the CoO/ TiO_2 composite resulted in considerable improvement in the photocatalytic degradation of methyl orange (MO) under UV irradiation. Nickel (Ni) oxide supported by TiO_2 has been of interest for decades in a wide range of catalytic reactions. NiO/TiO_2 composites under UV and visible light can degrade methyl orange (MO) [97]. The band gaps of CuO and Cu_2O are $\square 1.4$ and $\square 2.2$ eV, respectively, which makes both materials promising in the conversion of solar radiation [98]. Cu_2O/TiO_2 composites containing copper (I) oxide structure are thermodynamically unstable but the Copper (II) oxide (CuO) composite's structure is comparatively more stable thermodynamically. Stable oxides of copper with TiO_2 act as photocatalysts and for the production of hydrogen [99]. Additionally, the addition of nickel magnetic NPs to SiO_2/TiO_2 composite spheres (Ni–SiO_2/TiO_2) improved the composite's magnetic characteristics and increased the photocatalytic activity for the azo dye degradation [100]. Moreover, they asserted that Ni NPs might trap an e^- from TiO_2 and produce additional superoxide, increasing the magnetic composite's degrading efficacy [101].

Inner Transition Elements

TiO$_2$ photocatalyst composites with rare-earth metal (Ce, Er, La, and Yb) oxides show improvement in photocatalytic activity. These metals are hydrophobic and non-polar; they support photocatalysts in the degradation of polar and nonpolar compounds under UV light. For instance, herbicides (chloro-substituted dimethylurea) were degraded through photocatalyst-rare metal NCs in a reported kinetic study [102]. TiO$_2$ doped with lanthanoid metals (Ce, Eu, and Er) forms composites that act as active and eco-friendly agents for photo-degradation of water contaminants. The main focus of this chapter is to study the metal doping energy gap and photoactive electron-hole pairs that help in electron transfer and optical properties of the composites [103].

PHOTOCATALYTIC WATER PURIFICATION BY TIO$_2$-NON-METAL COMPOSITES

TiO$_2$ carbon-based NCs processed potential and sustainable solution for photocatalysis and water treatment applications (Table **3**). The NCs were designed to address the challenges of wastewater treatment from various industrial effluents and an experimental study was also performed on the photocatalytic degradation of ortho-nitro phenol (O-NP) [104]. In a wastewater treatment study, the nanomaterial TiO$_2$ was blended with biopolymer CA (cellulose acetate) and LDH to form a membrane of the composite. The performance of NC was evaluated through photocatalytic degradation experiments of organic pollutants using UV light. Excellent photosensitizing qualities are possessed by heterogeneous composites of vanadium porphyrin [VO (TPP)] and TiO$_2$, which effectively degrade organic contaminants [105]. The NCs exhibit good photocatalytic activity, credited to the cooperative effects between TiO$_2$ and carbon materials, enabling effective degradation and mineralization of pollutants. Furthermore, the NCs were tested for water treatment application also using a membrane for pure water flux determination. They examined the photodegradation and magnetic properties of the microbead composite after removing methylene blue (MB) dye and rhomanie-B (Rh-B) from the water system [106]. Therefore, α-Fe$_2$O$_3$@C@SiO$_2$/TiO$_2$ magnetic NCs were successfully synthesized and the nature of the surface sites that could be responsible for the major photocatalytic activity was elucidated. Kinetics and thermodynamics studies for photocatalytic degradation of RYD were carried out using the Langmuir-Hinshelwood (L-H) model.

Table 3. TiO$_2$ with the polymer of an organic compound with a band gap under light.

S.N.	TiO$_2$ with Organic Compound	Band Gap (eV)	Light Source Used	Refs.
1.	TiO$_2$ + Polyaniline	1.5 – 2.2 eV	Black 8W UV tube	[107]
2.	TiO$_2$ + Polystyrene	-	-	[108]
3.	TiO$_2$ + Poly-butylene succinate		20 W black light bulb	[109]
4.	TiO$_2$ + Polypyrrole	2.78- 2.95 eV	-	[110]
5.	TiO$_2$ + Poly-o phenylenediamine		1000W Xe lamp	[111]

A composite $Fe_3O_4/TiO_2/CuO$ NPs system was suggested to degrade methylene blue (MB) in an aqueous solution when exposed to sunshine [112]. They have reported that the removal rate targets pollutants through $TiO_2@SiO_2@Fe_3O_4@Ho$ core-shell NPs that were created as a new catalyst to degrade the target pollutants (MO and RhB) [113]. Without Holonium (Ho), the above-mentioned catalyst activity decomposed MB 71.2% after 150 min, and the other pollutant RhB decomposed almost 86.6% after 120 minutes. However, with Ho, the above-mentioned catalyst activity enhanced, MB decomposed 78.4% after 150 min, and the other pollutant RhB decomposed almost 92.1% after 120 min respectively. The synergistic properties of the above-mentioned composites possessed stability in the structure, magnetic nature, hydrophilicity, and excellent photocatalytic activity, which helps in the removal of reactive yellow dye (RYD) [114].

CONCLUSION

In this study, the TiO$_2$ nanomaterial and its composites with appropriate structural features for the possible application of the photodegradation of toxic organic compounds, dyes, and endocrine-disrupting chemicals under solar radiation in wastewater effluent have been discussed. It is clear from the discussion that POPs pose serious hazards to human health. The utilization of photosensitive semiconductors under solar radiation (UV or visible) to degrade persistent organic molecules into simple molecules such as (CO_2 and H_2O) is a significant contribution made by photo-degradation technology. Therefore, nanocomposite TiO$_2$ photocatalysts have the potential to accomplish the remediation of wastewater effluents for a sustainable future.

REFERENCES

[1]   Skocaj M, Filipic M, Petkovic J, Novak S. Titanium dioxide in our everyday life; is it safe? Radiology and Oology 2011; 45(4): 227-47.

[2]   Gázquez MJ, Moreno SMP, Bolívar JP. TiO$_2$ as white pigment and valorization of the waste coming from its production. Titanium Dioxide (Tio$_2$) and Its Applications.. Elsevier 2021; pp. 311-35.
[http://dx.doi.org/10.1016/B978-0-12-819960-2.00011-0]

[3]   Zhang W, Rhim JW. Titanium dioxide (TiO_2) for the manufacture of multifunctional active food packaging films. Food Packag Shelf Life 2022; 31: 100806.
[http://dx.doi.org/10.1016/j.fpsl.2021.100806]

[4]   Sengul AB, Asmatulu E. Toxicity of metal and metal oxide nanoparticles: a review. Environ Chem Lett 2020; 18(5): 1659-83.
[http://dx.doi.org/10.1007/s10311-020-01033-6]

[5]   Jin Y. Interaction between vinyl acetate-ethylene latex stabilized with polyvinyl alcohol and Portland cement 2016.

[6]   Jiang C, Chen Q. Effect of long-term low concentrations of TiO_2 nanoparticles on dewaterability of activated sludge and the relevant mechanism: the role of nanoparticle aging. Environ Sci Pollut Res Int 2022; 29(8): 12188-97.
[http://dx.doi.org/10.1007/s11356-021-16451-4] [PMID: 34562215]

[7]   Wu YN, Wuenschell JK, Fryer R, *et al.* Theoretical and experimental study of temperature effect on electronic and optical properties of TiO_2: comparing rutile and anatase. J Phys Condens Matter 2020; 32(40): 405705.
[PMID: 32544902]

[8]   Chen Q, Wang N, Zhu M, *et al.* TiO_2 nanoparticles cause mitochondrial dysfunction, activate inflammatory responses, and attenuate phagocytosis in macrophages: A proteomic and metabolomic insight. Redox Biol 2018; 15: 266-76.
[http://dx.doi.org/10.1016/j.redox.2017.12.011] [PMID: 29294438]

[9]   Jaeger A, Weiss DG, Jonas L, Kriehuber R. Oxidative stress-induced cytotoxic and genotoxic effects of nano-sized titanium dioxide particles in human HaCaT keratinocytes. Toxicology 2012; 296(1-3): 27-36.
[http://dx.doi.org/10.1016/j.tox.2012.02.016] [PMID: 22449567]

[10]   Dar G I, Saeed M, Wu A. Toxicity of TiO_2 NPs TiO_2 NPs: applications in nanobiotechnology and nanomedicine 2020; (): 67-103.

[11]   Kanan S, Moyet M A, Arthur R B, Patterson H H. Recent advances on TiO_2-based photocatalysts toward the degradation of pesticides and major organic pollutants from water bodies Catalysis Reviews 2019.

[12]   Rueda-Márquez JJ, Palacios-Villarreal C, Manzano M, Blanco E, Ramírez del Solar M, Levchuk I. Photocatalytic degradation of pharmaceutically active compounds (PhACs) in urban wastewater treatment plants effluents under controlled and natural solar irradiation using immobilized TiO_2. Sol Energy 2020; 208: 480-92.
[http://dx.doi.org/10.1016/j.solener.2020.08.028]

[13]   Kumar J, Bansal A. Photocatalysis by NPs of titanium dioxide for drinking water purification: a coeptual and state-of-art review. Materials Sciee Forum. 764: 130-50.
[http://dx.doi.org/10.4028/www.scientific.net/MSF.764.130]

[14]   Xie W, Pakdel E, Liang Y, Liu D, Sun L, Wang X. Natural melanin/TiO_2 hybrids for simultaneous removal of dyes and heavy metal ions under visible light. J Photochem Photobiol Chem 2020; 389: 112292.
[http://dx.doi.org/10.1016/j.jphotochem.2019.112292]

[15]   Lee HJ, Kang DW, Chi J, Lee DH. Degradation kinetics of recalcitrant organic compounds in a decontamination process with UV/H2O2 and UV/H2O2/TiO_2 processes. Korean J Chem Eng 2003; 20(3): 503-8.
[http://dx.doi.org/10.1007/BF02705556]

[16]   Athanasekou CP, Likodimos V, Falaras P. Recent developments of TiO_2 photocatalysis involving advanced oxidation and reduction reactions in water. J Environ Chem Eng 2018; 6(6): 7386-94.
[http://dx.doi.org/10.1016/j.jece.2018.07.026]

[17]   Cosa G, Galletero MS, Fernández L, Márquez F, García H, Scaiano JC. Tuning the photocatalytic activity of titanium dioxide by encapsulation inside zeolites exemplified by the cases of thianthrene photooxygenation and horseradish peroxidase photodeactivation. New J Chem 2002; 26(10): 1448-55.
[http://dx.doi.org/10.1039/B201397E]

[18]   Cheng Q, Yuan YJ, Tang R, *et al.* Rapid hydroxyl radical generation on (001)-facet-exposed ultrathin anatase TiO$_2$ nanosheets for enhaed photocatalytic lignocellulose-to-H2 conversion. ACS Catal 2022; 12(3): 2118-25.
[http://dx.doi.org/10.1021/acscatal.1c05713]

[19]   Kapilashrami M, Zhang Y, Liu YS, Hagfeldt A, Guo J. Probing the optical property and electronic structure of TiO$_2$ nanomaterials for renewable energy applications. Chem Rev 2014; 114(19): 9662-707.
[http://dx.doi.org/10.1021/cr5000893] [PMID: 25137023]

[20]   Pant B, Saud PS, Park M, Park SJ, Kim HY. General one-pot strategy to prepare Ag–TiO$_2$ decorated reduced graphene oxide nanocomposites for chemical and biological disinfectant. J Alloys Compd 2016; 671: 51-9.
[http://dx.doi.org/10.1016/j.jallcom.2016.02.067]

[21]   Singh P, Borthakur A. A review on biodegradation and photocatalytic degradation of organic pollutants: A bibliometric and comparative analysis. J Clean Prod 2018; 196: 1669-80.
[http://dx.doi.org/10.1016/j.jclepro.2018.05.289]

[22]   Gaur N, Dutta D, Singh A, Dubey R, Kamboj DV. Recent advances in the elimination of persistent organic pollutants by photocatalysis. Front Environ Sci 2022; 10: 872514.
[http://dx.doi.org/10.3389/fenvs.2022.872514]

[23]   Fawzi Suleiman Khasawneh O, Palaniandy P. Removal of organic pollutants from water by Fe2O3/TiO$_2$ based photocatalytic degradation: A review. Environmental Technology & Innovation 2021; 21: 101230.
[http://dx.doi.org/10.1016/j.eti.2020.101230]

[24]   Chiou CH, Wu CY, Juang RS. Photocatalytic degradation of phenol and m-nitrophenol using irradiated TiO$_2$ in aqueous solutions. Separ Purif Tech 2008; 62(3): 559-64.
[http://dx.doi.org/10.1016/j.seppur.2008.03.009]

[25]   Yamaguchi S, Fujita S, Nakajima K, *et al.* Support-boosted nickel phosphide nanoalloy catalysis in the selective hydrogenation of maltose to maltitol. ACS Sustain Chem& Eng 2021; 9(18): 6347-54.
[http://dx.doi.org/10.1021/acssuschemeng.1c00447]

[26]   Shibamoto T, Yasuhara A, Katami T. Dioxin formation from waste incineration Reviews of Environmental Contamination and Toxicology: Continuation of Residue Reviews 2007; 1-41.

[27]   Pant B, Park M, Park SJ. Recent advaes in TiO$_2$ films prepared by sol-gel methods for photocatalytic degradation of organic pollutants and antibacterial activities. Coatings 2019; 9(10): 613.
[http://dx.doi.org/10.3390/coatings9100613]

[28]   Sribenja S, Wahanthuek R. 2023.The development of TiO$_2$ based catalyst for poisoning organic molecules removal

[29]   Mousavi SE, Younesi H, Bahramifar N, Tamunaidu P, Karimi-Maleh H. A novel route to the synthesis of α-Fe2O3@C@SiO2/TiO$_2$ nanocomposite from the metal-organic framework as a photocatalyst for water treatment. Chemosphere 2022; 297: 133992.
[http://dx.doi.org/10.1016/j.chemosphere.2022.133992] [PMID: 35247450]

[30]   Arefi-Oskoui S, Khataee A, Jabbarvand Behrouz S, *et al.* Development of MoS2/O-MWCNTs/PES blended membrane for efficient removal of dyes, antibiotic, and protein. Separ Purif Tech 2022; 280: 119822.
[http://dx.doi.org/10.1016/j.seppur.2021.119822]

[31]   Sohrabi S, Keshavarz Moraveji M, Iranshahi D, Karimi A. Microfluidic assisted low-temperature and

speedy synthesis of TiO_2/ZnO/GOx with bio/photo active cites for amoxicillin degradation. Sci Rep 2022; 12(1): 15488.
[http://dx.doi.org/10.1038/s41598-022-19406-y] [PMID: 36109536]

[32] Rajendran S, Mani SS, Nivedhitha TR, *et al.* Facile One-Pot Synthesis of Cu_x O/TiO_2 Photocatalysts by Regulating Cu Oxidation State for Efficient Solar H_2 Production. ACS Appl Energy Mater 2024; 7(1): 104-16.
[http://dx.doi.org/10.1021/acsaem.3c02272]

[33] Ismail M, Akhtar K, Khan MI, *et al.* Pollution, toxicity and carcinogenicity of organic dyes and their catalytic bio-remediation. Curr Pharm Des 2019; 25(34): 3645-63.
[http://dx.doi.org/10.2174/1381612825666191021142026] [PMID: 31656147]

[34] Rojviroon T, Rojviroon O, Sirivithayapakorn S, Angthong S. Application of TiO_2 nanotubes as photocatalysts for decolorization of synthetic dye wastewater. Water Resour Ind 2021; 26: 100163.
[http://dx.doi.org/10.1016/j.wri.2021.100163]

[35] Pacheco-Álvarez M, Fuentes-Ramírez R, Brillas E, Peralta-Hernández JM. Assessing the electrochemical degradation of reactive orange 84 with Ti/IrO_2–SnO_2–Sb_2O_5 anode using electrochemical oxidation, electro-Fenton, and photoelectro-Fenton under UVA irradiation. Chemosphere 2023; 339: 139666.
[http://dx.doi.org/10.1016/j.chemosphere.2023.139666] [PMID: 37532204]

[36] Lian P, Qin A, Liao L, Zhang K. Progress on the nanoscale spherical TiO_2 photocatalysts: Mechanisms, synthesis and degradation applications. Nano Select 2021; 2(3): 447-67.
[http://dx.doi.org/10.1002/nano.202000091]

[37] Al-Hajji LA, Ismail AA, Bumajdad A, *et al.* Photodegradation of powerful five estrogens collected from waste water treatment plant over visible-light-driven Au/TiO_2 photocatalyst. Environmental Technology & Innovation 2021; 24: 101958.
[http://dx.doi.org/10.1016/j.eti.2021.101958]

[38] Wee SY, Aris AZ. Occurrence and public-perceived risk of endocrine disrupting compounds in drinking water. npj Clean Water 2019; 2(1): 4.
[http://dx.doi.org/10.1038/s41545-018-0029-3]

[39] Gao X, Kang S, Xiong R, Chen M. Environment-friendly removal methods for endocrine disrupting chemicals. Sustainability (Basel) 2020; 12(18): 7615.
[http://dx.doi.org/10.3390/su12187615]

[40] Tang C, Huang X, Wang H, Shi H, Zhao G. Mechanism investigation on the enhanced photocatalytic oxidation of nonylphenol on hydrophobic TiO_2 nanotubes. J Hazard Mater 2020; 382: 121017.
[http://dx.doi.org/10.1016/j.jhazmat.2019.121017] [PMID: 31446350]

[41] Pironti C, Ricciardi M, Proto A, Bianco PM, Montano L, Motta O. Endocrine-disrupting compounds: An overview on their occurree in the aquatic environment and human exposure. Water 2021; 13(10): 1347.
[http://dx.doi.org/10.3390/w13101347]

[42] Robinson JA, Ma Q, Staveley JP, Smolenski WJ, Ericson J. Degradation and transformation of 17α-estradiol in water–sediment systems under controlled aerobic and anaerobic conditions. Environ Toxicol Chem 2016; 36(3): 621-9.
[http://dx.doi.org/10.1002/etc.3383] [PMID: 26801177]

[43] Caron E, Sheedy C, Farenhorst A. Development of competitive ELISAs for 17β-estradiol and 17β-estradiol +estrone+estriol using rabbit polyclonal antibodies. J Environ Sci Health B 2010; 45(2): 145-51.
[http://dx.doi.org/10.1080/03601230903472090] [PMID: 20390944]

[44] Voutsa D. Analytical methods for determination of bisphenol A. Plastics in Dentistry and Estrogenicity: A Guide to Safe Practice. Berlin, Heidelberg: Springer Berlin Heidelberg 2013; pp. 51-77.

[45] Durán I, Beiras R. Acute water quality criteria for polycyclic aromatic hydrocarbons, pesticides, plastic additives, and 4-Nonylphenol in seawater. Environ Pollut 2017; 224: 384-91.
[http://dx.doi.org/10.1016/j.envpol.2017.02.018] [PMID: 28222980]

[46] González-Mariño I, Ares L, Montes R, *et al.* Assessing population exposure to phthalate plasticizers in thirteen Spanish cities through the analysis of wastewater. J Hazard Mater 2021; 401: 123272.
[http://dx.doi.org/10.1016/j.jhazmat.2020.123272] [PMID: 32645544]

[47] Zhu Q, Jia J, Zhang K, Zhang H, Liao C, Jiang G. Phthalate esters in indoor dust from several regions, China and their implications for human exposure. Sci Total Environ 2019; 652: 1187-94.
[http://dx.doi.org/10.1016/j.scitotenv.2018.10.326] [PMID: 30586805]

[48] Zhang W, Li Y, Su Y, Mao K, Wang Q. Effect of water composition on TiO_2 photocatalytic removal of endocrine disrupting compounds (EDCs) and estrogenic activity from secondary effluent. J Hazard Mater 2012; 215-216: 252-8.
[http://dx.doi.org/10.1016/j.jhazmat.2012.02.060] [PMID: 22436342]

[49] Gwenzi W, Kanda A, Danha C, Muisa-Zikali N, Chaukura N. Occurree, human health risks, and removal of pharmaceuticals in aqueous systems: Current knowledge and future perspectives Applied water sciee. Fundamentals and Applications 2021; 1: pp. 63-101.

[50] Kumar A, Sharma V, Kumar A, Krishnan V. Nanomaterials for Photocatalytic Decomposition of Endocrine Disruptors in Water Nanostructured Materials for Environmental Applications 2021; 299-320.

[51] Brossault DFF, McCoy TM, Routh AF. Self-assembly of $TiO_2/Fe_3O_4/SiO_2$ microbeads: A green approach to produce magnetic photocatalysts. J Colloid Interface Sci 2021; 584: 779-88.
[http://dx.doi.org/10.1016/j.jcis.2020.10.001] [PMID: 33139018]

[52] He Z, Hong T, Chen J, Song S. A magnetic TiO_2 photocatalyst doped with iodine for organic pollutant degradation. Separ Purif Tech 2012; 96: 50-7.
[http://dx.doi.org/10.1016/j.seppur.2012.05.005]

[53] Donga C, Mishra SB, Abd-El-Aziz AS, Mishra AK. Advaes in graphene-based magnetic and graphene-based/TiO_2 NPs in the removal of heavy metals and organic pollutants from industrial wastewater. J Inorg Organomet Polym Mater 2021; 31(2): 463-80.
[http://dx.doi.org/10.1007/s10904-020-01679-3]

[54] Barone P, Stranges F, Barberio M, Renzelli D, Bonanno A, Xu F. Study of band gap of silver NPs—titanium dioxide NCs. J Chem. 2014;2014:1–6.
[http://dx.doi.org/10.1155/2014/589707]

[55] Manzoor T, Pandith AH. Enhancing the photoresponse by CdSe-Dye-TiO_2-based multijunction systems for efficient dye-sensitized solar cells: A theoretical outlook. J Comput Chem 2019; 40(28): 2444-52.
[http://dx.doi.org/10.1002/jcc.26019] [PMID: 31290168]

[56] Xiong Z, Ma J, Ng WJ, Waite TD, Zhao XS. Silver-modified mesoporous TiO_2 photocatalyst for water purification. Water Res 2011; 45(5): 2095-103.
[http://dx.doi.org/10.1016/j.watres.2010.12.019] [PMID: 21215983]

[57] Janczarek M, Kowalska E. Defective dopant-free TiO_2 as an efficient visible light-active photocatalyst. Catalysts 2021; 11(8): 978.
[http://dx.doi.org/10.3390/catal11080978]

[58] Hu X, Hu X, Tang C, *et al.* Mechanisms underlying degradation pathways of microcystin-LR with doped TiO_2 photocatalysis. Chem Eng J 2017; 330: 355-71.
[http://dx.doi.org/10.1016/j.cej.2017.07.161]

[59] Mogal SI, Mishra M, Gandhi VG, Tayade RJ. Metal doped titanium dioxide: synthesis and effect of metal ions on physico-chemical and photocatalytic properties.

[60]   Wan Z, Cai R, Jiang S, Shao Z. Nitrogen- and TiN-modified Li4Ti5O12: one-step synthesis and electrochemical performance optimization. J Mater Chem 2012; 22(34): 17773-81.
[http://dx.doi.org/10.1039/c2jm33346e]

[61]   Lazar M, Varghese S, Nair S. Photocatalytic water treatment by titanium dioxide: recent updates. Catalysts 2012; 2(4): 572-601.
[http://dx.doi.org/10.3390/catal2040572]

[62]   Rockafellow EM, Stewart LK, Jenks WS. Is sulfur-doped TiO_2 an effective visible light photocatalyst for remediation? Appl Catal B 2009; 91(1-2): 554-62.
[http://dx.doi.org/10.1016/j.apcatb.2009.06.027]

[63]   Wu Z, Ye Y, Cai T, *et al.* Synergistic Effect of Self-Doped TiO_2 Nanotube Arrays and Ultraviolet (UV) on Enhanced Disinfection of Rainwater. Water Air Soil Pollut 2022; 233(10): 416.
[http://dx.doi.org/10.1007/s11270-022-05868-3]

[64]   Chen D, Cheng Y, Zhou N, *et al.* Photocatalytic degradation of organic pollutants using TiO_2-based photocatalysts: A review Journal of Cleaner Production 2020; 268: 121725.

[65]   Olivo-Alanís D, García-González A, Mueses MA, García-Reyes RB. Generalized kinetic model for the photocatalytic degradation processes: Validation for dye wastewater treatment in a visible-LED tubular reactor. Appl Catal B 2022; 317: 121804.
[http://dx.doi.org/10.1016/j.apcatb.2022.121804]

[66]   Burri DR, Choi KM, Han SC, Burri A, Park SE. Selective conversion of ethylbenzene into styrene over K2O/TiO_2-ZrO2 catalysts: Unified effects of K2O and CO2. J Mol Catal Chem 2007; 269(1-2): 58-63.
[http://dx.doi.org/10.1016/j.molcata.2006.12.021]

[67]   Stanković M, Pavlović S, Marinković D, Tišma M, Gabrovska M, Nikolova D. Solid green biodiesel catalysts derived from coal fly ash. Resources, Challenges and Applications 2020; 185.

[68]   Bekena F, Kuo DH. 10 nm sized visible light TiO_2 photocatalyst in the presence of MgO for degradation of methylene blue. Mater Sci Semicond Process 2020; 116: 105152.
[http://dx.doi.org/10.1016/j.mssp.2020.105152]

[69]   Doudin N, Collinge G, Persaud RR, *et al.* Binding and stability of MgO monomers on anatase TiO_2(101). J Chem Phys 2021; 154(20): 204703.
[http://dx.doi.org/10.1063/5.0047521] [PMID: 34241167]

[70]   Wang W, Yang J, Gong Y, Hong H. Tunable synthesis of TiO_2/SrO core/shell nanowire arrays with enhanced photocatalytic activity. Mater Res Bull 2013; 48(1): 21-4.
[http://dx.doi.org/10.1016/j.materresbull.2012.09.067]

[71]   Imagawa H, Tanaka T, Takahashi N, Matsunaga S, Suda A, Shinjoh H. Synthesis and characterization of Al_2O_3 and ZrO_2–TiO_2 nano-composite as a support for NO storage–reduction catalyst. J Catal 2007; 251(2): 315-20.
[http://dx.doi.org/10.1016/j.jcat.2007.08.002]

[72]   Gomathi Devi LN, Krishnamurthy G. Photocatalytic degradation of the herbicide pendimethalin using NPs of BaTiO3/TiO_2 prepared by gel to crystalline conversion method: A kinetic approach. *Journal of Environmental Sciee and Health.* Part B 2008; 43(7): 553-61.
[PMID: 18803109]

[73]   Shelimov BN, Tolkachev NN, Baeva GN, Stakheev AY, Kazanskii VB. Photocatalytic removal of nitrogen oxides from air on TiO_2 modified with bases and platinum. Kinet Catal 2011; 52(4): 518-24.
[http://dx.doi.org/10.1134/S002315841104015X]

[74]   Hao H, Hao S, Hou H, *et al.* A novel label-free photoelectrochemical immunosensor based on CdSe quantum dots sensitized Ho^{3+}/Yb^{3+}-TiO_2 for the detection of Vibrio parahaemolyticus. Methods 2019; 168: 94-101.
[http://dx.doi.org/10.1016/j.ymeth.2019.06.005] [PMID: 31181257]

[75] Li W, Li D, Lin Y, *et al.* Evidee for the active species involved in the photodegradation process of methyl orange on TiO_2. J Phys Chem C 2012; 116(5): 3552-60.
[http://dx.doi.org/10.1021/jp209661d]

[76] Yin M, Li Z, Kou J, Zou Z. Mechanism investigation of visible light-induced degradation in a heterogeneous TiO_2/eosin Y/rhodamine B system. Environ Sci Technol 2009; 43(21): 8361-6.
[http://dx.doi.org/10.1021/es902011h] [PMID: 19924970]

[77] Li M, Zhang S, Lv L, Wang M, Zhang W, Pan B. A thermally stable mesoporous $ZrO2–CeO2–TiO_2$ visible light photocatalyst. Chem Eng J 2013; 229: 118-25.
[http://dx.doi.org/10.1016/j.cej.2013.05.106]

[78] Tao Q, Jursich G, Takoudis C. Structural and dielectric characterization of atomic layer deposited $HfO2$ and TiO_2 as promising gate oxides. In 2010 IEEE/SEMI Advanced Semiconductor Manufacturing Conference (ASMC). 2010; pp. 17-22.

[79] Jahdi M, Mishra SB, Nxumalo EN, Mhlanga SD, Mishra AK. Smart pathways for the photocatalytic degradation of sulfamethoxazole drug using F-Pd co-doped TiO_2 nanocomposites. Appl Catal B 2020; 267: 118716.
[http://dx.doi.org/10.1016/j.apcatb.2020.118716]

[80] Safajou H, Khojasteh H, Salavati-Niasari M, Mortazavi-Derazkola S. Enhaed photocatalytic degradation of dyes over graphene/Pd/TiO_2 NCs: TiO_2 nanowires versus TiO_2 NPs. J Colloid Interface Sci 2017; 498: 423-32.
[http://dx.doi.org/10.1016/j.jcis.2017.03.078] [PMID: 28349885]

[81] Wetchakun N, Chainet S, Phanichphant S, Wetchakun K. Efficient photocatalytic degradation of methylene blue over BiVO4/TiO_2 nanocomposites. Ceram Int 2015; 41(4): 5999-6004.
[http://dx.doi.org/10.1016/j.ceramint.2015.01.040]

[82] Almulhem N, Awada C, Shaalan NM. Photocatalytic degradation of phenol red in water on Nb (x)/TiO_2 NCs. Crystals (Basel) 2022; 12(7): 911.
[http://dx.doi.org/10.3390/cryst12070911]

[83] Ismail AA, Abdelfattah I, Atitar MF, *et al.* Photocatalytic degradation of imazapyr using mesoporous Al2O3–TiO_2 nanocomposites. Separ Purif Tech 2015; 145: 147-53.
[http://dx.doi.org/10.1016/j.seppur.2015.03.012]

[84] Karunakaran C, Magesan P, Gomathisankar P, Vinayagamoorthy P. Photocatalytic degradation of dyes by Al2O3-TiO_2 and ZrO2-TiO_2 NCs.

[85] Nguyen TB, Huang CP, Doong R. Photocatalytic degradation of bisphenol A over a $ZnFe_2O_4$/TiO_2 nanocomposite under visible light. Sci Total Environ 2019; 646: 745-56.
[http://dx.doi.org/10.1016/j.scitotenv.2018.07.352] [PMID: 30064101]

[86] Mohammad A, Khan ME, Cho MH, Yoon T. Fabrication of binary SnO2/TiO_2 nanocomposites under a sonication-assisted approach: Tuning of band-gap and water depollution applications under visible light irradiation. Ceram Int 2021; 47(11): 15073-81.
[http://dx.doi.org/10.1016/j.ceramint.2021.02.065]

[87] Orozco-Hernández G, Durán PG, Aperador W. Tribocorrosion evaluation of Nb2O5, TiO_2, and Nb2O5+ TiO_2 coatings for medical applications. Lubricants 2021; 9(5): 49.
[http://dx.doi.org/10.3390/lubricants9050049]

[88] Bai S, Liu H, Sun J, *et al.* Improvement of TiO_2 photocatalytic properties under visible light by WO3/TiO_2 and MoO3/TiO_2 composites. Appl Surf Sci 2015; 338: 61-8.
[http://dx.doi.org/10.1016/j.apsusc.2015.02.103]

[89] Al-Ahmed A. Metal doped TiO_2 photocatalysts for CO2 photoreduction.

[90] Secordel X, Berrier E, Capron M, *et al.* TiO_2-supported rhenium oxide catalysts for methanol oxidation: Effect of support texture on the structure and reactivity evidenced by an operando Raman

study. Catal Today 2010; 155(3-4): 177-83.
[http://dx.doi.org/10.1016/j.cattod.2010.01.003]

[91] Malinga NN, Jarvis ALL. Synthesis, characterization and magnetic properties of Ni, Co and FeCo nanoparticles on reduced graphene oxide for removal of Cr(VI). J Nanostructure Chem 2020; 10(1): 55-68.
[http://dx.doi.org/10.1007/s40097-019-00328-7]

[92] Linley S, Leshuk T, Gu FX. Magnetically separable water treatment technologies and their role in future advaed water treatment: a patent review. Clean (Weinh) 2013; 41(12): 1152-6.
[http://dx.doi.org/10.1002/clen.201100261]

[93] Leonel AG, Mansur AA, Mansur HS. Magnetic Iron Oxide NPs and Nanohybrids for Advaed Water Treatment Technology. Handbook of Magnetic Hybrid Nanoalloys and Their NCs. Cham: Springer International Publishing 2022; pp. 1-24.

[94] Roy DRD, Yadav A, Singh K, Pandey G. Green Synthesized TiO_2-SnO2 NC for the Photocatalytic Degradation of Methylene Blue Dye. Jordan Journal of Physics 2023; 16(2): 181-94.

[95] Tarcea CI, Pantilimon CM, Matei E, Predescu AM, Berbecaru AC, Rapa M, *et al.* Photocatalytic degradation of methylene blue dye using TiO_2 and Fe3O4/SiO2/TiO_2 as photocatalysts. In IOP Conferee Series: Materials Sciee and Engineering. IOP Publishing. 2020; 877: p. (1)012008.

[96] Gopalan Sibi M, Verma D, Kim J. Magnetic core–shell nanocatalysts: promising versatile catalysts for organic and photocatalytic reactions. Catal Rev, Sci Eng 2020; 62(2): 163-311.
[http://dx.doi.org/10.1080/01614940.2019.1659555]

[97] Chu L, Li M, Cui P, Jiang Y, Wan Z, Dou S. The study of NiO/TiO_2 photocatalytic activity for degradation of methylene orange. Energy and Environment Focus 2014; 3(4): 371-4.
[http://dx.doi.org/10.1166/eef.2014.1125]

[98] Kumar S, Bhawna , Gupta A, *et al.* New insights into Cu/Cu2O/CuO NC heterojution facilitating photocatalytic generation of green fuel and detoxification of organic pollutants. J Phys Chem C 2023; 127(15): 7095-106.
[http://dx.doi.org/10.1021/acs.jpcc.2c08094]

[99] Jung M, Hart JN, Scott J, Ng YH, Jiang Y, Amal R. Exploring Cu oxidation state on TiO_2 and its transformation during photocatalytic hydrogen evolution. Appl Catal A Gen 2016; 521: 190-201.
[http://dx.doi.org/10.1016/j.apcata.2015.11.013]

[100] Mahesh KPO, Kuo DH. Synthesis of Ni nanoparticles decorated SiO2/TiO_2 magnetic spheres for enhanced photocatalytic activity towards the degradation of azo dye. Appl Surf Sci 2015; 357: 433-8.
[http://dx.doi.org/10.1016/j.apsusc.2015.08.264]

[101] Chen CC, Jaihindh D, Hu SH, Fu YP. Magnetic recyclable photocatalysts of Ni-Cu-Zn ferrite@SiO2@TiO_2@Ag and their photocatalytic activities. J Photochem Photobiol Chem 2017; 334: 74-85.
[http://dx.doi.org/10.1016/j.jphotochem.2016.11.005]

[102] Rusek J, Baudys M, Paušová Š, Paz Y, Krýsa J. Composite TiO_2-SiO2-REOs photocatalysts for water treatment: Degradation kinetics of monuron and its intermediates. J Photochem Photobiol Chem 2023; 445: 115025.
[http://dx.doi.org/10.1016/j.jphotochem.2023.115025]

[103] Cerrato E, Gaggero E, Calza P, Paganini MC. The role of Cerium, Europium and Erbium doped TiO_2 photocatalysts in water treatment: A mini-review. Chemical Engineering Journal Advances 2022; 10: 100268.
[http://dx.doi.org/10.1016/j.ceja.2022.100268]

[104] Afzal S. 2019.

[105] Yadav V, Verma P, Negi H, Singh RK, Saini VK. Efficient degradation of 4-nitrophenol using VO(TPP) impregnated TiO_2 photocatalyst: Insight into kinetics and mechanism. J Mater Res 2023;

38(1): 237-47.
[http://dx.doi.org/10.1557/s43578-022-00856-z]

[106] Ansari MAH, Khan ME, Mohammad A, Baig MT, Chaudary A, Tauqeer M. Application of nanocomposites in wastewater treatment. In: Khan ME, Aslam J, Verma C, editors. Nanocomposites—Advanced materials for energy and environmental aspects. Woodhead Publishing Series in Composites Science and Engineering. Cambridge: Woodhead Publishing; 2023. p. 297-319.
[http://dx.doi.org/10.1016/B978-0-323-99704-1.00025-4]

[107] Rahman KH, Kar AK. Effect of band gap variation and sensitization process of polyaniline (PANI)-TiO$_2$ p-n heterojunction photocatalysts on the enhancement of photocatalytic degradation of toxic methylene blue with UV irradiation. J Environ Chem Eng 2020; 8(5): 104181.
[http://dx.doi.org/10.1016/j.jece.2020.104181]

[108] S D, T SJ, C R. Solid-phase photodegradation of polystyrene by nano TiO$_2$ under ultraviolet radiation. Environ Nanotechnol Monit Manag 2019; 12: 100229.
[http://dx.doi.org/10.1016/j.enmm.2019.100229]

[109] Miyauchi M, Li Y, Shimizu H. Enhanced degradation in nanocomposites of TiO$_2$ and biodegradable polymer. Environ Sci Technol 2008; 42(12): 4551-4.
[http://dx.doi.org/10.1021/es800097n] [PMID: 18605585]

[110] Wang D, Wang Y, Li X, Luo Q, An J, Yue J. Sunlight photocatalytic activity of polypyrrole–TiO$_2$ nanocomposites prepared by '*in situ*' method. Catal Commun 2008; 9(6): 1162-6.
[http://dx.doi.org/10.1016/j.catcom.2007.10.027]

[111] Yang C, Zhang M, Dong W, Cui G, Ren Z, Wang W. Highly efficient photocatalytic degradation of methylene blue by PoPD/TiO$_2$ nanocomposite. PLoS One 2017; 12(3): e0174104.
[http://dx.doi.org/10.1371/journal.pone.0174104] [PMID: 28329007]

[112] Kianfar AH, Arayesh MA. Synthesis, characterization and investigation of photocatalytic and catalytic applications of Fe3O4/TiO$_2$/CuO nanoparticles for degradation of MB and reduction of nitrophenols. J Environ Chem Eng 2020; 8(1): 103640.
[http://dx.doi.org/10.1016/j.jece.2019.103640]

[113] Manikandan A, Thanrasu K, Dinesh A, *et al.* Photocatalytic applications of magnetic hybrid nanoalloys and their NCs. Handbook of Magnetic Hybrid Nanoalloys and their NCs. Cham: Springer International Publishing 2022; pp. 1193-224.
[http://dx.doi.org/10.1007/978-3-030-90948-2_59]

[114] Fatima SK, Ceesay AS, Khan MS, *et al.* Visible Light-Induced Reactive Yellow 145 Discoloration: Structural and Photocatalytic Studies of Graphene Quantum Dot-Incorporated TiO $_2$. ACS Omega 2023; 8(3): 3007-16.
[http://dx.doi.org/10.1021/acsomega.2c05805] [PMID: 36713734]

CHAPTER 13

A Comparative Study of Different Types of Nanomaterial-Based Carbon-di-oxide Sensors and Their Diverse Applications

Ratindra Gautam[1,*], Shivani Chaudhary[2], Karnica Srivastava[3], C. K. Kaithwas[1], U. B. Singh[2] and A. K. Srivastava[1]

[1] *Department of Applied Science, Institute of Engineering and Technology, Dr. Rammanohar Lohia Avadh University, Ayodhya 224000, Uttar Pradesh, India*

[2] *Department of Physics, Deen Dayal Upadhyay Gorkhpur University, Gorakhpur, Uttar Pradesh, India*

[3] *Department of Physics, Isabella Thoburn College, Lucknow 226007, Uttar Pradesh, India*

Abstract: Carbon dioxide is one of the greenhouse gases created by human activities like burning fossil fuels for power generation, oil refining, production of natural gas for transportation, and many other such processes. It is a colorless, relatively inert, and highly oxidizing gas; its concentration has a big effect on the world's climate resulting in sea level rise, global warming, the greenhouse effect, and the possible development of subtropical deserts. Thus, both qualitative and quantitative CO_2 detection is crucial for many industries, including food and beverage packaging, air quality, biotechnology, health and medical research, marine and environmental science, and industrial monitoring. This chapter majorly focuses on the different types of CO_2 nano-sensors and their comparison based on their performances.

Keywords: Carbon dioxide sensors, Detection, Environment, Industrial, Monitoring, Medical research.

INTRODUCTION

In the world of technology and scientific instruments, nano-sensors are crucial components that allow the transformation of physical events into data that can be measured and understood [1]. These devices are engineered to identify particular stimuli from the surroundings and convert them into electrical signals or other measurable outputs. The fundamental principle behind sensors is their capacity to

[*] **Corresponding author Ratindra Gautam:** Department of Applied Science, Institute of Engineering and Technology, Dr. Rammanohar Lohia Avadh University, Ayodhya 224000, Uttar Pradesh, India; E-mail: ratindra03@gmail.com

Neha Agarwal, Vijendra Singh Solanki, Neetu Singh & Maulin P. Shah (Eds.)

connect the physical realm with digital data, enabling a more profound comprehension and control of our environment [2, 3].

Fundamentally, a sensor functions by converting physical alterations, like changes in temperature, pressure, light, or chemical composition, into signals that are ready for processing and examination [2]. Sensors are essential due to the restrictions of human senses and the requirement for exactness, precision, and immediate monitoring in different fields. Our human senses have inherent limitations in their range and precision, which can make it difficult to detect small changes or differences in our surroundings [4]. With their advanced mechanisms and technologies, nano-sensors can effectively surpass these limitations by providing a dependable and accurate way to collect data. Essentially, sensors are crucial for improving our capacity to observe, measure, and control physical phenomena. Whether utilized in industrial automation, environmental monitoring, healthcare, or consumer electronics, sensors play a crucial role in advancing technology. They provide a constant flow of data that guides decision-making, enhances system dependability, and encourages creativity. There are several important categories to consider when it comes to the necessity of nano-sensors [5, 6]:

- With nano-sensors, measurements can be made with a level of precision and accuracy that exceeds what humans can achieve, allowing for detailed parameter readings.
- When it comes to automated systems, nano-sensors play a crucial role by detecting changes in the environment and adjusting system parameters to enable the autonomous operation of machinery and processes [7, 8].
- When it comes to safety and monitoring, nano-sensors play a vital role in various applications. They are essential for detecting toxic gases, monitoring structural integrity, and overseeing critical parameters in healthcare settings.
- Efficiency is enhanced through the continuous monitoring and optimization of processes, leading to reduced resource consumption in various sectors such as industrial, agricultural, and energy.
- At the cutting edge of technological advancement, nano-sensors are propelling the progress of smart devices, Internet of Things (also known as IoT) applications, and advanced scientific research [9].

Within the constantly changing realm of environmental monitoring and safety, the identification and quantification of carbon dioxide (CO_2) are crucial. Carbon dioxide, a transparent and scentless gas, is a natural element of Earth's atmosphere [10]. The increasing CO_2 levels, resulting from human activities like industrial production, transportation, and energy generation, highlight the critical importance of thorough monitoring and mitigation strategies [11]. Confronting the concerning CO_2 emission ratios requires a deep understanding of the complex relationship between CO_2 production, its negative impact on the environment, and the crucial role of carbon dioxide nano-sensors in tackling these issues [7].The continuous increase in CO_2 emissions from human activities has disturbed the fragile equilibrium of our atmosphere, leading to global warming and climate change.

Nevertheless, the rising levels of human-made activities have resulted in a significant increase in CO_2 emissions, sparking worries about its effects on climate change and indoor air quality. Researchers are working diligently to understand and address these impacts, making advanced carbon dioxide nano-sensors crucial for progress.

Exploring the complex realm of carbon dioxide sensing technologies, this chapter investigates a wide range of nano-sensors created to measure and analyze CO_2 levels in different scenarios. Whether in industrial settings, commercial buildings, or environmental monitoring stations, the demand for precise and dependable CO_2 nano-sensors is crucial. CO_2 nano-sensors have wide applications in various fields as shown in Fig. (**1**).

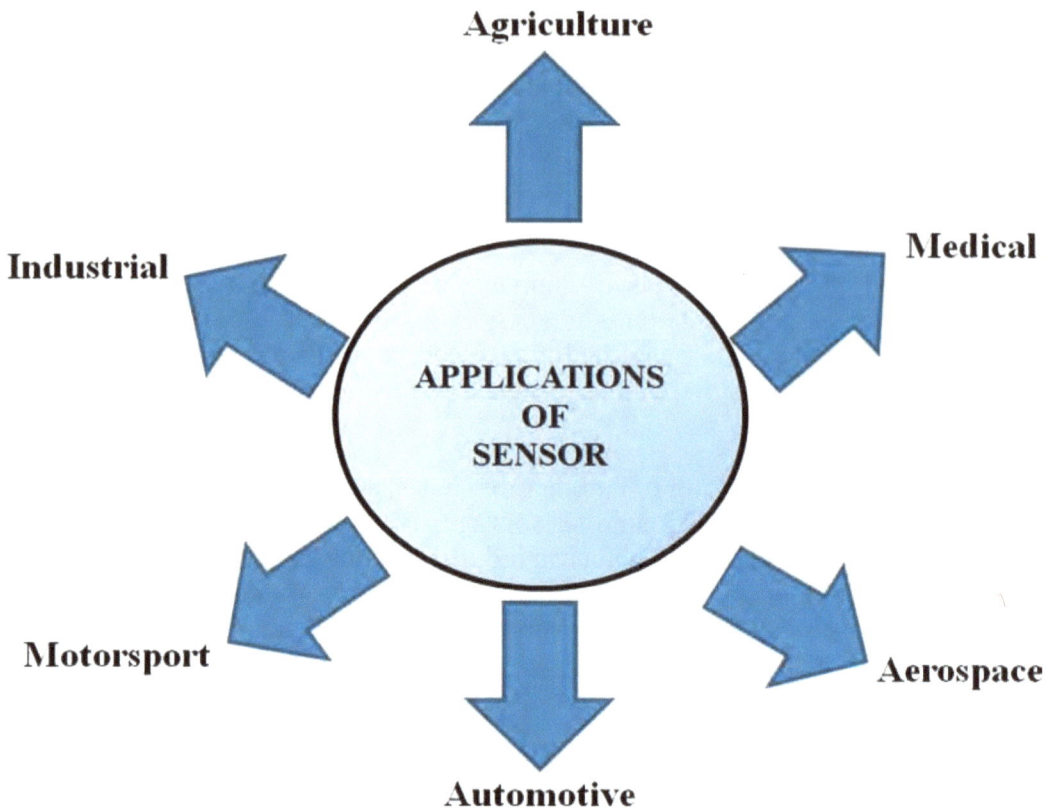

Fig. (1). Major applications of sensors.

As we delve into this chapter, we will explore the complexities of various carbon dioxide nano-sensors, illuminating their operational principles, advantages, drawbacks, and the exciting advancements propelling this field.

From NDIR nano-sensors to chemical and electrochemical nano-sensors, each type has its own unique benefits and difficulties. This investigation seeks to provide researchers, engineers, and enthusiasts with a thorough grasp of the intricacies involved in carbon dioxide sensing technologies. Exploring the details of these nano-sensors is our way of adding to the overall understanding that will help create better, precise, and eco-friendly methods for tracking and reducing the effects of carbon dioxide on our surroundings and health. Exploring the world of carbon dioxide nano-sensors involves delving into the intricate mechanisms that drive their functionality and the advancements leading us towards a more knowledgeable and eco-friendly future.

THE ANATOMY OF CO₂ GAS NANO-SENSORS

Carbon dioxide nano-sensors are made up of four key components: the sampling area, transducer, signal processing electronics, and a signal display unit [12] (Fig. **2).** The chapter delves into the specific roles of each component in the overall functionality of a CO_2 sensor. The area where sampling takes place is where surface chemistry occurs and triggers the sensor's reaction to CO_2. The transducer plays a vital role in converting the chemical recognition event into a signal that can be measured [10, 12]. Signal processing electronics and the display unit work together to convert the transduced signal into useful data.

Fig. (2). CO₂ Gas Sensor.

Sensing Techniques for CO₂ Measurements

Potentiometric nano-sensors, such as the Severinghaus type, which depend on changes in the pH of a liquid solution, are one of the CO_2 detecting systems discussed in the theory [13]. A plastic membrane lets CO_2 through but prevents water and electrolytes from passing through these nano-sensors, which include a glass electrode with a bicarbonate solution in water. The fact that the sensor monitors CO_2 in its ionic state rather than directly is its primary drawback. There are a few more negatives, such as the high cost of upkeep and the possibility of interaction with volatile acids or basic gasses [12, 13].

Gas chromatography (GC) is a technique that uses a column to selectively adsorb and elute gas molecules [14]. Mass spectrometry (MS) separates the sample's molecules into their component parts and sorts them by mass and charge. While gas chromatographs (GCs) are less expensive but take more time to examine materials, mass spectrometers (MSs) provide fast results but are expensive [14].

In response to these difficulties, research is underway to create field-portable miniaturized GC-MS systems that overcome obstacles in instrumentation, miniaturization, deployment, and sampling. For precise chemical identification, a purity level of 75% is required [12].

APPLICATIONS OF CARBON DIOXIDE NANO-SENSORS IN DIFFERENT AREAS

Carbon Dioxide Sensing—Biomedical Applications to Human Subjects: Illuminating Respiratory Insights

Advancements in measuring carbon dioxide (CO_2) levels in the human body have transformed it from a specialized area of medical research to a crucial element in numerous biomedical applications. The technology for sensing carbon dioxide has been incredibly valuable in comprehending respiratory dynamics, assisting in medical diagnostics, and improving patient care [15]. Exploring the significant impact of carbon dioxide sensing in the field of biomedical applications, this article highlights its crucial role in monitoring and improving respiratory health in humans [15, 16].

- Carbon dioxide detection is widely used in respiratory monitoring, enabling healthcare providers to monitor ventilation changes and evaluate gas exchange efficiency. For conditions like chronic obstructive pulmonary disease (COPD), asthma, and sleep apnea, ongoing monitoring of CO_2 levels provides important information for diagnosis and treatment strategies. Enhancing diagnostic precision and enabling personalized medical interventions by correlating CO_2 levels with respiratory patterns [17].
- One significant use of carbon dioxide sensing is in the biomedical field of anesthesia. Capnography is a crucial method for monitoring patients' safety during surgery by measuring end-tidal carbon dioxide ($EtCO_2$) levels. Through ongoing monitoring of *etc*O_2, anesthesiologists can quickly identify alterations in ventilation, endotracheal tube positioning, and overall respiratory condition, ultimately averting complications and promoting favorable patient results [18].
- Carbon dioxide sensing is extremely valuable in critical care settings and emergency medicine for managing patients experiencing respiratory distress or failure. By continuously monitoring CO_2 levels, healthcare providers can evaluate the impact of ventilation support, quickly adjust ventilator settings, and detect early indications of respiratory compromise. Having access to immediate feedback is essential for making timely interventions, particularly in critical scenarios involving compromised respiratory function [19, 20].
- Thanks to technological progress, portable and non-invasive carbon dioxide sensing devices have been created, expanding their use beyond clinical

environments. Individuals with chronic respiratory conditions can now take advantage of at-home monitoring, offering healthcare professionals important data to enhance treatment plans and enhance patient outcomes. These devices provide a way to monitor respiratory health at home, bridging the gap between hospital visits.

The field of carbon dioxide sensing in biomedical applications is constantly evolving due to ongoing innovations. Advancements in miniaturized nano-sensors, wearable device integration, and data analytics are creating exciting opportunities for personalized healthcare. Monitoring and analyzing CO_2 levels in real-time has the potential to enhance patient outcomes, improve telemedicine capabilities, and promote a proactive approach to respiratory health management [15].

Carbon Dioxide (CO_2) Sensors for the Agri-food Industry: Enhancing Quality and Sustainability

Accuracy and efficiency are of utmost importance in the agri-food business, which is situated at the crossroads of agricultural techniques and technology innovation. Within this framework, the incorporation of CO_2 nano-sensors has become a game-changer, revolutionizing the way different processes in the agri-food supply chain are monitored, managed, and optimized. The significance of carbon dioxide (CO_2) nano-sensors in the food and agriculture sector, including an examination of their uses, advantages, and critical function in guaranteeing product quality and longevity as [12]-

- Controlled environment storage is a key use case for CO_2 nano-sensors in the agricultural and food processing industries. During transit and storage, perishable foods like fruits and vegetables may spoil. Adjusting controlled atmospheres to prolong the shelf-life of food is made possible by CO_2 nano-sensors, which precisely monitor gas concentrations inside storage settings. The industry can guarantee that customers obtain fresh and high-quality goods by maintaining ideal CO_2 levels, which mitigates the danger of early ripening and reduces spoiling [21].
- The field of precision agriculture relies heavily on CO_2 nano-sensors, especially for greenhouse management. The levels of carbon dioxide (CO_2) are an important variable in the equation used to optimize the circumstances for crop development in controlled greenhouses. In order to improve crop yields and make photosynthesis easier, nano-sensors are used to monitor and regulate CO_2 levels. By reducing the need for fertilizers and other inputs, this level of accuracy helps to enhance output while also contributing to resource efficiency [22, 23].

- Strict quality standards must be maintained in the food processing industry. Processing facilities use CO_2 nano-sensors to keep an eye on the air quality and make sure it is safe for food to be processed. In the packaging industry, for instance, too much carbon dioxide gas might change the taste or reduce the product's freshness. Using CO_2 nano-sensors for real-time monitoring enables quick modifications, which reduces the chance of quality deterioration and guarantees that final goods fulfill both consumer and regulatory standards [24].
- Economic and environmental consequences stem from post-harvest losses, which are a major worry for the agri-food sector. The use of CO_2 nano-sensors allows for proactive monitoring and control, which in turn helps to reduce post-harvest losses. Distributors and farmers may optimise resource use and reduce losses by making well-informed choices about storage conditions, transportation logistics, and distribution schedules based on accurate measurements of goods' respiration rates [7, 23].
- At a time when sustainability is at the forefront of people's minds, CO_2 nano-sensors are crucial for the advancement of environmentally friendly methods in the food and agriculture sector. These nano-sensors help with resource efficiency, waste reduction, and sustainable agriculture by optimizing controlled atmospheres, greenhouse conditions, and storage settings. The capacity to adjust environmental factors in response to real-time data permits the prudent use of water, electricity, and fertilizers, in line with overarching sustainability objectives.

With the rapid advancement of technology, there are exciting possibilities for the future of CO_2 nano-sensors in the agri-food sector. To make CO_2 monitoring systems even more accurate and responsive, they may be integrated with IoT platforms, data analytics, and AI. These innovations allow agri-food management to become more data-driven and networked, which strengthens its ability to withstand changing market and environmental conditions.

Carbon Dioxide Sensor Module Based on NDIR Technology: Precision in Gas Sensing for a Sustainable Future

Non-dispersive infrared (NDIR) technology has emerged as a cornerstone in the search for precise and reliable gas detection, leading to a revolution in carbon dioxide (CO_2) sensing technology in recent years. A giant step forward in the area, the "Carbon Dioxide Sensor Module Based on NDIR Technology" has the potential to revolutionize many different sectors with its exceptional accuracy, adaptability, and use [25]. The ideas underlying near-infrared (NDIR) technology, its benefits, and the revolutionary potential it has for a sustainable future. The underlying premise of near-field infrared (NDIR) technology is that various gases absorb infrared light at different wavelengths. When it comes to carbon dioxide, the gas has a unique infrared absorption band at around 4.26 micrometers [26]. Using this basic premise, the NDIR Technology-based Carbon Dioxide Sensor Module makes use of an infrared light source, an infrared detector, and a gas sample container [25]. The sensor gives a proportionate and precise reading of the CO_2 content in the room by measuring the quantity of infrared light absorbed by CO_2 molecules. There is a plethora of benefits to using NDIR technology in CO_2 detecting modules. Its remarkable precision is the first and most important quality. When pinpoint accuracy is of the utmost importance, such as in environmental monitoring, industrial operations, or indoor air quality evaluation, NDIR nano-sensors are the way to go because of their ability to detect CO_2 concentrations. In addition, NDIR nano-sensors have a fast reaction time, which allows for continuous monitoring and easy adaptation to new circumstances. A wide range of sectors may benefit from the Carbon Dioxide Sensor Module that is based on NDIR Technology due to its adaptability [26, 27]. These nano-sensors are useful for measuring emissions of greenhouse gases and for studying the effects of climate change in environmental monitoring. To ensure occupational safety and monitor operations where CO_2 levels need to be tightly managed, NDIR-based sensors are essential in industrial settings. The module is also useful for assessing the interior air quality, which may provide light on the effectiveness of ventilation systems and any hazards to human health. In the fight for environmental friendliness and reduced energy consumption, NDIR-based CO_2 sensors play a crucial role. By adjusting ventilation rates based on real-time CO_2 concentrations, the nano-sensors contribute to smart HVAC systems in buildings. Eliminating the need for superfluous ventilation not only creates a healthy interior atmosphere but also helps to minimize energy usage [27]. The module's accuracy enables focused interventions, which is in line with the worldwide movement towards sustainable energy management methods. An NDIR-based carbon dioxide sensor module is more than just cutting-edge innovation; it represents the gas sensing industry's bright future. New opportunities will emerge as a result of combining NDIR nano-sensors with data analytics, Internet of Things (IoT) platforms, and other emerging technologies [9]. More responsive and integrated CO_2 level management is possible in a number of settings thanks to real-time monitoring, predictive analytics, and smooth connection with smart systems.

Costal Network: A Low-Cost Carbon Dioxide Monitoring System for Coastal and Estuarine Sensor Networks

It is essential for ecosystems and populations that coastal and estuarine areas remain healthy. These fragile ecosystems are very sensitive to changes in atmospheric carbon dioxide (CO_2) levels, which have far-reaching effects on aquatic life, water chemistry, and environmental health generally [28]. The Costal Network's accessibility and low price are two of its main benefits. The high price and complicated equipment of traditional CO_2 monitoring systems in maritime conditions prevent their broad implementation. The Costal Network provides an affordable answer to these problems without lowering standards for data dependability or quality [28]. With the widespread availability of monitoring tools, more people than ever before. Researchers and environmentalists now have new tools at their disposal thanks to the proliferation of affordable carbon dioxide monitoring equipment, which will allow them to better study and protect these crucial ecosystems. It is essential to monitor the levels of CO_2 in these environments for various important purposes. First and foremost, carbon dioxide impacts the pH of seawater, which has consequences for the well-being of marine creatures like corals, shellfish, and plankton [29]. Moreover, fluctuations in CO_2 levels can influence carbon sequestration mechanisms and lead to ocean acidification, which can endanger biodiversity and ecosystem stability. Through the observation of CO_2 levels, scientists can gather valuable information about the condition of coastal ecosystems and monitor shifts in the environment as time progresses. With the Costal Network's real-time data, adaptive management plans may be informed and resource managers and lawmakers can make evidence-based decisions. The potential for the Costal Network to achieve beneficial environmental results and promote sustainability in coastal and estuarine areas globally is great, thanks to its emphasis on cooperation and knowledge exchange. In our efforts to protect the Earth's natural resources, projects such as the Costal Network stand out as symbols of creativity and optimism, motivating united efforts to protect and maintain our coastal legacy.

Revealing the Subsurface Dynamics with an Integrated Probe System for Soil Carbon Dioxide Concentration Measurements

Delving into the complex interactions of carbon dioxide (CO_2) within the soil is crucial for deciphering the intricacies of terrestrial ecosystems. Soil is more than just a static material; it's a vibrant ecosystem with complex activities such as microbial processes, root respiration, and decomposition of organic matter [30]. Soil CO_2 concentrations are a crucial indicator of the fundamental processes at play. Tracking fluctuations in soil CO_2 levels offers valuable information on nutrient cycling, microbial activity, and the general well-being of the soil ecosystem [30]. This information is crucial for sustainable agriculture, ecosystem management, and our overall comprehension of the global carbon cycle. This system introduces a new method for exploring below the surface. Conventional soil sampling techniques can be labor-intensive, time-consuming, and have the potential to disrupt the natural soil composition. This system addresses these limitations by providing real-time, in-situ measurements. This probe comes with state-of-the-art nano-sensors that can penetrate the soil, enabling researchers to monitor changes in CO_2 concentrations at different depths without disturbing the soil profile. The integrated probe system's functionality is based on precision and efficiency. Arranged along the length of the probe are nano-sensors that enable measurements at various depths. When the probe is inserted into the soil, the nano-sensors collect real-time data on soil CO_2 concentrations. The system seamlessly captures and stores these measurements, offering researchers a detailed overview of subsurface CO_2 dynamics [31]. By adopting this method, measurements become more precise while also reducing disruption to the soil composition. Researchers in ecology and environmental science find the integrated probe system valuable for understanding soil ecosystems in depth. Through ongoing monitoring of soil CO_2 concentrations, we are able to evaluate the effects of climate change, land-use alterations, and disturbances on soil health. This tool helps forecast changes in ecosystem functioning, grasp carbon sequestration capabilities, and execute successful conservation and restoration plans. Continual advancements may involve incorporating data analytics, remote sensing technologies, and connecting to broader environmental monitoring networks. These improvements would help in gaining a better grasp of soil dynamics on larger spatial and temporal scales.

A Vision *via* Optical Ingenuity: Technologies for Measuring Dissolved Carbon Dioxide

Optical approaches for estimating quantities of dissolved carbon dioxide (CO_2) are at the forefront of this breakthrough in measurement technology, which has been encouraged by the desire for a better knowledge of aquatic habitats. Techniques based on the interaction of light and matter have been developed for the purpose of optically detecting dissolved CO_2. These methods make use of the unique absorption and fluorescence characteristics of carbon dioxide molecules. The unique imprint that light leaves behind as it passes through water and interacts with dissolved gases, such as CO_2, may be studied in great detail. Absorption spectroscopy involves scrutinizing the absorption lines of CO_2 molecules, providing a direct and quantifiable measure of their concentration [32]. On the other hand, fluorescence spectroscopy capitalizes on the unique fluorescence emissions exhibited by certain molecules, offering a highly sensitive and selective approach to CO_2 measurement [32]. Thanks to this optical brilliance, scientists can now understand the movements of carbon underneath the ocean's surface. Advancements in fiber optics technology have propelled optical sensing into new dimensions. Compact optical probes with fiber optic nano-sensors are essential for conducting measurements in difficult conditions. With these probes, you can achieve precise measurements and easily explore a wide range of aquatic environments and depths with remarkable accuracy. One remarkable aspect of optical methods is their capability to offer instantaneous tracking of dissolved CO_2 levels. Having quick access to data can completely alter the game, particularly in settings with rapidly changing conditions. Our ability to capture dissolved CO_2 in real time helps us understand short-term variations and the factors affecting aquatic carbon dynamics more comprehensively. Optical methods for measuring dissolved CO_2 are being explored for their potential in various aquatic environments, including freshwater lakes and coastal ecosystems [30]. However, challenges such as interference from other dissolved species, variations in water properties, and the need for accurate calibration are being addressed. Current research focuses on refining sensor selectivity, developing robust calibration strategies, and enhancing signal processing algorithms to ensure the reliability and accuracy of optical CO_2 measurements. The future of optical methods in measuring dissolved CO_2 is marked by innovation and collaboration between environmental scientists, engineers, and technologists, aiming to refine sensor designs, optimize data analysis techniques, and integrate optical platforms with emerging technologies like artificial intelligence.

The Role of Metal-Organic Frameworks (MOFs) and Porous Materials in Gas Sensing for Vehicle Use: Towards a More Sustainable Tomorrow

Understanding gas levels in automotive applications is essential for monitoring and regulating emissions produced during combustion. With their intricate network of voids and open spaces, porous materials provide improved sensitivity and selectivity in gas sensing applications. These materials offer a significant surface area for gas molecules to engage with, aiding in effective adsorption and desorption processes [33]. Metal-organic frameworks (MOFs) and porous polymers are highly appealing for gas sensing applications because of their flexibility and design versatility.

Understanding gas detection in porous materials is essential for reducing pollution and managing emissions in automotive systems. Through the incorporation of these nano-sensors into exhaust systems, vehicles are able to monitor and control pollutants such as NOx, CO, and hydrocarbons, guaranteeing adherence to emission standards and minimizing the presence of harmful pollutants [34]. Gas sensors using porous materials help enhance fuel efficiency by analyzing exhaust gas content. This allows for smart feedback systems to regulate fuel-air ratios instantly, resulting in decreased fuel usage and reduced greenhouse gas emissions [34]. Understanding gas sensing is crucial to guarantee the safety of vehicles and passengers by detecting dangerous gases, thus avoiding accidents and ensuring the well-being of occupants.MOFs are perfect for CO_2 sensing because of their unique ability to selectively interact with gasses. To improve the sensitivity and selectivity of MOFs, their pore diameters and chemical capabilities may be engineered to preferentially absorb CO_2 molecules [35]. Environment monitoring, industrial procedures, and indoor air quality evaluation all rely on this sensitivity.

With MOF-based CO_2 nano-sensors, one can track changes in CO_2 levels in real-time and with great accuracy. Sensors help reduce the effects of increasing CO_2 levels and provide important data for climate research. Integrating MOF-based nano-sensors into industrial processes allows for the monitoring and optimization of CO_2 levels, which promotes sustainable practices and ensures compliance with environmental standards. Potential applications for CO_2 nano-sensors based on MOFs include carbon capture and storage, safety monitoring, and medical diagnostics. Still, fixing issues with stability, repeatability, and scalability is necessary to have it used by everyone [11]. Improved stability and robustness in MOFs, novel synthesis methods, and sensor design optimization are the current areas of attention in the field of study. With their potential for enhanced stability, broader applicability, and better performance, MOFs have a promising future in CO_2 sensing. By using new technology such as data analytics and the Internet of Things (IoT), it is possible to build a system of smart nano-sensors that might provide larger-scale, linked insights into the dynamics of CO_2 in real-time.

Flexing Innovation: The Era of Continuous Carbon Dioxide Monitoring with Flexible Sensor Patches

An innovative technique that makes flexible sensor patches useful and adaptable for a range of applications is their ability to conform to curved surfaces. They provide live, ongoing monitoring of carbon dioxide levels, offering a detailed view of CO_2 levels as they change. This information is essential for making well-informed decisions in different situations. One of the most thrilling uses of flexible sensor patches involves incorporating them into wearable technology [36]. A small patch can be discreetly attached to clothing or accessories to continuously monitor CO_2 levels in the wearer's environment.

These patches are well-suited for use in dynamic environments where conventional nano-sensors may not be sufficient. These devices are handy for keeping track of CO_2 levels in indoor areas with changing numbers of people or for evaluating air quality in vehicles. With their flexibility, they can easily blend into different environments, offering essential information for improving ventilation, controlling surroundings, and safeguarding the people inside [36].

These patches help in implementing timely interventions to address elevated CO_2 levels in industries, workplaces, and public spaces. The insights derived from the data collected by flexible sensor patches help in making informed decisions to create healthier and more sustainable environments. When conducting scientific research, flexible sensor patches offer detailed information on the impact of environmental factors, human activities, and spatial configurations on CO_2 concentrations. By combining wireless communication technologies, data analytics, and artificial intelligence, a seamless interconnected network of nano-sensors could be developed to monitor CO_2 levels collectively and autonomously. This could usher in an era of smart and responsive environmental management.

Solid-State Electrochemical Carbon Dioxide Sensors: Decoding Fundamentals, Exploring Materials, and Embracing Diverse Applications

Electrochemical sensing, the foundation of solid-state electrochemical carbon dioxide nano-sensors, is based on gas-electrode interactions that generate detectable electrical signals. These nano-sensors can monitor concentrations in real time by responding to certain electrochemical processes triggered by CO_2. Because of their selectivity, sensitivity, and general performance, these nano-sensors rely significantly on the materials chosen for them, which may include catalysts, polymers, and metal oxides [37].

Particularly in varied settings, it is difficult to achieve high CO_2 sensing sensitivity and selectivity. Improvements in sensitivity and selectivity in sensor materials and designs are the subject of continuing research, with the goal of providing more reliable readings in a wider range of environments. Environmental monitoring, industrial operations, and wearable technology are just a few of the many sectors that might benefit from solid-state electrochemical CO_2 nano-sensors. Air quality assessments, greenhouse gas emissions tracking, and climate research data collection all rely on them [38]. They ensure that operations are optimized and that emissions rules are followed in industrial environments. For applications where real-time CO_2 readings are crucial for informed decision-making, such as personal health, workplace safety, and wearable solutions, solid-state electrochemical CO_2 nano-sensors are a great choice due to their small size. Constant innovation and collaboration with other new technologies like the IoT, data analytics, and AI are shaping the future of these nano-sensors, which have

the potential to radically alter the way CO_2 emissions throughout the world are tracked and controlled.

Plastic Film Nano-sensors: A Thin Layer of Innovation Transforming Sensing Technologies

A novel kind of sensing device, plastic film nano-sensors make use of the adaptability and ease of use of very thin plastic films. The ability of these nano-sensors to adapt to different surfaces opens up exciting possibilities for their use in environmental and healthcare monitoring. While there is some variation in the material composition of these nano-sensors, polymers are often used. In order to develop materials with tailored properties, researchers are investigating a wide range of options, such as conductive polymers, piezoelectric polymers, and materials with intrinsic sensing capabilities [39].

The incorporation of plastic film nano-sensors into garments, accessories, or even the skin allows for the monitoring of physiological data; this is one of the key uses for wearable technology [39]. In environmental monitoring, they may be used to track air quality, humidity, or pollutant levels; they can be attached to non-standard surfaces, placed in unusual places, or even embedded in clothing.

Remote patient monitoring, tailored treatment, and healthcare delivery are all seeing breakthroughs in the healthcare sector because of Plastic Film Sensors [40]. Their production technique is often less expensive than conventional ways of fabricating nano-sensors, which makes them a practical and affordable option for widespread use.

Problems with calibration, stability over time, and durability plague plastic film nano-sensors. These issues, as well as the exploration of new materials and the improvement of production methods, are the subject of ongoing research into ways to make these nano-sensors more effective and reliable. Integrating new features, improving energy efficiency, and increasing sensitivity are all possible outcomes of future advances.

CONCLUSION AND FUTURE OUTLOOK

Numerous fields have been profoundly impacted by carbon dioxide sensing, including biomedicine, agriculture, food safety, environmental tracking, and wearable electronics. It started out as specialized research but has now become an essential tool for tracking respiratory health, connecting patients with their doctors between appointments, and giving them control over their chronic diseases. Carbon dioxide nano-sensors have caused a paradigm change in the agri-food sector, leading to more sustainable methods, less waste, and better use of available resources. A foundational component in environmental monitoring, industrial safety, and sustainable energy management, the incorporation of Non-Dispersive Infrared (NDIR) technology into carbon dioxide sensor modules represents a watershed moment in the accuracy of gas sensing. The integrated probe system provides useful information on soil dynamics, and the Costal Network's inexpensive CO_2 monitoring equipment satisfies the urgent requirement to track CO_2 levels in the estuary and coastal regions. The measurement of dissolved carbon dioxide in aquatic ecosystems has been made possible by optical inventiveness, while gas sensing in automotive applications shows promise thanks to Metal-Organic Frameworks (MOFs) and porous materials. Dynamic approaches to continuous carbon dioxide monitoring are provided by flexible sensor patches, while real-time monitoring in varied situations is offered by solid-state electrochemical carbon dioxide nano-sensors. Wearable tech, environmental monitoring, and healthcare are some of the places you could see plastic film nano-sensors used.

REFERENCES

[1]   Betty C, *et al.* Reliability studies of highly sensitive and specific multi-gas sensor based on nanocrystalline SnO_2 film 2014; 193(): 484-91.
[http://dx.doi.org/10.1016/j.snb.2013.11.118]

[2]   Molina A, Escobar-Barrios V, Oliva JJSM. A review on hybrid and flexible CO_2 gas sensors 2020; 270: 116602.
[http://dx.doi.org/10.1016/j.synthmet.2020.116602]

[3]   S.C.e., A New Conventional method for Synthesizing Nanomaterials: Green Synthesis/ Biosynthesis. Emerging Domains of Material Science. Thanuj International Publisers Tamil Nadu, India 2021; 1(1): 5.

[4]   Chandramouli M, *et al.* Based Carbon Dioxide Sensors: Past. Present, and Future Perspectives 2021; 12: 2353-60.

[5]   Abdelkarem K, *et al.* Design of high-sensitivity La-doped ZnO sensors for CO_2 gas detection at room temperature 2023; 13(1): 18398.
[http://dx.doi.org/10.1038/s41598-023-45196-y]

[6]   Elrashidi A, Traversa E, Elzein BJFieR. Highly sensitive ultra-thin optical CO_2 gas sensors using nanowall honeycomb structure and plasmonic nanoparticles 2022; 10: 909950.
[http://dx.doi.org/10.3389/fenrg.2022.909950]

[7]   Li YongWei. Automatic carbon dioxide enrichment strategies in the greenhouse: a review 2018.

[8]   Kumar U, *et al.* Micro and Nanofibers-Based Sensing Devices. Smart Nanostructure Materials and Sensor Technology. Springer 2022; pp. 97-112.
[http://dx.doi.org/10.1007/978-981-19-2685-3_5]

[9]   Potyrailo RAJCr. Multivariable sensors for ubiquitous monitoring of gases in the era of internet of things and industrial internet 2016; 116(19): 11877-923.
[http://dx.doi.org/10.1021/acs.chemrev.6b00187]

[10]   Blitz-Raith AH, *et al.* Separation of cobalt (II) from nickel (II) by solid-phase extraction into Aliquat 336 chloride immobilized in poly (vinyl chloride) 2007; 71(1): 419-23.

[11]   Zhang J, *et al.* Voltammetric ion-selective electrodes for the selective determination of cations and anions 2010; 82(5): 1624-33.
[http://dx.doi.org/10.1021/ac902296r]

[12]   Neethirajan S, *et al.* Carbon dioxide (CO_2) sensors for the agri-food industry—a review 2009; 2: 115-21.

[13]   Severinghaus JW. Electrodes for Blood pO_2 and pCO_2 Determination The Journal of the American Society of Anesthesiologists. The American Society of Anesthesiologists 1959.

[14]   Sipior J, *et al.* Phase fluorometric optical carbon dioxide gas sensor for fermentation off-gas monitoring 1996; 12(2): 266-71.
[http://dx.doi.org/10.1021/bp960005t]

[15]   Dervieux E, Théron M, Uhring WJS. Carbon dioxide sensing—biomedical applications to human subjects 2021; 22(1): 188.

[16]   Scheer BV, Perel A, Pfeiffer UJJCc. Clinical review: complications and risk factors of peripheral arterial catheters used for haemodynamic monitoring in anaesthesia and intensive care medicine 2002; 6: 1-7.

[17]   Guyenet PG, Bayliss DAJN. Neural control of breathing and CO_2 homeostasis 2015; 87(5): 946-61.

[http://dx.doi.org/10.1016/j.neuron.2015.08.001]

[18]   Campbell RL, Saxen MAJAp. Respiratory effects of a balanced anesthetic technique--revisited fifteen years later 1994; 41(1): 1.

[19]   Tsuboi T, *et al.* Importance of the PaCO$_2$ from 3 to 6 months after initiation of long-term non-invasive ventilation 2010; 104(12): 1850-7.
[http://dx.doi.org/10.1016/j.rmed.2010.04.027]

[20]   Lumb AB, Nunn JF. Nunn's applied respiratory physiology. Butterworth-Heinemann 2005.

[21]   Rego R, Mendes AJS, Chemical AB. Chemical, Carbon dioxide/methane gas sensor based on the permselectivity of polymeric membranes for biogas monitoring 2004; 103(1-2): 2-6.

[22]   Jayas D, *et al.* Evaluation of a computer-controlled ventilation system for a potato storage facility 2001; 43: 5.5-5.12.

[23]   Lee D-D, Lee D-SJIsj. Environmental gas sensors 2001; 1(3): 214-24.

[24]   Tan ES, *et al.* Freeze damage detection in oranges using gas sensors 2005; 35(2): 177-82.
[http://dx.doi.org/10.1016/j.postharvbio.2004.07.008]

[25]   Zhou L, *et al.* Carbon dioxide sensor module based on NDIR technology 2021; 12(7): 845.
[http://dx.doi.org/10.3390/mi12070845]

[26]   Gibson D, MacGregor CJS. A novel solid state non-dispersive infrared CO$_2$ gas sensor compatible with wireless and portable deployment 2013; 13(6): 7079-103.
[http://dx.doi.org/10.3390/s130607079]

[27]   Ishihara T, *et al.* Application of mixed oxide capacitor to the selective carbon dioxide sensor: I. Measurement of carbon dioxide sensing characteristics 1991; 138(1): 173-.

[28]   Bresnahan PJ, *et al.* A Low-Cost Carbon Dioxide Monitoring System for Coastal and Estuarine Sensor Networks 2023; 36(1): 14-5.
[http://dx.doi.org/10.5670/oceanog.2023.s1.4]

[29]   Northcott D, *et al.* Impacts of urban carbon dioxide emissions on sea-air flux and ocean acidification in nearshore waters 2019; 14(3): e0214403-.
[http://dx.doi.org/10.1371/journal.pone.0214403]

[30]   Hassan S, *et al.* Integrated Probe System for Measuring Soil Carbon Dioxide Concentrations 2023; 23(5): 2580.

[31]   Pumpanen J, *et al.* Comparison of different chamber techniques for measuring soil CO$_2$ efflux 2004; 123(3-4): 159-76.
[http://dx.doi.org/10.1016/j.agrformet.2003.12.001]

[32]   Gerlach G, Oelßner WJCDSF. Principles and Applications, Opto-Chemical CO$_2$ Sensors . 2019; p. 133.

[33]   Zhang C, *et al.* Metal oxide resistive sensors for carbon dioxide detection 2022; 472: 214758-.
[http://dx.doi.org/10.1016/j.ccr.2022.214758]

[34]   Fleming L, *et al.* Reducing N$_2$O induced cross-talk in a NDIR CO$_2$ gas sensor for breath analysis using multilayer thin film optical interference coatings 2018; 336: 9-16.

[35]   Lei J, *et al.* Design and sensing applications of metal–organic framework composites 2014; 58: 71-8.
[http://dx.doi.org/10.1016/j.trac.2014.02.012]

[36]   Hetzler Z, *et al.* Flexible sensor patch for continuous carbon dioxide monitoring 2022; 10: 983523.
[http://dx.doi.org/10.3389/fchem.2022.983523]

[37]   Mulmi S, Thangadurai VJJTES, Eds. Choice—review—solid-state electrochemical carbon dioxide sensors: fundamentals, materials and applications . 2020; 167: p. (3)037567.

[38]   Gorbova E, *et al.* Fundamentals and principles of solid-state electrochemical sensors for high

temperature gas detection 2021; 12(1): 1.
[http://dx.doi.org/10.3390/catal12010001]

[39]   Elanjeitsenni VP, Vadivu KS, Prasanth BMJMRE. A review on thin films, conducting polymers as sensor devices 2022; 9(2): 022001.
[http://dx.doi.org/10.1088/2053-1591/ac4aa1]

[40]   Liu Z, *et al.* A thin-film temperature sensor based on a flexible electrode and substrate 2021; 7(1): 42.
[http://dx.doi.org/10.1038/s41378-021-00271-0]

CHAPTER 14

Application of Nanotechnology in Air Remediation

Anjali Mehta[1], **Tanisha Kathuria**[1] and **Sudesh Kumar**[2,*]

[1] *Department of Chemistry, Banasthali Vidyapith, Rajasthan 304022, India*

[2] *Department of Chemistry, National Council of Educational Research and Training (NCERT), New Delhi 110016, India*

Abstract: The world's persistent daily development continues to cause unceasing damage to the air. As reported by the World Health Organization, over six million people worldwide lost their lives due to residing and working in environments affected by air pollution in 2016. Despite the effectiveness of traditional techniques such as desulfurization, denitrification, and dust removal in reducing emissions from the sources of stationary combustion, they have not proven successful in reducing the frequency of atmospheric haze conditions. Current research globally urges the advancement of technologies to create nanomaterials (NMs) capable of efficiently and intelligently trapping CO_2, CO, and other harmful gases from the air. Diverse NMs play pivotal roles as nano adsorbents, nanocatalysts, nanofilters, and nanosensors, showcasing the versatility and effectiveness of nanotechnological applications in this field. This technology facilitates air pollution remediation by treating volatile organic compounds, greenhouse gases, and bioaerosols through adsorption, photocatalytic degradation, thermal decomposition, and air filtration processes. This chapter specifically delves into the practical use of a range of NMs for air pollution remediation applications.

Keywords: Air pollution, Air purification, Nanotechnology, Nanosensors, Nanofilters.

INTRODUCTION

Air pollution comprises a broad spectrum of volatile organic compounds (VOCs) and damaging gases that originate from different natural and human-related sources. Industrial processes like manufacturing, power generation, and refineries release a wide range of pollutants such as nitrogen oxides (NO_x), volatile organic compounds (VOCs), carbon monoxide (CO), sulfur dioxide (SO_2), and particulate matter (PM) as a result of burning fossil fuels and conducting different production processes [1]. Furthermore, the contribution of agricultural practices to air pollution through activities like livestock farming, fertilizer use, and biomass

---

* **Corresponding author Sudesh Kumar:** Department of Chemistry, National Council of Educational Research and Training (NCERT), New Delhi 110016, India; E-mail: chem.ambi@gmail.com

Neha Agarwal, Vijendra Singh Solanki, Neetu Singh & Maulin P. Shah (Eds.)

burning illustrates the need to reduce their environmental impact by transitioning to cleaner energy sources. NOx has a strong photochemical reactivity in the atmosphere, as it can cause acid precipitation, haze, secondary particle generation, and negative health impacts [2]. Approximately 90% of the total emissions of NO_x occur from coal-fired power plants, industrial facilities, such as industrial boilers, iron-steel plants, cement, and motor vehicles [3]. The combination of these sources results in a diverse range of pollutants in the air, which not only harms the environment but also plays a significant role in driving climate change. The global concern about the adverse impact of air pollution on human health continues to grow. As per the World Health Organization (WHO), almost 80% of urban populations are exposed to inadequate air quality, which increases the risks of heart disease, stroke, respiratory illnesses, and lung cancer [4]. Recent studies have shown that in 2015, poor air quality contributed to an increased number of premature deaths, affecting 6.4 million individuals. Out of these, 4.2 million deaths were associated with ambient air quality, while 2.8 million were attributable to indoor air quality [5, 6].

Hence, there is a demand for technology capable of monitoring, detecting, and ideally purifying air contaminants. Innovations in nanotechnology offer a novel approach to environmental cleanup and enhance the effectiveness of conventional methods by minimizing emissions or preventing the formation of pollutants [7]. In theory, nanostructured materials are employed for green chemistry and environmental remediation for the reasons: (1) These materials have textile flexibility properties and an increased number of reactive edges, leading to inherently higher surface reactivity; (2) In comparison to bulk materials, they have huge surface-to-bulk ratios and large specific surface areas (SSAs);(3) Their chemical characteristics, including their oxidation and reduction potentials as well as their Lewis acid and Lewis base qualities can be adjusted or customized for specific reactions [8]. Thus, nanotechnology presents promising solutions for efficient air purification through the use of nano adsorbents, sensors, membranes/nanofilters, and nanocatalysts [9, 10]. This chapter delves into the specific applications of NMs in remediating air pollution.

NANOTECHNOLOGY FOR ENVIRONMENTAL POLLUTION REMEDIATION

Nanotechnology refers to the exploration of technology, engineering, and science at the nanoscale, typically ranging from 1 to 100 nanometers. This scale enables innovative applications across diverse disciplines including physics, medicine, chemistry, biology, electronics, and engineering. It is noteworthy that materials sized around 1000 nanometers are also categorized as NMs [11]. Within the nanoworld, nanoparticles (NPs) stand out for their exceptional characteristics, attributed to their significant surface-to-volume ratio, making them inherently more reactive than larger forms of the same materials. Such NMs are being extensively utilized across various environmental sectors, from air pollution control and water purification to soil remediation and renewable energy enhancement. These materials play crucial roles in making renewable energy sources more affordable and efficient. Specifically, they are employed in processes such as particle filtration, gas adsorption, and sensor technologies. NMs are integral components of purification devices aimed at reducing PM, gaseous pollutants like NO_x and VOCs, as well as toxic substances such as formaldehyde (HCHO), as illustrated in Fig. (**1**).

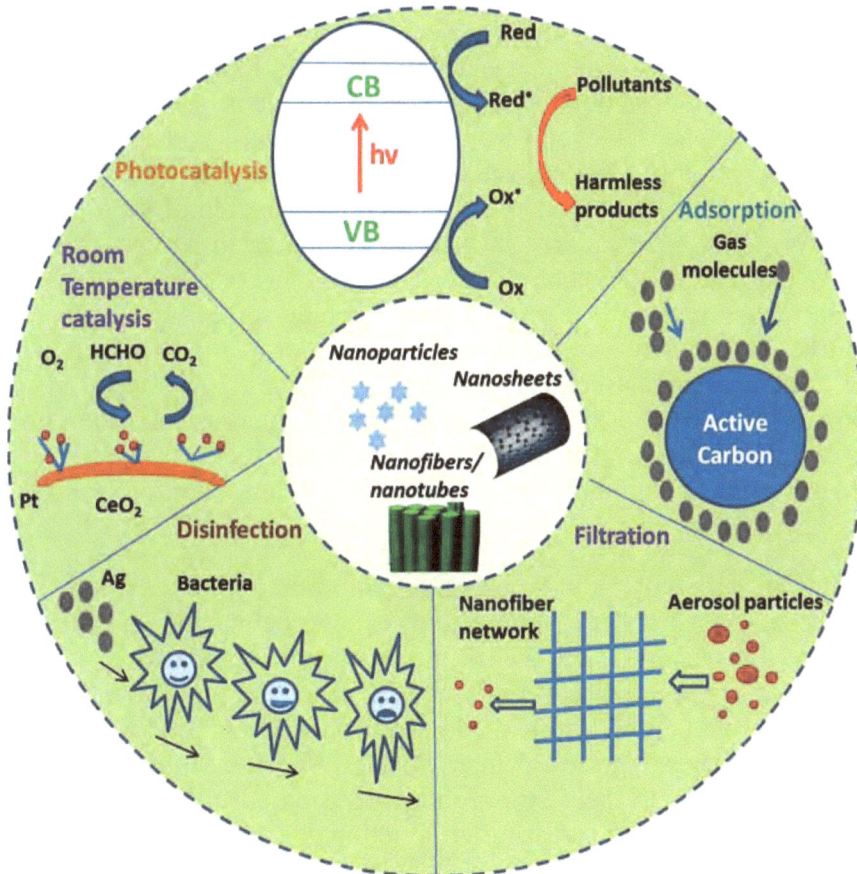

Fig. (1). Various NM-based air remediation technologies [14].

Some applications of nanotechnology are nearing commercialization, including nanosensors and nanoscale coatings that aim to replace thicker polymer coatings, which are less efficient and more wasteful in preventing corrosion. Nanosensors are also being developed for detecting aquatic toxins, and nanostructured metals are being investigated for their ability to break down hazardous organic compounds at room temperature. These advancements in nanotechnology present promising solutions for addressing environmental challenges by improving the efficiency and selectivity of air purification systems. NMs such as carbon-based and metal-based variants are under extensive exploration for their potential in air pollution control [12, 13].

NANOSENSORS FOR AIR QUALITY MONITORING

Nanosensors have become invaluable in monitoring air quality due to the growing global concern about air pollution. The emergence of rapid and precise sensors capable of detecting pollutants at the molecular level has the potential to empower humans to protect the environment's sustainability and human wellness. Greater advancements in environmental decision-making, ecosystem monitoring and process control can be achieved through the adoption of contaminant detection technology that is highly sensitive and cost-effective. One of the technologies that are highly desired is a continuous monitoring tool capable of rapidly providing detailed information, particularly regarding pollutants, in a very short analysis time [15]. Currently, nanotechnology presents an opportunity to make a substantial impact on pollutant sensing by enhancing sensors to be more specific and sensitive, particularly for air monitoring purposes. Nanosensors stand out for their key characteristics, such as fast response times, high specificity, quick detection capabilities, and the ability to identify numerous toxic compounds at levels as low as parts per million (ppm) and parts per billion (ppb) [16, 17]. Table **1,** provides a concise summary of various nanosensors commonly used for air quality monitoring.

Various carbon materials including carbon nano tubes (CNTs), grapheme oxide (GO), charcoal, and graphene among others, have demonstrated their utility as chemical and biosensors. Their sensitivity can be tailored effectively through simple chemical treatments, making them versatile tools for sensing applications [18]. CNTs have gained widespread adoption as the foundation for developing a variety of gas sensor categories, primarily due to their exceptional sensitivity to atoms and molecules and their extraordinarily strong adsorptive ability [19]. For example, by incorporating a carboxyl functional group (-COOH) into CNTs within a gas sensor, it is possible for carbonyl to achieve a detection rate of 1 ppm [20]. Recently, epitaxial graphene has been employed for detecting NO_2 gas at parts per billion (ppb) levels. It was observed that single-layer graphene is significantly more effective than bilayer graphene in terms of carrier concentration response for this purpose. In a research investigation, Ovsianytskyi and co-workers [21] developed a graphene-based gas sensor. The sensor was designed to detect hydrogen sulfide and was functionalized using silver NPs along with charged impurities, as reported in their work. Chemical vapour deposition was used to synthesize graphene, and then a wet-chemical process was used to add contaminants and silver NPs to the graphene.

Table 1. Overview of the most significant sensors based on nanostructured materials for air quality monitoring.

Nanosensors	Analytes	Detection Limit	Response Time	Refs.
Graphene	Trinitrotoluene	0.01–0.1 mg mL^{-1}	1 min	[26]
SnO$_2$ Ag-RGO	NO$_2$	5ppm	49 s	[27]
In-doped ZnO NPs	Trinitrotoluene	9×10^{-9} M	<6.3 s	[28]
MoS$_2$	NO$_2$ and NH$_3$ molecules	1.5 to 50 ppm	–	[29]
Graphene-SnO$_2$ Nanocomposites	NO$_2$	1–5 ppm	1 ppm: 175 s 3 ppm: 138 s 5 ppm: 129 s	[30]
ZnO Nanorods	NO$_2$	1 ppm to 100 ppm	1 ppm: 48 s 10 ppm: 42 s 50 ppm: 39 100 ppm: 35 s	[31]
ZnO BNW's	NO$_2$	1 ppm to 100 ppm	1 ppm: 26 s 10 ppm: 22 s 50 ppm: 20 s 100 ppm: 17 s	[31]
Cu NPs decorated SWCNTs Nanotube	H$_2$S	5 ppm to 150 ppm	5 ppm H$_2$S:10 s	[32]

Moreover, in gas-sensing applications, titanium dioxide (TiO$_2$) is an extensively examined semiconducting oxide. It is characterized by a wide band gap ranging from approximately 3.2 to 3.35 eV. Furthermore, the modification of TiO$_2$ with WO$_3$ leads to a unique heterojunction that can be used as a humidity sensor and a photocatalyst. According to observations, TiO$_2$/WO$_3$ heterojunction provides selectivity to a variety of reducing and oxidising gases, a faster reaction, and a longer depletion layer. The study of using a sol-gel approach, Kathirvelan and colleagues [22] have developed a selective ethylene sensing nanocomposite made of TiO$_2$-WO$_3$. The solvothermal approach was also used by Epifani and co-workers [23] to create TiO$_2$-WO$_3$ nanocomposites for acetone vapor detection. Similarly, Song and co-workers [24] have reported employing gold-modified porous α-Fe$_2$O$_3$ nanorods to generate a room temperature, and low-ppm-level triethylamine sensor. Their observations include a lower detection limit (1 ppm), faster response time (17-50 ppm), and enhanced selectivity, particularly towards triethylamine. Three mechanisms are primarily responsible for gas sensing in semiconductor metal oxides: charge transfer, gas molecule adsorption, and gas molecule desorption [25].

NANOCATALYSTS FOR AIR POLLUTION REMEDIATION

This NM, as its name suggests, possesses catalytic properties. Its high surface-t--volume ratio contributes to its exceptional reactivity, making it stand out in catalysis applications. There are several ways that the NPs can be used to control air pollution. One approach is the photocatalytic remediation, which involves using semiconducting materials, wherein exposure to light with an energy equivalent to the material's band gap creates an electron-hole pair [33]. A NM's active surface, where the reaction occurs, is one of its most important catalytic properties.

The active surface of the NM, where reaction occurs, is one of its most important catalytic features [34]. Comparing this nanocatalyst system to other common catalysts, it enables faster and more selective chemical transformation with a higher product yield and the capacity to recover the catalyst. Different nanocatalysts can be used to eliminate the release of air pollutants and harmful gases under different treatment applications as shown in Table **2**. Many different kinds of NMs have been employed in catalytic air cleanup methods. TiO_2 NMs exhibit utility in reducing CO_2 greenhouse gases into synthetic gas and hydrocarbons as illustrated in Fig. (**2**). The process that turns CO_2 into useful energy is intricate and includes two steps: first, CO_2 is directly reduced to carbon and the reactive oxygen species (ROS) and then the photocatalyst reacts with ROS to produce syngas' constituent parts [35].

Table 2. Application of nanocatalysts in addressing greenhouse gases and various air pollutants.

Target contaminant	Photocatalyst	synthesis	Performance	Refs.
HCHO	Au-TiO$_2$	Calcination in air	Conversion rate: 83.3%	[40]
CO$_2$	Nanostructured titanium dioxide	Sol-gel method	High conversion of CO$_2$ to H$_2$, CO, CH$_2$OH, and CH$_4$	[35]
NO$_x$, VOCs	Nanosilver–TiO$_2$ nanofibers	The sol-gel reaction is followed by an electrospinning process.	Decomposition efficiency was 21% for NO$_x$ and 30% for VOCs	[41]

(Table 2) cont.....

Target contaminant	Photocatalyst	synthesis	Performance	Refs.
Gas-phase methanol	Graphene-TiO$_2$ composite mats	Hydrothermal reaction and Electrospinning method	Removal efficiency: 100%	[42]
Benzene	BiVO$_4$ loaded TiO$_2$	Simple coupling method	Degradation: 66.8%	[37]
SO$_2$	Mn/Copper Slag	Impregnation method	SO$_2$ removal accomplished is almost 99%.	[39]
NO	Carbon nanodots/ ZnFe$_2$O$_4$	Hydrothermal reaction	Removal efficiency: 38.0%	[43]

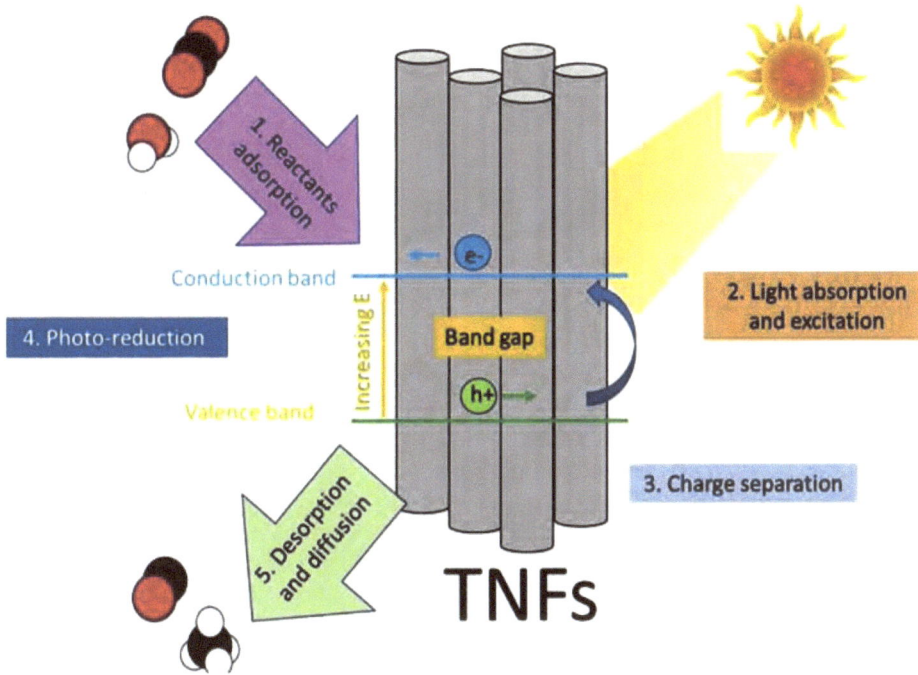

Fig. (2). Undoped and Cu-doped TiO$_2$ nanofibers for photocatalytic CO$_2$ reduction [38].

Yu and co-workers [36] evaluated the rate of formaldehyde breakdown over titanium dioxide-reduced graphene oxide under visible light produced by hydrothermally combining graphite powder with titanium tetrabutoxide. After 100 minutes of radiation, the formaldehyde was removed with removal efficiencies of 60.0%, 20.0%, and 10.0%, utilizing titanium dioxide-reduced graphene oxide, titanium dioxide, and graphene oxide respectively. To degrade gaseous benzene under visible light (450 < λ < 900 nm), Hu and co-workers [37] developed a

$BiVO_4/TiO_2$ heterojunction with easy charge separation. According to the experimental results, benzene was mineralized with 52% efficiency and converted into CO_2 over 50 hours without any evident signs of catalyst deactivation.

Cu-doped TiO_2 nanofibers and TiO_2 nanofibers (TNFs) are also used for photocatalytic CO_2 reduction with water vapor. The CO_2 reduction rates of all undoped and Cu-doped TNFs are found to be greater than those of commercial catalysts (P-25). Only carbon monoxide and methane are reaction products during the photocatalytic reduction of CO_2 (as indicated in Figure 2) [38]. Rabiee and co-workers [39] proposed the idea of applying a unique catalytic oxidation technique to the photocatalytic oxidation of SO_2 from simulated flue gas under ultra-violet irradiation, employing manganese supported on Copper Slag nanocatalyst. The test findings obtained have the potential to be successfully applied in the industrialized usage of flue Gas desulfurization systems for the removal of SO_2, hence improving the management of sulfur dioxide pollution. These nanocatalysts have great promise for air remediation because of their capacity to lower air pollutants.

AIR FILTRATION

The most popular method for effectively and economically eliminating submicrometer aerosol particles from a gas stream in an environment with low dust concentration is the use of a fibrous filter. Fibrous filters in practical applications do not function as sieves due to their tendency to quickly become clogged, leading to a significant increase in pressure drop [44]. The process of sieving small aerosol particles through a fibrous filter before an accumulation forms on the filter's surface may be the least significant mechanism. The high-pressure drop across the filter will cause the filter's service life to shorten quickly once the cake forms due to sieving [45].

One of the most important novel NMs for air filtration applications is nanofibers. Le and co-workers [46] showed that 91.6% of butanol was destroyed in 55 minutes and 69% of bacteria and 63% of fungus were eliminated in 6 hours after the air was passed through the air cleaner apparatus, which was fitted with four photocatalytic filters, four UV-A lamps, and a pre-filter coated with nanosilver. According to Zhang and co-workers [47], metal-organic frameworks (MOFs) nanofibrous filters have good removal efficiency for PM 2.5 (particulate matter with a diameter of 2.5 micrometres or less), PM10 (diameter of 10 micrometres or less), and toxic SO_2. Silver NP filters have emerged as one of the most widely utilized NMs in air filtration systems in recent years. Indoor air can be cleaned more efficiently with this type of air filtration device in hospitals [48].

Nanostructured Membranes

Nanotechnology offers numerous methods to purify the air. A filter's porous characteristic, which permits gas to pass through, keeps particles in the filter. Three methods are used by the membrane to remove particles from the filter: direct particle contact with the filter structure, the application of inertia force when the gas direction is changed, and the electric charge interaction between the particle and the filter structure. The primary drawback of filters is that they cause a pressure decrease, which must be overcome with a lot of energy. Compared to ordinary filters, nanofilters are more efficient because their holes range from 1 to 10 nm, which allows them to efficiently remove a variety of bacteria, viruses, and organic pollutants [49].

Several researchers have reported that the characteristics of carbon NMs include chemical stability, low density, structural diversity, and large-scale production accountability. Moreover, carbon NM structures possess numerous adsorptive sites on their surface. Additionally, their surface is amphoteric (hydroxyl functional groups, -OH), allowing for deprotonation (resulting in a negative charge) or protonation (resulting in a positive charge). This has the advantage of increasing the number of oxygen-containing functional groups that may be added to the carbon NMs surface, increasing their chemical adsorption activity [50].

Carbon nano tubes (CNTs)

CNTs can trap gases at a rate that is 100 times more rapid than traditional approaches. Consequently, they are well-suited for large-scale separation processes. Conventional membranes suffer from the negative connection between gas separation quality and gas flow rate; carbon nanotube-based membranes do not exhibit these drawbacks. Research has demonstrated that noncovalent forces such as van der Waals forces, π–π stacking, hydrophobic interactions, hydrogen bonding, and electrostatic forces are among how CNTs may interact with organic molecules [51]. CNTs offer high specificity, assessable adsorption sites, variable surface chemistry, and a short intraparticle distance. CNTs have been used to capture methane and carbon dioxide emissions from automobiles and industrial chimneys. Moreover, CNTs can capture greenhouse gases from coal mines and power plants. Because of their improved characteristics, particularly their strong affinity for selective adsorption of contaminants, these materials can separate air pollutants as nano adsorbents.

Furthermore, CNTs have large pore diameters, pore volumes, and surface areas, which improve the reactive sites connected to particular air pollutants [50]. The adsorptive action of these NMs can be enhanced by the insertion of functional groups that amplify the reactive sites [52]. Two distinct varieties of CNTs were synthesized, symbolizing the differences between single-walled and multi-walled CNTs (SWCNTs, MWCNTs) in terms of their shape and porosity. In terms of structure, CNTs are formed from a single graphitic carbon sheet that has been rolled into a hollow, cylindrical form with a diameter of one nanometer. The SWCNTs only have a single, cylindrical layer of graphitic carbon, but the MWCNTs include several concentric cylindrical single layers [53]. There are four distinct places in SWCNT bundles where adsorption can occur: inside the nanotube pores, outside the bundles in the grooves separating adjacent tubes, and in the interstitial channels separating the tubes within the bundles [54]. As MWCNT aggregates do not adopt a bundle structure, their adsorptive sites are found on both the outer and inner surfaces of the pores [55]. Therefore, SWCNT bundles adsorb small molecules like gases, whereas MWCNTs adsorb large biomolecules like bacteria and viruses through their interstitial channels [56]. Although the shape, number, and size of aromatic molecules determine the increased number of aromatic rings inside MWCNTs considerably improves the sorption affinity of phenolic compounds on MWCNTs, leading to better π–π interactions [57]. CO_2 emissions from power plants are filtered using CNT membranes. In addition to CNTs' high-energy binding sites, they are ideal sorbents based on their adsorptive capability to improve air remediation, particularly that of dangerous air pollutants [58].

NANOTECHNOLOGY: A REVOLUTIONARY APPROACH FOR ADSORBING TOXIC GASES

NO_2 Adsorption

Ion exchange zeolites, activated carbon, and FeOOH distributed on active carbon fiber are common adsorbents used to remove NO_x at low temperatures. Research by Long and Yang [59] suggests that CNTs can be used as an adsorbent for the removal of NO. The absorption of NO_x was almost 78 mg/g of CNTs. When NO and O_2 move through CNTs, NO is oxidized to NO_2, which then binds to the surface as nitrate species. As NO and O_2 flow within CNTs, NO undergoes oxidation to NO_2, subsequently adhering to the surface as nitrate species. While CO_2 is far less adsorbed on CNTs than NO_2 or NO, SO_2 may also be adsorbed on them, but at an unpromising adsorption rate.

CO_2 Capture

Numerous CO_2 collection technologies, such as cryogenic, membrane, adsorption, absorption, and others, have been studied. Single-walled NPs, zeolite, activated carbon, and nanoporous silica-based molecular baskets are some of these adsorbents. The chemical alteration of CNTs will allow the trapping of CO_2 emissions [60]. The effectiveness of CO_2 adsorption on modified CNTs generally rises with relative humidity but falls with temperature [61, 62].

Isopropyl Alcohol Adsorption

Isopropyl alcohol (IPA) is widely used as a solvent and in the manufacturing of semiconductors and optoelectronic devices. The absence of air pollution management results in the untreated discharge of IPA vapor into the atmosphere. The oxidation of SWNTs by an HNO_3 and NaClO solution, which was utilized as an adsorbent to absorb IPA vapor, was studied by Hsu and Lu. The most effective combination for adsorbing IPA was SWNTs/NaClO, followed by SWNTs/HNO_3. IPA is attracted by chemical and physical interactions during the adsorption process. Van der Waals forces among adsorbates and adsorbents induce physical adsorption, while chemical adsorption is the consequence of adsorbate molecules interacting chemically with adsorbent surface functional groups. Physical forces primarily cause the IPA vapor adsorption process on SWNTs and SWNTs/NaClO at relatively low intake concentrations of IPA; at relatively high intake concentration intake, however, both physical and chemical forces can produce the IPA vapor adsorption mechanism on SWNTs/NaClO [63].

CONCLUSION

Nanotechnology plays a key role in improving air quality by offering innovative approaches to air pollution. NMs, including nanosensors, nanofilters, and nanocatalysts, have demonstrated the potential for use in the removal of air pollutants. The effectiveness of using nanosensors to detect harmful gases including hydrogen sulfide (H_2S), sulfur dioxide (SO_2), and nitrogen dioxide (NO_2) was examined in this work. Several studies have also been conducted on the design and optimization of nanostructured membranes for capturing a range of gas contaminants. Additionally, by reducing air pollutants, the nanocatalysts showed great promise in air remediation. These substances have excellent selectivity, efficacy, and affordability, which make them useful components of air filtration systems. As pollutants are removed, air quality is improved, and health risks are reduced through the use of these NMs. Moreover, nanotechnology has led to innovative approaches to addressing and monitoring air pollution due to rapid advancements in the field.

REFERENCES

[1]   Mohamed EF, Awad G. Nanotechnology and nanobiotechnology for environmental remediation Magnetic Nanostructures: Environmental and Agricultural Applications 2019; 77-93.
       [http://dx.doi.org/10.1007/978-3-030-16439-3_5]

[2]   Perring AE, Pusede SE, Cohen RC. An observational perspective on the atmospheric impacts of alkyl and multifunctional nitrates on ozone and secondary organic aerosol. Chem Rev 2013; 113(8): 5848-

70.
[http://dx.doi.org/10.1021/cr300520x] [PMID: 23614613]

[3]   Li M, Liu H, Geng G, *et al.* Anthropogenic emission inventories in China: a review. Natl Sci Rev 2017; 4(6): 834-66.
[http://dx.doi.org/10.1093/nsr/nwx150]

[4]   Srivastava AK. Air pollution: Facts, causes, and impacts. Asian Atmospheric Pollution. Elsevier 2022; pp. 39-54.
[http://dx.doi.org/10.1016/B978-0-12-816693-2.00020-2]

[5]   Gakidou E, Afshin A, Abajobir AA, *et al.* Global, regional, and national comparative risk assessment of 84 behavioural, environmental and occupational, and metabolic risks or clusters of risks, 1990–2016: a systematic analysis for the Global Burden of Disease Study 2016. Lancet 2017; 390(10100): 1345-422.
[http://dx.doi.org/10.1016/S0140-6736(17)32366-8] [PMID: 28919119]

[6]   Landrigan PJ, Fuller R, Acosta NJR, *et al.* The Lancet Commission on pollution and health. Lancet 2018; 391(10119): 462-512.
[http://dx.doi.org/10.1016/S0140-6736(17)32345-0] [PMID: 29056410]

[7]   Adeleye AS, Conway JR, Garner K, Huang Y, Su Y, Keller AA. Engineered nanomaterials for water treatment and remediation: Costs, benefits, and applicability. Chem Eng J 2016; 286: 640-62.
[http://dx.doi.org/10.1016/j.cej.2015.10.105]

[8]   Ibrahim RK, Hayyan M, AlSaadi MA, Hayyan A, Ibrahim S. Environmental application of nanotechnology: air, soil, and water. Environ Sci Pollut Res Int 2016; 23(14): 13754-88.
[http://dx.doi.org/10.1007/s11356-016-6457-z] [PMID: 27074929]

[9]   Handojo L, Nursanto EB, Indarto A. Progress of NMs application in environmental concerns. Nanohybrids in Environmental & Biomedical Applications. CRC Press 2019; pp. 189-205.
[http://dx.doi.org/10.1201/9781351256841-8]

[10]   Nguyen-Tri P, Nguyen TA, Nguyen TV. NM for air remediation: an introduction. NMs for Air Remediation. Elsevier 2020; pp. 3-8.
[http://dx.doi.org/10.1016/B978-0-12-818821-7.00001-4]

[11]   Nasrollahzadeh M, Sajadi SM, Sajjadi M, Issaabadi Z. An introduction to nanotechnology Interface science and technology. 2019; 28: pp. 1-27.
[http://dx.doi.org/10.1016/B978-0-12-813586-0.00001-8]

[12]   Mehndiratta P, Jain A, Srivastava S, Gupta N. Environmental pollution and nanotechnology. Environ Pollut 2013; 2(2): 49.
[http://dx.doi.org/10.5539/ep.v2n2p49]

[13]   Mohamed EF. Nanotechnology: future of environmental air pollution control. J Environ Sustain 2017; 6(2): 429.

[14]   Cao J, Huang Y, Zhang Q. 14 Cao JJ, Huang Y, Zhang Q. Ambient air purification by nanotechnologies: from theory to application. Catalysts 2021; 11(11): 1276.
[http://dx.doi.org/10.3390/catal11111276]

[15]   Yunus IS, Harwin , Kurniawan A, Adityawarman D, Indarto A. Nanotechnologies in water and air pollution treatment. Environ Technol Rev 2012; 1(1): 136-48.
[http://dx.doi.org/10.1080/21622515.2012.733966]

[16]   Kim J, Choi SW, Lee JH, Chung Y, Byun YT. Gas sensing properties of defect-induced single-walled carbon nanotubes. Sens Actuators B Chem 2016; 228: 688-92.
[http://dx.doi.org/10.1016/j.snb.2016.01.094]

[17]   Zhou R, Hu G, Yu R, Pan C, Wang ZL. Piezotronic effect enhanced detection of flammable/toxic gases by ZnO micro/nanowire sensors. Nano Energy 2015; 12: 588-96.
[http://dx.doi.org/10.1016/j.nanoen.2015.01.036]

[18]   Gupta VK, Saleh TA. Sorption of pollutants by porous carbon, carbon nanotubes and fullerene- An overview. Environ Sci Pollut Res Int 2013; 20(5): 2828-43.
[http://dx.doi.org/10.1007/s11356-013-1524-1] [PMID: 23430732]

[19]   Zaporotskova IV, Boroznina NP, Parkhomenko YN, Kozhitov LV. Carbon nanotubes: Sensor properties. A review. Mod Electron Mater 2016; 2(4): 95-105.
[http://dx.doi.org/10.1016/j.moem.2017.02.002]

[20]   Fu D, Lim H, Shi Y, *et al.* Differentiation of gas molecules using flexible and all-carbon nanotube devices. J Phys Chem C 2008; 112(3): 650-3.
[http://dx.doi.org/10.1021/jp710362r]

[21]   Ovsianytskyi O, Nam YS, Tsymbalenko O, Lan PT, Moon MW, Lee KB. Highly sensitive chemiresistive H2S gas sensor based on graphene decorated with Ag nanoparticles and charged impurities. Sens Actuators B Chem 2018; 257: 278-85.
[http://dx.doi.org/10.1016/j.snb.2017.10.128]

[22]   Kathirvelan J, Vijayaraghavan R, Thomas A. Ethylene detection using TiO_2–WO_3 composite sensor for fruit ripening applications. Sens Rev 2017; 37(2): 147-54.
[http://dx.doi.org/10.1108/SR-12-2016-0262]

[23]   Epifani M, Comini E, Díaz R, *et al.* Acetone sensing with TiO2-WO3 nanocomposites: an example of response enhancement by inter-oxide cooperative effects. Procedia Eng 2014; 87: 803-6.
[http://dx.doi.org/10.1016/j.proeng.2014.11.676]

[24]   Song X, Xu Q, Zhang T, Song B, Li C, Cao B. Room-temperature, high selectivity and low-ppm-level triethylamine sensor assembled with Au decahedrons-decorated porous α-Fe2O3 nanorods directly grown on flat substrate. Sens Actuators B Chem 2018; 268: 170-81.
[http://dx.doi.org/10.1016/j.snb.2018.04.096]

[25]   Tomer VK, Duhan S, Malik R, Nehra SP, Devi S. A novel highly sensitive humidity sensor based on ZnO/SBA-15 hybrid nanocomposite. J Am Ceram Soc 2015; 98(12): 3719-25.
[http://dx.doi.org/10.1111/jace.13836]

[26]   Avaz S, Roy RB, Mokkapati VRSS, *et al.* Graphene based nanosensor for aqueous phase detection of nitroaromatics. RSC Advances 2017; 7(41): 25519-27.
[http://dx.doi.org/10.1039/C7RA03860G]

[27]   Wang Z, Zhang Y, Liu S, Zhang T. Preparation of Ag NPs-SnO2 NPs-reduced graphene oxide hybrids and their application for detection of NO_2 at room temperature. Sens Actuators B Chem 2016; 222: 893-903.
[http://dx.doi.org/10.1016/j.snb.2015.09.027]

[28]   Ge Y, Wei Z, Li Y, Qu J, Zu B, Dou X. Highly sensitive and rapid chemiresistive sensor towards trace nitro-explosive vapors based on oxygen vacancy-rich and defective crystallized In-doped ZnO. Sens Actuators B Chem 2017; 244: 983-91.
[http://dx.doi.org/10.1016/j.snb.2017.01.092]

[29]   Cho B, Hahm MG, Choi M, *et al.* Charge-transfer-based gas sensing using atomic-layer MoS2. Sci Rep 2015; 5(1): 8052.
[http://dx.doi.org/10.1038/srep08052] [PMID: 25623472]

[30]   Kim HW, Na HG, Kwon YJ, *et al.* Microwave-assisted synthesis of graphene–SnO_2 nanocomposites and their applications in gas sensors. ACS Appl Mater Interfaces 2017; 9(37): 31667-82.
[http://dx.doi.org/10.1021/acsami.7b02533] [PMID: 28846844]

[31]   Navale YH, Navale ST, Ramgir NS, *et al.* Zinc oxide hierarchical nanostructures as potential NO2 sensors. Sens Actuators B Chem 2017; 251: 551-63.
[http://dx.doi.org/10.1016/j.snb.2017.05.085]

[32]   Asad M, Sheikhi MH, Pourfath M, Moradi M. High sensitive and selective flexible H2S gas sensors based on Cu nanoparticle decorated SWCNTs. Sens Actuators B Chem 2015; 210: 1-8.

[http://dx.doi.org/10.1016/j.snb.2014.12.086]

[33]   Theerthagiri J, Duraimurugan K, Kim HS, Madhavan J. Graphitic Carbon Nitride-Based Nanostructured Materials for Photocatalytic Applications. Photocatalytic Functional Materials for Environmental Remediation 2019; pp. 291-307.
[http://dx.doi.org/10.1002/9781119529941.ch10]

[34]   Tai XH, Lai CW, Juan JC, Lee KM. Nanocatalyst-based catalytic oxidation processes. NMs for air remediation. Elsevier 2020; pp. 133-50.
[http://dx.doi.org/10.1016/B978-0-12-818821-7.00007-5]

[35]   Akhter P, Hussain M, Saracco G, Russo N. Novel nanostructured-TiO2 materials for the photocatalytic reduction of CO2 greenhouse gas to hydrocarbons and syngas. Fuel 2015; 149: 55-65.
[http://dx.doi.org/10.1016/j.fuel.2014.09.079]

[36]   Yu L, Wang L, Sun X, Ye D. Enhanced photocatalytic activity of rGO/TiO$_2$ for the decomposition of formaldehyde under visible light irradiation. J Environ Sci (China) 2018; 73: 138-46.
[http://dx.doi.org/10.1016/j.jes.2018.01.022] [PMID: 30290862]

[37]   Hu Y, Li D, Zheng Y, *et al.* BiVO4/TiO2 nanocrystalline heterostructure: A wide spectrum responsive photocatalyst towards the highly efficient decomposition of gaseous benzene. Appl Catal B 2011; 104(1-2): 30-6.
[http://dx.doi.org/10.1016/j.apcatb.2011.02.031]

[38]   Camarillo R, Rizaldos D, Jiménez C, Martínez F, Rincón J. Enhancing the photocatalytic reduction of CO2 with undoped and Cu-doped TiO2 nanofibers synthesized in supercritical medium. J Supercrit Fluids 2019; 147: 70-80.
[http://dx.doi.org/10.1016/j.supflu.2019.02.013]

[39]   Rabiee F, Mahanpoor K. Photocatalytic oxidation of SO2 from flue gas in the presence of Mn/copper slag as a novel nanocatalyst: optimizations by Box-Behnken design. Iran J Chem Chem Eng 2019; 38(3): 69-85.

[40]   Zhu X, Jin C, Li XS, *et al.* Photocatalytic formaldehyde oxidation over plasmonic Au/TiO2 under visible light: moisture indispensability and light enhancement. ACS Catal 2017; 7(10): 6514-24.
[http://dx.doi.org/10.1021/acscatal.7b01658]

[41]   Srisitthiratkul C, Pongsorrarith V, Intasanta N. The potential use of nanosilver-decorated titanium dioxide nanofibers for toxin decomposition with antimicrobial and self-cleaning properties. Appl Surf Sci 2011; 257(21): 8850-6.
[http://dx.doi.org/10.1016/j.apsusc.2011.04.083]

[42]   Roso M, Boaretti C, Pelizzo MG, Lauria A, Modesti M, Lorenzetti A. Nanostructured photocatalysts based on different oxidized graphenes for VOCs removal. Ind Eng Chem Res 2017; 56(36): 9980-92.
[http://dx.doi.org/10.1021/acs.iecr.7b02526]

[43]   Huang Y, Liang Y, Rao Y, *et al.* Environment-friendly carbon quantum dots/ZnFe2O4 photocatalysts: characterization, biocompatibility, and mechanisms for NO removal. Environ Sci Technol 2017; 51(5): 2924-33.
[http://dx.doi.org/10.1021/acs.est.6b04460] [PMID: 28145696]

[44]   Chen CY. Filtration of aerosols by fibrous media. Chem Rev 1955; 55(3): 595-623.
[http://dx.doi.org/10.1021/cr50003a004]

[45]   Samuel Sunday Ogunsola, Mayowa Ezekiel Oladipo, Peter Olusakin Oladoye, Mohammed Kadhom, Carbon nanotubes for sustainable environmental remediation: A critical and comprehensive review, Nano-Structures & Nano-Objects, Volume 37, 2024, 101099, ISSN 2352-507X.
[http://dx.doi.org/10.1016/j.nanoso.2024.101099]

[46]   Le TS, Dao TH, Nguyen DC, Nguyen HC, Balikhin IL. Air purification equipment combining a filter coated by silver NPs with a nano-TiO2 photocatalyst for use in hospitals. Adv Nat Sci: Nanosci Nanotechnol 2015; 6(1): 015016.

[47]   Zhang Y, Yuan S, Feng X, Li H, Zhou J, Wang B. Preparation of nanofibrous metal-organic framework filters for efficient air pollution control. J Am Chem Soc 2016; 138(18): 5785-8.
        [http://dx.doi.org/10.1021/jacs.6b02553] [PMID: 27090776]

[48]   Lee BU, Yun SH, Jung JH, Bae GN. Effect of relative humidity and variation of particle number size distribution on the inactivation effectiveness of airborne silver nanoparticles against bacteria bioaerosols deposited on a filter. J Aerosol Sci 2010; 41(5): 447-56.
        [http://dx.doi.org/10.1016/j.jaerosci.2010.02.005]

[49]   Sutherland KS, Chase G. Filters and filtration handbook. Elsevier 2011.

[50]   Gupta VK, Saleh TA. Sorption of pollutants by porous carbon, carbon nanotubes and fullerene- An overview. Environ Sci Pollut Res Int 2013; 20(5): 2828-43.
        [http://dx.doi.org/10.1007/s11356-013-1524-1] [PMID: 23430732]

[51]   Rengaraj S, Yeon JW, Kim Y, Kim WH. Application of Mg-mesoporous alumina prepared by using magnesium stearate as a template for the removal of nickel: kinetics, isotherm, and error analysis. Ind Eng Chem Res 2007; 46(9): 2834-42.
        [http://dx.doi.org/10.1021/ie060994n]

[52]   Kyzas GZ, Matis KA. Nanoadsorbents for pollutants removal: A review. J Mol Liq 2015; 203: 159-68.
        [http://dx.doi.org/10.1016/j.molliq.2015.01.004]

[53]   Zhao YL, Stoddart JF. Noncovalent functionalization of single-walled carbon nanotubes. Acc Chem Res 2009; 42(8): 1161-71.
        [http://dx.doi.org/10.1021/ar900056z] [PMID: 19462997]

[54]   Ren X, Chen C, Nagatsu M, Wang X. Carbon nanotubes as adsorbents in environmental pollution management: A review. Chem Eng J 2011; 170(2-3): 395-410.
        [http://dx.doi.org/10.1016/j.cej.2010.08.045]

[55]   Machado FM, Bergmann CP, Fernandes THM, *et al.* Adsorption of Reactive Red M-2BE dye from water solutions by multi-walled carbon nanotubes and activated carbon. J Hazard Mater 2011; 192(3): 1122-31.
        [http://dx.doi.org/10.1016/j.jhazmat.2011.06.020] [PMID: 21724329]

[56]   Upadhyayula VKK, Deng S, Mitchell MC, Smith GB. Application of carbon nanotube technology for removal of contaminants in drinking water: A review. Sci Total Environ 2009; 408(1): 1-13.
        [http://dx.doi.org/10.1016/j.scitotenv.2009.09.027] [PMID: 19819525]

[57]   Chen W, Duan L, Zhu D. Adsorption of polar and nonpolar organic chemicals to carbon nanotubes. Environ Sci Technol 2007; 41(24): 8295-300.
        [http://dx.doi.org/10.1021/es071230h] [PMID: 18200854]

[58]   Gangupomu RH, Sattler ML, Ramirez D. CNTs for air pollutant control *via* adsorption: A review. Reviews in Nanoscience and Nanotechnology 2014; 3(2): 149-60.
        [http://dx.doi.org/10.1166/rnn.2014.1048]

[59]   Long RQ, Yang RT. CNTs as a superior sorbent for nitrogen oxides. Ind Eng Chem Res 2001; 40(20): 4288-91.
        [http://dx.doi.org/10.1021/ie000976k]

[60]   Gui MM, Yap YX, Chai SP, Mohamed AR. Multi-walled carbon nanotubes modified with (3-aminopropyl)triethoxysilane for effective carbon dioxide adsorption. Int J Greenh Gas Control 2013; 14: 65-73.
        [http://dx.doi.org/10.1016/j.ijggc.2013.01.004]

[61]   White CM, Strazisar BR, Granite EJ, Hoffman JS, Pennline HW. Separation and capture of CO2 from large stationary sources and sequestration in geological formations--coalbeds and deep saline aquifers. J Air Waste Manag Assoc 2003; 53(6): 645-715.
        [http://dx.doi.org/10.1080/10473289.2003.10466206] [PMID: 12828330]

[62] Aaron D, Tsouris C. Separation of CO2 from flue gas: a review. Sep Sci Technol 2005; 40(1-3): 321-48.
[http://dx.doi.org/10.1081/SS-200042244]

[63] Hsu S, Lu C. Modification of single-walled CNTs for enhancing isopropyl alcohol vapor adsorption from air streams. Sep Sci Technol 2007; 42(12): 2751-66.
[http://dx.doi.org/10.1080/01496390701515060]

CHAPTER 15

Applications of Nanotechnology for Remediation of Soil

Reenu Gill[1,*] and **Mithlesh Kumar**[1]

[1] *Department of Physics, Rajkeeya Mahavidhyalaya Todarpur, Hardoi 241125, Uttar Pradesh, India*

Abstract: Soil is a valuable natural resource that favors the growth and development of plants, microbes, and other organisms living in both terrestrial and aquatic ecosystems. At present due to various anthropogenic causes like accelerated urbanization, industrialization, and ever-increasing population, the soil is becoming heavily contaminated due to industrial wastes, mining, excess use of agrochemicals like fertilizers and pesticides, and several other pollutants like toxic heavy metals and poly aromatic hydrocarbons. Due to these contaminants, plants, and soil microbes face various types of stresses which adversely affect the growth of plants and microorganisms. So, the remediation of such valued soil resources is essential to fulfill the need for food grains and to ensure the food security of growing populations throughout the world. Nanotechnology is the recent and most advanced technology that provides efficient, cost-effective, and environment-friendly ways for the remediation of contaminated soils. Various nanoparticles like zinc oxide (ZnO), titanium oxide (TiO_2), silver nanoparticles, nZVI (Nano zero-valent iron), silicon oxide (SiO_2), and aluminium oxide (Al_2O_3), *etc.* are used for remediation purposes of such degraded lands. The application of nanotechnology-based methods has great potential to restore degraded land to its optimal forms that are suitable for the growth of plants and microbes. The use of nanotechnology provides innovative techniques for the remediation of degraded soil due to their reactivity and versatility. So, the promotion of efficient and sustainable use of nanomaterials (NMs) can enhance the productivity and fertility of such soils. It is the necessity of the present time to provide sustainable remediation approaches that ensure a safe and healthy environment without degrading natural resources.

Keywords: Agrochemicals, Contaminants, Fertility, Microorganism, Nanoparticle, Nanotechnology, Nanomaterials.

INTRODUCTION

Soil is one of the most significant components of the environment which provides a critical terrestrial ecosystem in which plants and microorganisms survive and

---

[*] **Corresponding author Reenu Gill:** Department of Physics, Rajkeeya Mahavidhyalaya Todarpur, Hardoi 241125, Uttar Pradesh, India; E-mail: reenugill@yahoo.com

Neha Agarwal, Vijendra Singh Solanki, Neetu Singh & Maulin P. Shah (Eds.)

acquire all essential nutrients, water, and oxygen [1]. Soil with better and optimal physico-chemical properties is a prerequisite for optimum growth of plants and different microorganisms. Soil also helps in the regulation of the temperature of Earth. The important components of soil are soil separation (sand, silt, and clay), organic materials, minerals, water, and air. Due to rapid population growth, urbanization, and industrialization, this valuable natural resource is being contaminated due to the addition of various harmful substances. Soil, as a natural resource supports the human food production system and the cultivation of vegetation for fuel, fiber, and fodder. Natural resource soil is very important because it is used to meet the needs of living organisms for their survival [2]. Some important functions of the soils are listed as:

- It supports the root system.
- It provides the roots with minerals and nutrients.
- It helps in the exchange of gases and oxygen and protects against erosion.
- It filters various properties of the soil, retains water, and decomposes organic matter.

Soil is the living surface of the crust of the earth, which consists of numerous materials for example minerals, organic material, soil air, and ground water. Soils vary in thickness, texture, configuration, and genetic processes. The creation of different soil types depends on soil, climate, source rocks, organisms, time, and human influence [3]. The aim of this chapter is to explore the possibilities for remediation of contaminated soil having pesticides and their residues, heavy metals, and persistent organic pollutants (POP) using nanotechnology and its role in improving bioremediation and phytoremediation.

CONTAMINATION OF SOIL

Topsoil is deeply affected by environmental toxic waste often considered a "Universal sink". Contaminants usually pass into the soil as sewage, garbage, accidental discharges, or by-products and as remains from the fabrication of different materials. Such pollution of soils can cause unwanted changes in their physical, chemical, and biotic properties, all of which can contribute to changes in soil productivity and potency levels. Any substance found in the soil that exceeds naturally occurring levels and poses a risk to human health is a soil pollutant. For example, Arsenic occurs naturally in some soils, but spraying certain pesticides in the yard can lead to soil contamination, which causes deterioration or loss of one or more functions of soil [1]. Soil contaminants may be of different types such as inorganic and inorganic contaminants or particulate pollutants [5]. Domestic and industrial waste pollutes the soil the most (37%), followed by the industry/commercial sector (33%). Heavy metals and mineral oil are the main pollutants responsible for approximately 60% of soil pollution [6]. There are many ways to contaminate the soil such as:

• Solid waste seepage and waste yards.
• Ejection from industrialized leftovers into the soil.
• Filtration of polluted water into the soil.
• Surplus use of herbicides, pesticides, and fertilizers.

The widespread compounds that are responsible for soil contamination are petroleum hydrocarbons, pesticides, Heavy metals (HMs), and solvents. Soil contaminated with these chemicals has a higher risk of contaminating the food chain on account of the bioaccumulation potential of these contaminants. Cultivation efficiency is severely hampered by both the burden of producing more food and the challenge of stopping more soil deprivation. Nanopowered soil remediation can give a sustainable way to revitalize degraded soil properties. Applications that are based on Nanotechnology are cost-effective, easy to use, and involve adequate treatment and remediation approaches that can significantly diminish soil contamination [7].

SOIL REMEDIATION

Soil remediation is widely regarded as an effective method for mitigating soil contamination. The following measures have been proposed to control soil contamination. We can limit construction in sensitive areas to prevent soil erosion. In general, we would need fewer fertilizers and pesticides if we could all adopt the three R's: reduce, reuse, and recycle. This would reduce solid waste [8]. Some of the soil remediation methods are as follows:

• Extraction and separation techniques
• Thermal methods
• Chemical methods
• Microbial treatment methods
• Reducing the use of chemical fertilizers and pesticide
• Reforesting
• Solid waste treatments.

The main pollutants that soil remediation techniques target are organic composites like pesticides, polychlorinated biphenyls (PCBs), and PAH (polycyclic aromatic hydrocarbons); heavy metals like zinc, lead, arsenic, and chromium; and an extensive variety of combined toxins. Engineered NMs, coupled with different nanotechnology devices and systems, bring forth innovative and enhanced strategies for the restoration of contaminated soil.

Conventional Techniques for Soil Remediation

Among the conventional techniques of soil remediation; bioremediation, chemical, physical, and thermal treatments are the most commonly used techniques. The right soil remediation strategy is chosen by initially identifying the kind of soil and its physical characteristics, the type of contamination present and how feasible it is to isolate and remove it from the soil, the intensity required for handling, and cost-effectiveness.

Bioremediation

Living microorganisms, including fungi and bacteria, play a crucial role in the process of bioremediation, which aims to break down organic pollutants present in soil. *in situ* techniques involve the utilization of plants capable of accumulating contaminants or specific bacteria to reduce the contamination. On the other hand, ex-situ methods treat slurries or solids by combining them with contaminated soil. Although the treatment process may span several months or even years, bioremediation projects generally prove to be more cost-effective compared to alternative approaches.

Chemical processes

Contaminated soil is turned into harmless soil by chemical processes namely oxidation and reduction; such as stabilization. Once stabilization has been completed, a dry solid is typically generated by blending chemicals to chemically diminish the toxicity and mobility of pollutants present in the soil. Liquid and dry reagents are applied to the soil using specialized machinery, pump and treat systems, or direct injection as part of the chemical oxidation process. Through chemical reactions, the process transforms contaminants into less toxic, more stable compounds.

Physical processes

Three physical treatments are available for contaminated soil: encapsulation, soil washing, and thermal desorption.

The process of encapsulation involves the placement of layers of concrete, clay caps, lime, or synthetic textiles over the pollution source to control the movement of contaminants and prevent their migration from the confined area. Nevertheless, encapsulated soil is not conducive for growing crops, hence it is typically utilized as a last resort for soil remediation.

Soil washing utilizes surfactants and water to eliminate impurities from the soil by suspending or dissolving pollutants in the wash solution and segregating the soil based on particle size.

The thermal desorption soil remediation method uses heat to make contaminants more volatile and help them separate from the soil. After that, the volatilized pollutants are either gathered or thermally destroyed [9].

NANOTECHNOLOGY IN SOIL REMEDIATION

The remediation of contaminated soil presents several technical and financial difficulties due to the heterogeneous and complex soil-water interface. During *in situ* monitoring, soil remediation specialists specifically struggle to achieve effective mixing and exchanges among the remediation agents and the target chemicals. The time and energy requirements, lack of financial support, and environmental unfriendliness of many newly developed technologies, like sequestration, immobilization, and bioremediation, further limit their applicability in soil remediation scenarios. Comparatively, lesser size, large specific surface area, suitable reactivity, and versatility are associated with nanotechnology and ENMs [10]. Academicians have discovered that the addition of ENMs to conventional *in situ* approaches can enable the simultaneous elimination of several pollutants, improving the combined soil remediation techniques, because of the inherent benefits of these technologies [10, 11]. Therefore, to eliminate obstinate pollutants from complicated environmental media like soil, ENMs are the perfect instruments [12]. The efficiency of nano-phytoremediation in eradicating heavy metals (HMs) from polluted sites is significantly influenced by the level of impurity present, the bioavailability of metals, and the plant's capacity to accumulate these metals [13, 14].

It has been documented that a range of ENMs, encompassing carbon NMs, cellulose NMs as well as metal and metal oxide NMs, possess the capability to enhance agricultural crop production. This is achieved through their dual role as nutrient stimulants (as observed in metal and oxide nanoparticles) or as carriers/platforms for the efficient delivery of said nutrients across seeds, roots, and leaves [15].

APPLICATIONS OF NANOTECHNOLOGY IN SOIL REMEDIATION

Immobilization by NMs

Immobilization is one of the prominent ways of remediation. The effectiveness, affordability, and environmental friendliness of this kind of remediation technique are some of its most noteworthy benefits for removing metal pollutants from the soils. A variety of ENMs, such as carbon NMs and metal oxide NMs for example titanium oxide (TiO) and ferric oxide (Fe_2O_3), Carbon Nanotubes, and many other nanocomposites, have been used to immobilize soil contaminants. For instance, FeO NMs have a remarkable ability to absorb as well as immobilize heavy metals from various media samples, including arsenic and cadmium [12].

Photocatalytic Degradation by ENMs

Nano-photocatalysts are employed along with an ultraviolet light source like sunlight, during the photocatalytic degradation of soil contaminants to encourage the degradation of organic materials for example pesticides, persistent organic pollutants (POPs) like Polycyclic aromatic hydrocarbons (PAHs), and polychlorinated biphenyls (PCBs). The inherent qualities of the polluted soil samples, their acidity levels, and the presence or absence of organic matter all affect the effectiveness of this soil remediation technique. TiO is popularly used ENMs in this process of photocatalytic degradation because it can achieve up to 78 percent degradation rates in just five hours of treatment [15 - 17]. Fig. (**1**) depicts the photocatalytic application of NMs in the degradation of pollutants.

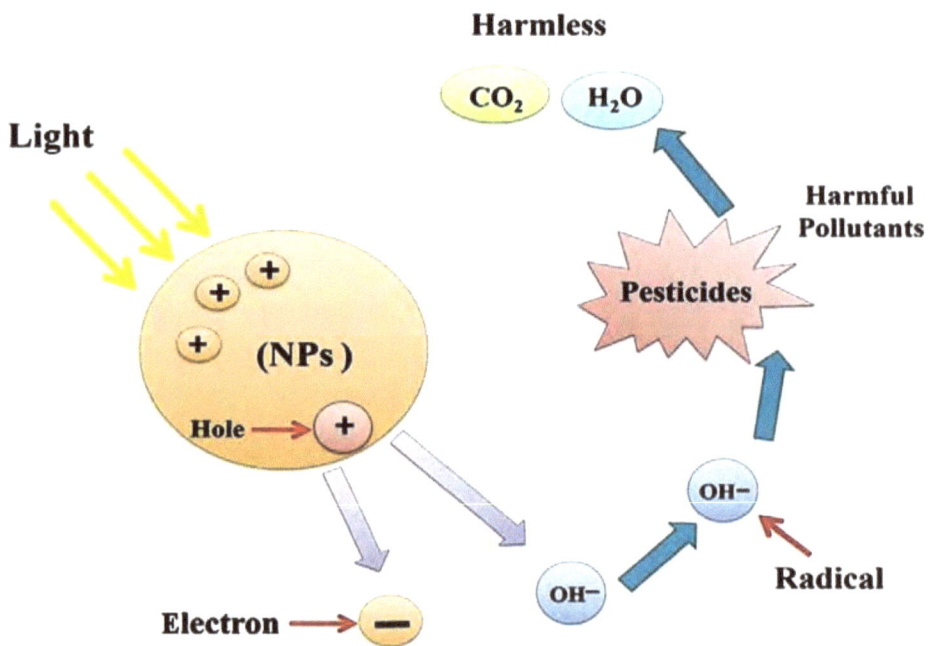

Fig. (1). Application of NMs in photocatalytic degradation [6].

The development of nanoparticles (NPs) may illuminate the environment-friendly and sustainable route to hasten the elimination of toxic elements from adulterated soils. Several efforts have been undertaken to enhance the efficiency of phytoremediation, *viz.* employing rhizobacteria, genetic modification, and chemical additives. In this regard, new avenues for improving remediation techniques have been opened up by the fusion of nanotechnology and bioremediation. Therefore, biological and nanotechnological remediation techniques are combined in advanced remediation approaches, where the regulation of nanoscale processes facilitates the adsorption and degradation of pollutants. Because of their special surface characteristics, NPs not only absorb a wide range of contaminants but also enhance the reaction rate by dropping the energy needed to break them down. Therefore, this method limits the spread of impurities from one medium to another while reducing their accumulation [18].

The degradation of organic contaminants like pesticides, PAHs, and PCBs in water treatment has been extensively used in photocatalytic degradation employing nano-photocatalysts exposed to UV light [19, 20]. Such a process has also been considered for practice in soil remediation in recent literature [21, 22]. One common ex-situ technique involves nonpolar solvents to wash contaminated soils and using photocatalysts to treat the leached water to remove harmful substances [23, 24]. However, the *in situ* approach would call for the direct addition of ENMs to the polluted soils along with a small amount of light irradiation [25]. Due to the poor light penetration in soil, this approach is severely constrained and would necessitate constant land plows [26, 27]. Overall, characteristics of the soil, pH level, and the occurrence of organic substances were found to be related to the efficiency of photocatalytic oxidation. These variables essentially controlled the number of hydroxyl radicals (OH·) produced in the soil environment by nano-photocatalysts and were important in the reasonable absorption of pollutants besides soil constituents [28, 29]. Since TiO_2 is the most widely used ENM in this field, it has been demonstrated that adding 50 mg of TiO_2 per soil g^{-1} to pyrene-polluted soils (red, quartz sand, and alluvial) can improve their quality.

Multifunctional ENMs and Combined Techniques

The development and creation of versatile ENMs, capable of simultaneously targeting various types of contaminants and exhibiting superior selectivity towards contaminants amidst intricate matrix components, signifies a notable progression. By concentrating and exposing pollutants to a comparatively high concentrated volume of reactants, the symbiotic role of ENMs in degradation and adsorption enhances the efficacy of removal and/or conversion. Here, for example, contaminants brought to the surface of ENMs can enhance the surface-driven degradation processes by facilitating the transfer of electrons among ENMs and target composites as well as the generation of free radicals. The encouragement of surface oxidation in the water and wastewater treatment process is essential, as it serves as a key component of advanced oxidation processes

(AOPs), like the Fenton reaction, as well as photo- and electro-catalytic oxidation processes. These processes produce OH⁻ radical over the catalyst or electron surfaces in real-time. Using CNT-supported TiO_2 as an example [30] or Pt/SiC single-atom [31] as catalysts has been investigated for surface-mediated reactions that break down resistant substances, like PFASs. Additionally, new nanocomposites have been suggested to enhance the durability of the sequestration effectiveness under complicated circumstances and to accomplish the simultaneous elimination of organic and inorganic substances [32, 33].

It is assumed that the best soil for a plant's growth and optimal health is one that is high in essential nutrients and micronutrients [34]. Man-made actions have presently contaminated the soil with various types of POPs compounds, which are HMs like Hg and Pb, volatile organic compounds (VOCs), PAHs, PCBs, agrochemicals (fertilizers, fungicides, and pesticides), and occasionally excess nutrients [35]. In addition to this, solid wastes, different kinds of chemicals, and solvents have been added to the environment and farming land as a result of urbanization and industrialization [36]. By using plants and microbes to break down polluting compounds, nano bioremediation is an affordable process for improving soil quality and decreasing contamination. The process can be able to eliminate, or reduce the volume of contaminants that appear by decomposing them in the soil [37, 38]. In the past, the application of biotechnology and chemical additives has been instrumental in researching and enhancing the efficiency of bioremediation [39, 40], but nanotechnology added a new dimension to the process. Table **1** elaborates on the various NP-mediated methods for removing contaminants from contaminated media. When a procedure is carried out at the nanoscale, it is known as nano bioremediation, which combines nanotechnology and bioremediation. The distinctive physicochemical features of NPs, which also act as catalysts, cause the target pollutants to be adsorbed, broken down, or altered. The distinct physicochemical characteristics of the NPs cause the target pollutants to be adsorbed, degraded, or modified. They also function as catalysts, by reducing the activation energy, which is required to break down the compounds [41]. The most widely used NPs in the nanobioremediation process are carbon- and metal-based [42, 43]. The removal of long-chain hydrocarbons and POPs is another remarkable application of polymeric NPs in the form of nanospheres or nanocapsules [44]. However, when it comes to HMs, the situation is completely different because they are not biodegradable and are more likely to find their way into food chains and biological systems [45]. Traditional methods to remove HMs from polluted soils include biosorption and bioaccumulation using microbes and plants. Nonetheless, new research has shown that NPs can be used to remediate heavy metals with impressive results [46]. It has been reported that applying NPs to particular microbes simultaneously or sequentially has produced convincing results [47]. By serving as microbes' nanocarriers or microbial biosorbents, they might hasten the removal of HMs [48]. Fig. (**2**) shows a schematic diagram that illustrates the steps involved in nanobioremediation, particularly about biogenic nanoparticles.

Table 1. Elimination of various pollutants from contaminated soil by using different NPs [18].

NPs used	Contaminant(s) Removed	Efficiency and condition for Operation	Refs.
Iron oxide nanoparticles coated with PVP (Polyvinylpyrrolidone)	Cadmium and lead	Bioremediation process mediated *via* Halomonas sp. *Halomonas* sp. was inoculated for 48 Hours at 180 rpm, and a temperature 28°C. For lead, complete elimination was recorded after 24 hours, whereas for Cd, it took 48 hours.	[53]
Zero valent iron (nZVI) commercial suspension at two different doses *i.e* 1% and 10%.	Arsenic	pH value of nZVI suspension $= 12.2 \pm 0.1$ To avoid the aggregation of nZVI in the suspension, polyacrylic acid is used as a stabilizer. At 10% of nZVI, Maximum immobilization of As in brownfield soil was observed.	[54]
Titanium oxide NMs bonded chitosan nanolayer (NTiO$_2$ - NCh)	Metals like lead, cadmium, zinc, copper, and arsenic Metals in contaminated soil	Nzvi and nGOx are applied to the contaminated soils significantly impacting the availability of As and metals. Cd, Pb, and Cu were immobilized by nGOx were immobilized by nGOx, whereas mobilized P and As. In the case of nZVI, it immolized the Pb and As effectively, and poorly Cd whereas improved the availability of Cu. This study shows that applications of both Nanoparticles might act as schemes for the stabilization and immobilization that can be utilized for phytoremediation at a later stage.	[55]
Palladium (Pd), Pd NPs	Copper and cadmium	pH = 7.0 during the investigation. the microwave–enforced absorption approach was used for 60-70s heating to assist in the removal. The use of NTiO$_2$-NCh was found to eradicate Cd and Cu by 70.67% and 88.01% respectively.	[56]

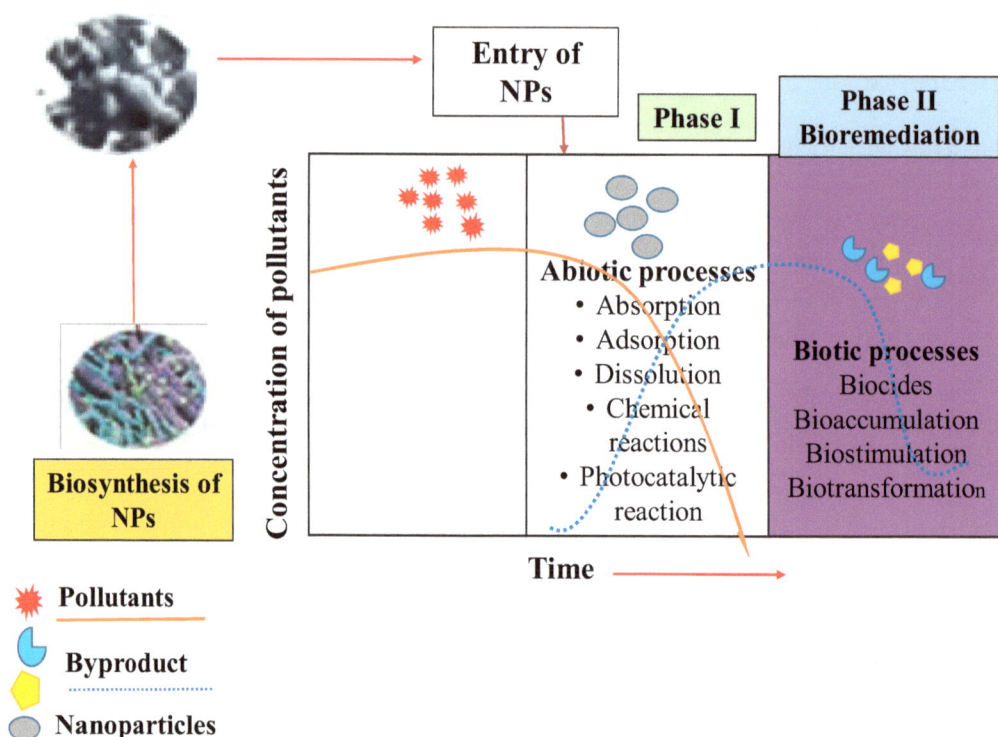

Fig. (2). Uses of biogenic NPs in the Nano bioremediation process [18].

The relationship between NPs and bacterial degradation has also drawn attention, but there are not many published studies available, so it is impossible to conduct a systematic review. Despite the efforts of numerous researchers, nanobioremediation remains too small to support a firm conclusion [16]. The incorporation of NPs with microbes for bioremediation involves a convergence of abiotic and biotic processes across two distinct phases [49]. After the introduction of NPs into the system, pollutants go through several physiochemical changes in the first phase of the process, which show abiotic processes like dissolution, adsorption, absorption, and chemical catalysis of photocatalytic reactions [50]. The second phase involves various biological processes *viz*. bioaccumulation, biocides, biotransformation, and biostimulation [51, 52]. These biotic processes are vital to the system's ability to eliminate pollutants.

NPs have additional beneficial characteristics which include more adjustable pore size, reduced temperature modification, and diverse surface chemistry [57]. Because of these characteristics, they can be excellent catalysts for both chemical reduction and the mitigation of the relevant pollutants. Therefore, highly toxic pollutants, such as HMs, chlorinated organic solvents, pesticides, PAHs, and PCBs have gained a lot of attention and are being seriously considered for the remediation of soil atmospheres Fig. (**3**).

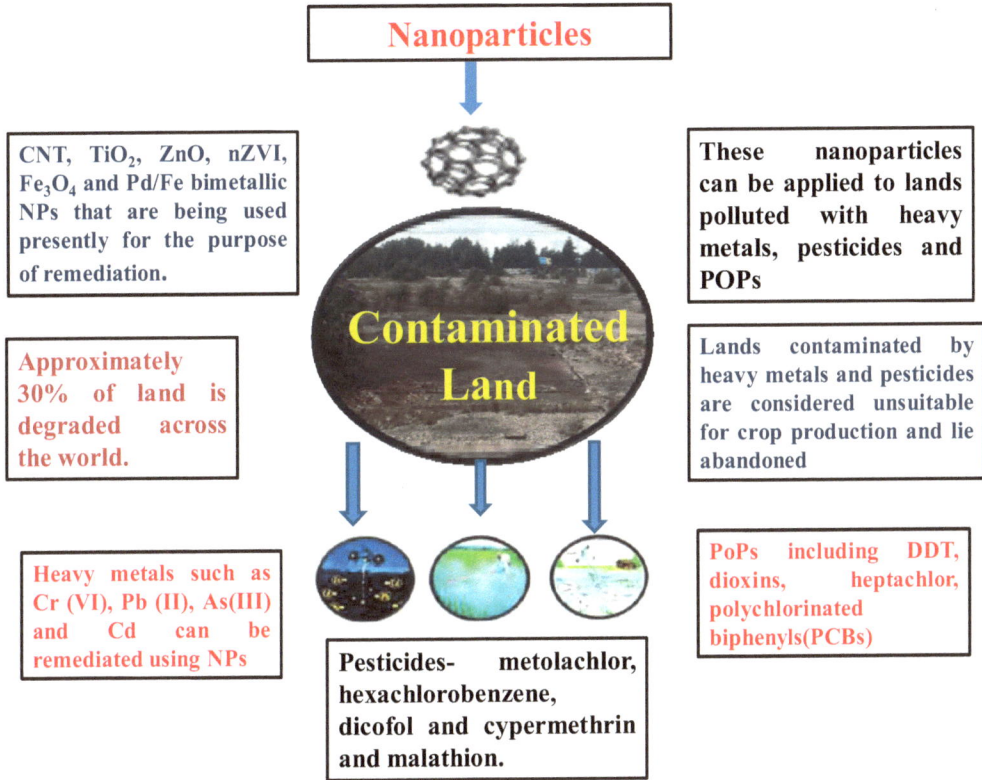

Fig. (3). NPs used for the removal of pesticides, HMs, and POPs from contaminated soil [7]..

The Role of NMs in the Remediation of HMs Contaminated Soil

Wang *et al.* used carboxymethyl cellulose (CMC) stabilized iron sulfide NPs (FeS NPs) to clean groundwater and Cr(VI) contaminated soil [58]. They found that FeS NPs efficiently immobilized Cr(VI) through reduction, coprecipitation, and adsorption. Another study reduced the motion of Cr and Mercury (Hg) in contaminated calcareous soil by using water treatment residual NPs (nWTR), which are generated through the purification of the water process and used as waste in the water industry [59]. When these nWTR were applied to the contaminated soil, the release of Hg and Cr from the soil was significantly reduced while the metal absorption process for both Cr and Hg was greatly increased. In the polluted soil, an elevated variation of Cr and Hg ions happened at the Nwtr application rate of 0.3 percent. Stable complexes of $Cr(OH)_2$ and Hg

$(OH)_2$ respectively, were formed by both Cr and Hg. A group of workers produced iron-based NPs stabilized by starch, containing FeS, magnetite (Fe_3O_4), and zerovalent iron (ZVI) [60]. When they used it to immobilize As, it was found that the bioaccessibility and leachability of As in soil were reduced as the Fe/As molar ratio increased. It was suggested that these NPs can be used for *in situ* immobilization of Ar in soil having higher concentrations of Ar and lower concentrations of Fe. Fe (II) Phosphate NPs synthesized by Liu and Zhao, decreased the bioaccessibility of soil-bound Cu(II) and Pb(II) [61].

By utilizing a novel variety of calcium phosphate NPs, the authors managed to significantly lower the TLCP leaching fraction of lead in soils from 66 percent to 10 percent [62]. Because of the small size of the particle, the NPs in both studies had higher mobility and reactivity in soils, and they demonstrated encouraging outcomes for *in situ* soil remediation. It was also predicted that these NPs would be useful in the remediation of radioactive nuclides and other dangerous HMs such as Cd, Cu, Zn, and U. Furthermore, they recommended using these NPs as nano fertilizer to meet crops' phosphorus requirements while resolving the eutrophication concern, which is often connected to the use of conventional fertilizers that comprise of phosphate [62]. Table **2** summarises various types of NMs that have been used to remediate HM-contaminated soil.

Table 2. Different types of NMs used to eradicate HMs [7].

NPs	Targeted heavy metals	Outcomes	Refs.
nZVI	Cr(VI)	Cr(VI) reduced to Cr(III)	[63]
nZVI	Zn and Pb	concentrations reduced in leachates and effective immobilization	[64]
nZVI	As, Pb Cr, Zn, Cd	metal availability reduced	[65]
CMC-nZVI	Cr (VI)	Reactive immobilization of Cr(VI)	[66]
CMC Stabilized	Cr (VI)	Reduction of Cr (VI) to Cr (III) in soil and water	[67]
CMC STABILIZED FeS NPs	Cr (VI)	Effective immobilization of Cr(VI) through reduction, coprecipitation and adsorption	[58]
Fe (II) phosphate	Pb(II)	Bioaccessibility and leachability of soil-bound Cu (II) and Pb(II) are reduced	[61]
Ca (II) Phosphate	Pb (II)	*in situ* immobilization of Pb (II) and possible	[62]

Reduction of POPs

POPs are very harmful to the environment and to living beings because they bioaccumulate in adipose tissue and may even act as carcinogens. The pesticide hexachlorocyclohexane (HCH) was reported to be degraded by a Gram-negative bacterial strain, NM05 of Sphinomonas [68, 69]. It was discovered that the experimental parameters (pH, temperature, HCH concentrations, *etc.*) had an impact on the degradation process [70]. The peroxymonosulphate (3×10^{-4} M) that assisted in the oxidation process for oxidizing PAHs in the sediments was activated in the study using LFBC *i.e* lignocellulosic fiber reinforce biodegradable composites at 0.75g/L and pH value 6.0. Although up to 90% of the total degradation was attained, specific levels of 4-ring PAHs (66%), 3-ring PAHs (61%), 5 rings PAHs (56%), 2-ring PAHs (52%), and 6- ring PAHs(29%) were noted [71]. The procedure also revealed increased sediment microbial diversity, with the phylum proteobacteria being the most commonly observed at first, but Hyphomonas predominated following the procedure [72]. The degradation of naphthalene in groundwater was carried out using 400 mg/L of synthesized CaO_2 NPs on naphthalene at an optimal concentration of 20 mg/L within a continuous-flow experimental system. This study shows that within 50 days, naphthalene can be completely remedied from column effluent in the presence of CaO_2 NPs and microbes (a large number of Coccibacilli) [73]. Applying NPs to the soil can enhance its microbial community, which is another method of lowering or eliminating the levels of harmful pollutants. Si NPs have been shown to enhance biomass and microbial colonization, particularly rhizospheric microbes that contribute to better soil health [74, 75]. Nonetheless, extended exposure to and buildup of these NPs in the soil may have an impact on the amount of organic matter and nutrients.

Restoration of Metalloid-Contaminated Soils through Nanotechnology-based Approaches

Metalloids are elements that possess characteristics of both metals and nonmetals. While they exist naturally, the industrial revolution has led to elevated concentrations that exceed safe thresholds [76]. Boron (B), Germanium (Ge), Antimony (Sb), Tellurium (Te), arsenic (As), and silicon (Si) are the six metalloids recognized in the periodic table [77]. The contamination of soil by metalloids is a worldwide issue, with these substances affecting human health, plant life, and animal populations. Even in trace amounts, metalloids can be hazardous [78]. Furthermore, metalloids can have a significant impact on the composition, abundance, and diversity of soil microbial communities, which leads to the altering of ecosystem dynamics. Because FeO NPs can absorb HMs, they are thought to be effective at remediating these contaminants from soils [79]. In this particular context, Zhang *et al.* demonstrated the efficiency of NPs based on iron, *i.e.* particles of FeS, Fe_3O_4, and ZVI to immobilize as in contaminated soils [80]. Fe_3O_4 NPs were found to be superior to other NPs in terms of immobilizing them. Other NPs are also considered as remediating soils contaminated with diverse metals and metalloids, in addition to oxides of iron [81]. Other well-known Fe NPs besides magnetite include nZVI, which is used to treat water that has been contaminated by metals or metalloids. The photolytic system was employed in a study to decrease the amount of toxic Cr^{6+} chromium and convert it to non-toxic Cr^{3+} in aqueous [82]. Up to 90% of the Cr^{6+} reduction efficiency was observed in the same study. Additionally, in certain instances, NPs were utilized to immobilize the trace elements present in the polluted soil. For instance, it was found that nanostructured TiO_2, ZnO, and MgO were effective adsorbents for removing Cr ions from the soil treated with leather factory waste [83]. Therefore, these aspects can be further examined to envisage how these NPs might be used to remove metalloids from contaminated soils. The ability of the nanoscale amorphous MnO to clean up soil contaminated with HMs like Cd, Cu, Zn, As, and Pb was also assessed [84, 85]. Certain NPs are used to eliminate HMs from other contaminated mediums, like water, so even though they aren't as frequently employed in metalloid remediation they can still be used in this context.

Pesticide Residue Detection using Biosensing Systems based on Nanotechnology

While the research on NPs and NMs-based sensors or nanosensor systems, is still in its early stages, there is much promise for these materials in the fields of agriculture and environmental remediation. NM-integrated biosensors are useful for monitoring and detecting a variety of substances, including pesticides, hazardous chemicals, pathogens, bacteria that cause odors, and other microbes. These applications can be found in both field and production sites [86]. Pesticide residue detection is just one of the many services that nanosensors can provide, and they can thus play a major role in enhancing soil health and productivity overall. Furthermore, food manufacturers are currently using nanosensors to detect chemicals, toxins, and pathogens in food, which helps prevent foodborne illnesses.

There are three primary categories of nanosensing techniques based on how they sense things. These are (a) Organophosphorus hydrolase (OPH) (b) immunoassays, and (c) inhibition of cholinesterase [87]. The majority of sensors currently under development for the detection of pesticide residues are based on Acetylcholinesterase (AChE) and utilize electrochemical principles. These sensors are particularly effective in detecting organophosphate and carbamate pesticides, both of which pose a significant threat due to their high toxicity to

AChE. AChE is a crucial enzyme responsible for the proper working of the human central nervous system and is extensively targeted by these pesticides in various nations. The particular molecular target of organophosphate and carbamate pesticides is the inhabitation of AChE functioning. Consequently, this inhibition has been employed to develop biological markers for the primary revealing of poisoning due to exposure to these pesticides [88].

To measure carbamate pesticides in apples, cabbage, and broccoli, Cesarino *et al.* (2012) created a sensor based on the immobilization of cholinesterase in a core-shell modified glassy electrode of MWCNTs (Multiwalled Carbon Nanotubes) and polyaniline [89]. Wu *et al.* also created a sensor to monitor dichlorvos, a hazardous organophosphate for AChE immobilization [90]. To detect methyl parathion, Dong *et al.* created a biosensor by immobilizing AChE on a glassy carbon electrode modified with chitosan nanocomposites and MWCNTs. This technique combined AChE inhibition with the electrochemical reduction of Ellman's reagent, DNTB. The detection point for methyl parathion was an inhibition effect on AChE activity, which was monitored by a change in the electrochemical reduction response of DTNB [91].

CHALLENGES AND FUTURE OF NANOTECHNOLOGY IN SOIL REMEDIATION

Even with the clear benefits of using ENMs for soil remediation, there are some safety and environmental health issues to take into account. The native soil ecosystem will unavoidably be altered by the introduction of ENMs. These materials may have an impact on the sprouting of plant seeds, the growth and development of plant roots as well as shoots, the growth and metabolism of soil microorganisms, and even the lives of specific invertebrate creatures that live in the soil, like earthworms, snails, and other insects. It should be remembered that, although the possible negative effects of ENMs must be taken into account, there are situations in which ENM introduction may benefit the soil ecosystem. For example, it has been demonstrated that metal and metal oxide NMs, cellulose NMs, and carbon NMs all function as nutrient stimulants or improve the transfer of particular nutrients to the soil *via* plant parts like roots, seeds, and leaves. Thus, these kinds of materials lead to a decrease in the occurrence of undesirable soil contaminants while also increasing crop yields and improving agricultural practices.

Recent reports have shown that there are significant consequences associated with the growing use of NPs in agriculture. For instance, the intentional application of NPs may cause them to accumulate or cause an increase in the concentration level of their constituents in the soil, which would alter the properties of the soil [92].

One of the most significant factors influencing the availability of soil nutrients, general soil health, microbial dynamics, and plant growth as well as development is soil pH, which is known to be transformed by the presence of NPs in soils [93].

CONCLUSION

The nanotechnology-based methods have great potential to restore contaminated land to its optimal forms, which facilitates the growth of plants and microbes. As discussed above, the efficient and sustainable use of nanomaterials (NMs) can enhance the productivity and fertility of contaminated soils by sustainable remediation approaches that ensure a safe and healthy environment. There is encouraging potential for nano remediation to reduce the overall cost and time needed for large-scale clean-up of polluted sites. Additionally, they work well for remediation that takes place on-site, which excludes the requirement for soil disposal, treatment, and transportation after remediation. Undoubtedly, nanotechnology has proven to be a good substitute for existing methods because of all the advantages it provides but researchers must also focus on any possible potential risks it might present to the environment. Before the widespread utilization of NPs, it is imperative to conduct in-depth ecosystem analyses and extensive long-term assessments to avert any potential detrimental environmental consequences.

REFERENCES

[1]   Midhat L, Ouazzani N, Hejjaj A, Ouhammou A, Mandi L. Accumulation of heavy metals in metallophytes from three mining sites (Southern Centre Morocco) and evaluation of their phytoremediation potential. Ecotoxicol Environ Saf 2019; 169: 150-60.
[http://dx.doi.org/10.1016/j.ecoenv.2018.11.009] [PMID: 30445246]

[2]   Parikh SJ, James BR. Soil: The foundation of agriculture. Natl Educ. Knowl 2012; p. 3.

[3]   De Deyn GB, Kooistra L. The role of soils in habitat creation, maintenance and restoration Philos Trans R Soc Lond B Biol Sci 2021; 376(1834): 20200170.

[4]   JRC.2014. Soil themes: soil contamination. Available from: http://eusoils.jrc.ec.europa.eu/library/themes/contamination

[5]   Mirsal IA. Soil Pollution: Origin, Monitoring & Remediation. 2nd ed., Berlin: Springer-Verlag 2008.

[6]   Panagos P, Van Liedekerke M, Yigini Y, Montanarella L, Montanarella L. Contaminated sites in Europe: review of the current situation based on data collected through a European network. J Environ Public Health 2013; 2013: 1-11.
[http://dx.doi.org/10.1155/2013/158764] [PMID: 23843802]

[7]   Bakshi M, Abhilash P C. Nano-Materials as Photocatalysts for Degradation of Environmental Pollutants: Challenges and Possibilities 2020; 345-70.

[8]   Ashraf MA, Maah MJ, Yusoff I. Soil Contamination. Risk Assessment and Remediation 2014.

[9]   Polyguard.   Encapsulation   of   contaminated   soil.   Available   from: https://polyguard.com/architectural/blog/encapsulation-of-contaminated-soil

[10]   Li Y, Xu R, Ma C, *et al.* Potential functions of engineered nanomaterials in cadmium remediation in soil-plant system: A review. Environ Pollut 2023; 336: 122340.

[http://dx.doi.org/10.1016/j.envpol.2023.122340] [PMID: 37562530]

[11] Albalawi F, Hussein MZ, Fakurazi S, Masarudin MJ. Engineered Nanomaterials: The Challenges and Opportunities for Nanomedicines. Int J Nanomedicine 2021; 16: 161-84.
[http://dx.doi.org/10.2147/IJN.S288236] [PMID: 33447033]

[12] Cuffari B. How is Nanotechnology Used in Soil Remediation? AZoNano 2020.

[13] Rajput V, Minkina T, Semenkov I, Klink G, Tarigholizadeh S, Sushkova S. Phylogenetic analysis of hyperaccumulator plant species for heavy metals and polycyclic aromatic hydrocarbons. Environ Geochem Health 2020; 16: 68-75.
[PMID: 32040786]

[14] Ghazaryan KA, Movsesyan HS, Khachatryan HE, *et al.* Copper phytoextraction and phytostabilization potential of wild plant species growing in the mine polluted areas of Armenia. Environ Geochem Health 2018; 19: 155-63.

[15] Aqeel M. Jamil Mohd, Yusoff I. Soil Contamination, Risk Assessment and Remediation. Environmental Risk Assessment of Soil Contamination. InTech 2014.
[http://dx.doi.org/10.5772/57287]

[16] Qian Y, Qin C, Chen M, Lin S. Nanotechnology in soil remediation – applications *vs.* implications. Ecotoxicol Environ Saf 2020; 201: 110815.
[http://dx.doi.org/10.1016/j.ecoenv.2020.110815] [PMID: 32559688]

[17] Wang A, Teng Y, Hu X, *et al.* Diphenylarsinic acid contaminated soil remediation by titanium dioxide (P25) photocatalysis: Degradation pathway, optimization of operating parameters and effects of soil properties. Sci Total Environ 2016; 541: 348-55.
[http://dx.doi.org/10.1016/j.scitotenv.2015.09.023] [PMID: 26410709]

[18] Rajput VD, Minkina T, Upadhyay SK, *et al.* Nanotechnology in the Restoration of Polluted Soil. Nanomaterials (Basel) 2022; 12(5): 769.
[http://dx.doi.org/10.3390/nano12050769] [PMID: 35269257]

[19] Zhu Y, Liu X, Hu Y, *et al.* Behavior, remediation effect and toxicity of nanomaterials in water environments. Environ Res 2019; 174: 54-60.
[http://dx.doi.org/10.1016/j.envres.2019.04.014] [PMID: 31029942]

[20] Osin OA, Yu T, Cai X, *et al.* Photocatalytic degradation of 4-nitrophenol by C, N-TiO2: degradation efficiency *vs.* embryonic toxicity of the resulting compounds. Front Chem 2018; 6: 192.
[http://dx.doi.org/10.3389/fchem.2018.00192] [PMID: 29915782]

[21] Rachna MR, Rani M, Shanker U. Degradation of tricyclic polyaromatic hydrocarbons in water, soil and river sediment with a novel TiO_2 based heterogeneous nanocomposite. J Environ Manage 2019; 248: 109340.
[http://dx.doi.org/10.1016/j.jenvman.2019.109340] [PMID: 31386991]

[22] Zuo, Fanjiao, Yameng Zhu, Tiantian Wu, Caixia Li, Yang Liu, Xiwei Wu, Jinyue Ma, Kaili Zhang, Huizi Ouyang, Xilong Qiu, and et al. 2024. "Titanium Dioxide Nanomaterials: Progress in Synthesis and Application in Drug Delivery" Pharmaceutics 16, no. 9: 1214.
[http://dx.doi.org/10.3390/pharmaceutics16091214]

[23] Fenoll J, Flores P, Hellín P, Hernández J, Navarro S. Minimization of methabenzthiazuron residues in leaching water using amended soils and photocatalytic treatment with TiO2 and ZnO. J Environ Sci (China) 2014; 26(4): 757-64.
[http://dx.doi.org/10.1016/S1001-0742(13)60511-2] [PMID: 25079405]

[24] Delgado-Balderas R, Hinojosa-Reyes L, Guzmán-Mar JL, Garza-González MT, López-Chuken UJ, Hernández-Ramírez A. Photocatalytic reduction of Cr(VI) from agricultural soil column leachates using zinc oxide under UV light irradiation. Environ Technol 2012; 33(23): 2673-80.
[http://dx.doi.org/10.1080/09593330.2012.676070] [PMID: 23437668]

[25] Zhang L, Li P, Gong Z, Li X. Photocatalytic degradation of polycyclic aromatic hydrocarbons on soil

surfaces using TiO2 under UV light. J Hazard Mater 2008; 158(2-3): 478-84.
[http://dx.doi.org/10.1016/j.jhazmat.2008.01.119] [PMID: 18372106]

[26] Gu J, Dong D, Kong L, Zheng Y, Li X. Photocatalytic degradation of phenanthrene on soil surfaces in the presence of nanometer anatase TiO_2 under UV-light. J Environ Sci (China) 2012; 24(12): 2122-6.
[http://dx.doi.org/10.1016/S1001-0742(11)61063-2] [PMID: 23534208]

[27] Dong D, Li P, Li X, *et al.* Photocatalytic degradation of phenanthrene and pyrene on soil surfaces in the presence of nanometer rutile TiO_2 under UV-irradiation. Chem Eng J 2010; 158(3): 378-83.
[http://dx.doi.org/10.1016/j.cej.2009.12.046]

[28] Miditana SR, Tirukkovalluri SR, Raju IM, Alim SA. Photocatalytic degradation of Orange-II by surfactant assisted Mn/Mg co-doped TiO_2 nanoparticles under visible light irradiation. Current Chemistry Letters 2024; 13(1): 265-76.
[http://dx.doi.org/10.5267/j.ccl.2023.6.003]

[29] Chang Chien SW, Chang CH, Chen SH, Wang MC, Madhava Rao M, Satya Veni S. Effect of sunlight irradiation on photocatalytic pyrene degradation in contaminated soils by micro-nano size TiO_2. Sci Total Environ 2011; 409(19): 4101-8.
[http://dx.doi.org/10.1016/j.scitotenv.2011.06.050] [PMID: 21762957]

[30] Song C, Chen P, Wang C, Zhu L, Madhava Rao M, Satya Veni S. Photodegradation of perfluorooctanoic acid by synthesized TiO_2–MWCNT composites under 365 nm UV irradiation. Chemosphere 2012; 86(8): 853-9.
[http://dx.doi.org/10.1016/j.chemosphere.2011.11.034] [PMID: 22172634]

[31] Dahong H. Glen Andrew de V, Chiheng C, Qianhong Z, Eli S, Jing M, Huolin Xi, Jacob A, Charles AS, Junfeng N, Gary L, Haller, Jae-Hong K. Single-atom Pt catalyst for effective C–F bond activation *via* hydro defluorination. ACS Catal 2018; 8: 9353-8.

[32] Liu J, Cheng S, Cao N, *et al.* Actinia-like multifunctional nanocoagulant for single-step removal of water contaminants. Nat Nanotechnol 2019; 14(1): 64-71.
[http://dx.doi.org/10.1038/s41565-018-0307-8] [PMID: 30478276]

[33] Huang Y, Fulton AN, Keller AA. Simultaneous removal of PAHs and metal contaminants from water using magnetic nanoparticle adsorbents. Sci Total Environ 2016; 571: 1029-36.
[http://dx.doi.org/10.1016/j.scitotenv.2016.07.093] [PMID: 27450251]

[34] McGrath JM, Spargo J, Penn CJ. Soil Fertility and Plant Nutrition. In: Van Alfen NK, Ed. Encyclopedia of Agriculture and Food Systems. Oxford, UK: Academic Press, Elsevier 2014; pp. 166-84.
[http://dx.doi.org/10.1016/B978-0-444-52512-3.00249-7]

[35] Gorovtsov A, Demin K, Sushkova S, *et al.* The effect of combined pollution by PAHs and heavy metals on the topsoil microbial communities of Spolic Technosols of the lake Atamanskoe, Southern Russia. Environ Geochem Health 2021; 1-17.
[PMID: 34528142]

[36] Rajput VD, Yadav AN, Jatav HS, Singh SK, Minkina T. Sustainable Management and Utilization of Sewage Sludge. Cham: Springer 2022.
[http://dx.doi.org/10.1007/978-3-030-85226-9]

[37] Ashraf S, Siddiqa A, Shahida S, Qaisar S. Titanium-based nanocomposite materials for arsenic removal from water: A review. Heliyon 2019; 5(5): e01577.
[http://dx.doi.org/10.1016/j.heliyon.2019.e01577]

[38] Chandra R, Kumar V, Tripathi S, Sharma P. Phytoremediation of industrial pollutants and life cycle assessment Phytoremediation of Environmental Pollutants. Boca Raton, FL, USA: CRC Press 2017; pp. 441-70.
[http://dx.doi.org/10.4324/9781315161549-18]

[39] Gerhardt KE, Gerwing PD, Greenberg BM. Opinion: Taking phytoremediation from proven

technology to accepted practice. Plant Sci 2017; 256: 170-85.
[http://dx.doi.org/10.1016/j.plantsci.2016.11.016] [PMID: 28167031]

[40]   Gong X, Huang D, Liu Y, *et al.* Remediation of contaminated soils by biotechnology with nanomaterials: bio-behavior, applications, and perspectives. Crit Rev Biotechnol 2018; 38(3): 455-68.
[http://dx.doi.org/10.1080/07388551.2017.1368446] [PMID: 28903604]

[41]   Mehndiratta P, Jain A, Srivastava S, Gupta N. Environmental pollution and nanotechnology. Environ Pollut 2013; 2(2): 49.
[http://dx.doi.org/10.5539/ep.v2n2p49]

[42]   Gong JL, Wang B, Zeng GM, *et al.* Removal of cationic dyes from aqueous solution using magnetic multi-wall carbon nanotube nanocomposite as adsorbent. J Hazard Mater 2009; 164(2-3): 1517-22.
[http://dx.doi.org/10.1016/j.jhazmat.2008.09.072] [PMID: 18977077]

[43]   Chen C, Tsyusko OV, McNear DH Jr, Judy J, Lewis RW, Unrine JM. Effects of biosolids from a wastewater treatment plant receiving manufactured nanomaterials on Medicago truncatula and associated soil microbial communities at low nanomaterial concentrations. Sci Total Environ 2017; 609: 799-806.
[http://dx.doi.org/10.1016/j.scitotenv.2017.07.188] [PMID: 28768212]

[44]   Chauhan R, Yadav HOS, Sehrawat N. Nanobioremediation: A new and a versatile tool for sustainable environmental clean up-Overview. J Mater Environ Sci 2020; 11: 564-73.

[45]   Briffa J, Sinagra E, Blundell R. Heavy metal pollution in the environment and their toxicological effects on humans. Heliyon 2020; 6(9): e04691.
[http://dx.doi.org/10.1016/j.heliyon.2020.e04691] [PMID: 32964150]

[46]   Misra M, Ghosh Sachan S. Nanobioremediation of heavy metals: Perspectives and challenges. J Basic Microbiol. 2021;61(12):1–16.

[47]   Abdi O, Kazemi M. A review study of biosorption of heavy metals and comparison between different biosorbents. J Mater Environ Sci 2015; 6: 1386-99.

[48]   Ayangbenro A, Babalola O. A new strategy for heavy metal polluted environments: A review of microbial biosorbents. Int J Environ Res Public Health 2017; 14(1): 94.
[http://dx.doi.org/10.3390/ijerph14010094] [PMID: 28106848]

[49]   Usman M, Farooq M, Wakeel A, *et al.* Nanotechnology in agriculture: Current status, challenges and future opportunities. Sci Total Environ 2020; 721: 137778.
[http://dx.doi.org/10.1016/j.scitotenv.2020.137778] [PMID: 32179352]

[50]   Abebe B, Murthy HCA, Amare E. Summary on adsorption and photocatalysis for pollutant remediation: Mini review. Journal of Encapsulation and Adsorption Sciences 2018; 8(4): 225-55.
[http://dx.doi.org/10.4236/jeas.2018.84012]

[51]   Desiante WL, Minas NS, Fenner K. Micropollutant biotransformation and bioaccumulation in natural stream biofilms. Water Res 2021; 193: 116846.
[http://dx.doi.org/10.1016/j.watres.2021.116846] [PMID: 33540344]

[52]   Filote C, Roşca M, Hlihor R, *et al.* Sustainable application of biosorption and bioaccumulation of persistent pollutants in wastewater treatment: Current practice. Processes (Basel) 2021; 9(10): 1696.
[http://dx.doi.org/10.3390/pr9101696]

[53]   Cao X, Alabresm A, Chen YP, Decho AW, Lead J. Improved metal remediation using a combined bacterial and nanoscience approach. Sci Total Environ 2020; 704: 135378.
[http://dx.doi.org/10.1016/j.scitotenv.2019.135378] [PMID: 31806322]

[54]   Gil-Díaz M, Diez-Pascual S, González A, *et al.* A nanoremediation strategy for the recovery of an As-polluted soil. Chemosphere 2016; 149: 137-45.
[http://dx.doi.org/10.1016/j.chemosphere.2016.01.106] [PMID: 26855217]

[55]   Baragaño D, Forján R, Welte L, Gallego JLR. Nanoremediation of As and metals polluted soils by

means of graphene oxide nanoparticles. Sci Rep 2020; 10(1): 1896.
[http://dx.doi.org/10.1038/s41598-020-58852-4] [PMID: 32024880]

[56]   Mahmoud ME, Abou Ali SAA, Elweshahy SMT. Microwave functionalization of titanium oxide nanoparticles with chitosan nanolayer for instantaneous microwave sorption of Cu(II) and Cd(II) from water. Int J Biol Macromol 2018; 111: 393-9.
[http://dx.doi.org/10.1016/j.ijbiomac.2018.01.014] [PMID: 29309870]

[57]   Tang WW, Zeng GM, Gong JL, *et al.* Impact of humic/fulvic acid on the removal of heavy metals from aqueous solutions using nanomaterials: A review. Sci Total Environ 2014; 468-469: 1014-27.
[http://dx.doi.org/10.1016/j.scitotenv.2013.09.044] [PMID: 24095965]

[58]   Wang T, Liu Y, Wang J, Wang X, Liu B, Wang Y. In-situ remediation of hexavalent chromium contaminated groundwater and saturated soil using stabilized iron sulfide nanoparticles. J Environ Manage 2019; 231: 679-86.
[http://dx.doi.org/10.1016/j.jenvman.2018.10.085] [PMID: 30391712]

[59]   Moharem M, Elkhatib E, Mesalem M. Remediation of chromium and mercury polluted calcareous soils using nanoparticles: Sorption –desorption kinetics, speciation and fractionation. Environ Res 2019; 170: 366-73.
[http://dx.doi.org/10.1016/j.envres.2018.12.054] [PMID: 30623883]

[60]   Zhang M, Wang Y, Zhao D, Pan G. Immobilization of arsenic in soils by stabilized nanoscale zero-valent iron, iron sulfide (FeS), and magnetite (Fe_3O_4) particles. Chin Sci Bull 2010; 55(4-5): 365-72.
[http://dx.doi.org/10.1007/s11434-009-0703-4]

[61]   Liu R, Zhao D. *in situ* immobilization of Cu(II) in soils using a new class of iron phosphate nanoparticles. Chemosphere 2007; 68(10): 1867-76.
[http://dx.doi.org/10.1016/j.chemosphere.2007.03.010] [PMID: 17462708]

[62]   Liu R, Zhao D. Synthesis and characterization of a new class of stabilized apatite nanoparticles and applying the particles to *in situ* Pb immobilization in a fire-range soil. Chemosphere 2013; 91(5): 594-601.
[http://dx.doi.org/10.1016/j.chemosphere.2012.12.034] [PMID: 23336925]

[63]   Singh R, Misra V, Singh RP. Removal of Cr(VI) by nanoscale zero-valent iron (nZVI) from soil contaminated with tannery wastes. Bull Environ Contam Toxicol 2012; 88(2): 210-4.
[http://dx.doi.org/10.1007/s00128-011-0425-6] [PMID: 21996721]

[64]   Gil-Díaz M, Ortiz LT, Costa G, *et al.* Immobilization and leaching of Pb and Zn in an acidic soil treated with zerovalent iron nanoparticles (nZVI): Physicochemical and toxicological analysis of leachates. Water Air Soil Pollut 2014; 225(6): 1990.
[http://dx.doi.org/10.1007/s11270-014-1990-1]

[65]   Gil-Díaz M, Pinilla P, Alonso J, Lobo MC. Viability of a nanoremediation process in single or multi-metal(loid) contaminated soils. J Hazard Mater 2017; 321: 812-9.
[http://dx.doi.org/10.1016/j.jhazmat.2016.09.071] [PMID: 27720472]

[66]   Madhavi V, Prasad TNVKV, Reddy BR, Reddy AVB, Gajulapalle M. Conjunctive effect of CMC–zero-valent iron nanoparticles and FYM in the remediation of chromium-contaminated soils. Appl Nanosci 2014; 4(4): 477-84.
[http://dx.doi.org/10.1007/s13204-013-0221-1]

[67]   Xu Y, Zhao D. Reductive immobilization of chromate in water and soil using stabilized iron nanoparticles. Water Res 2007; 41(10): 2101-8.
[http://dx.doi.org/10.1016/j.watres.2007.02.037] [PMID: 17412389]

[68]   Singh R, Manickam N, Mudiam MKR, Murthy RC, Misra V. An integrated (nano-bio) technique for degradation of γ-HCH contaminated soil. J Hazard Mater 2013; 258-259: 35-41.
[http://dx.doi.org/10.1016/j.jhazmat.2013.04.016] [PMID: 23692681]

[69]   Xie Y, He Y, Irwin PL, Jin T, Shi X. Antibacterial activity and mechanism of action of zinc oxide

nanoparticles against Campylobacter jejuni. Appl Environ Microbiol 2011; 77(7): 2325-31.
[http://dx.doi.org/10.1128/AEM.02149-10] [PMID: 21296935]

[70]   Hung CM, Huang CP, Chen CW, Dong CD. Degradation of organic contaminants in marine sediments by peroxymonosulfate over LaFeO₃ nanoparticles supported on water caltrop shell-derived biochar and the associated microbial community responses. J Hazard Mater 2021; 420: 126553.
[http://dx.doi.org/10.1016/j.jhazmat.2021.126553] [PMID: 34273879]

[71]   Gholami F, Shavandi M, Dastgheib SMM, Amoozegar MA. Naphthalene remediation from groundwater by calcium peroxide (CaO2) nanoparticles in permeable reactive barrier (PRB). Chemosphere 2018; 212: 105-13.
[http://dx.doi.org/10.1016/j.chemosphere.2018.08.056] [PMID: 30144671]

[72]   Rizwan M, Ali S, Zia ur Rehman M, *et al.* Alleviation of cadmium accumulation in maize (Zea mays L.) by foliar spray of zinc oxide nanoparticles and biochar to contaminated soil. Environ Pollut 2019; 248: 358-67.
[http://dx.doi.org/10.1016/j.envpol.2019.02.031] [PMID: 30818115]

[73]   Rajput VD, Minkina T, Feizi M, *et al.* Effects of silicon and silicon-based nanoparticles on rhizosphere microbiome, plant stress and growth. Biology (Basel) 2021; 10(8): 791.
[http://dx.doi.org/10.3390/biology10080791] [PMID: 34440021]

[74]   Srivastava S, Shukla A, Rajput VD, *et al.* Arsenic Remediation through Sustainable Phytoremediation Approaches. Minerals (Basel) 2021; 11(9): 936.
[http://dx.doi.org/10.3390/min11090936]

[75]   Gajić G, Djurdjević L, Kostić O, Jarić S, Mitrović M, Pavlović P. Ecological Potential of Plants for Phytoremediation and Ecorestoration of Fly Ash Deposits and Mine Wastes. Front Environ Sci 2018; 6: 124.
[http://dx.doi.org/10.3389/fenvs.2018.00124]

[76]   Fordyce FM, Everett PA, Bearcock JM, Lister TR. Soil metal/metalloid concentrations in the Clyde Basin, Scotland, UK: implications for land quality. Earth Environ Sci Trans R Soc Edinb 2017; 108(2-3): 191-216.
[http://dx.doi.org/10.1017/S1755691018000282]

[77]   Yazdi MH, Sepehrizadeh Z, Mahdavi M, Shahverdi AR, Faramarzi MA. Metal, metalloid, and oxide nanoparticles for therapeutic and diagnostic oncology. Nano Biomed Eng 2016; 8(4): 246-67.
[http://dx.doi.org/10.5101/nbe.v8i4.p246-267]

[78]   Martínez-Alcalá I, Bernal MP. Environmental Impact of Metals, Metalloids, and Their Toxicity. Metalloids in Plants. Hoboken, NJ, USA: John Wiley and Sons 2020; pp. 451-88.
[http://dx.doi.org/10.1002/9781119487210.ch21]

[79]   Kumpiene J, Lagerkvist A, Maurice C. Stabilization of As, Cr, Cu, Pb and Zn in soil using amendments--a review. Waste Manag 2008; 28(1): 215-25.
[http://dx.doi.org/10.1016/j.wasman.2006.12.012] [PMID: 17320367]

[80]   Kadhim Q. Jabbar, Azeez A. Barzinjy, Samir M. Hamad, Iron oxide nanoparticles: Preparation methods, functions, adsorption and coagulation/flocculation in wastewater treatment, Environmental Nanotechnology, Monitoring & Management, Volume 17, 2022, 100661, ISSN 2215-1532.
[http://dx.doi.org/10.1016/j.enmm.2022.100661]

[81]   Martínez-Fernández D, Vítková M, Michálková Z, Komárek M. Engineered NMs for Phytoremediation of Metal/Metalloid-Contaminated Soils: Implications for Plant Physiology. In: Ansari AA, Gill SS, Gill R, Lanza GR, Newman L, Eds. Phytoremediation: Management of Environmental Contaminants. Springer 2017; 5: pp. 369-403.
[http://dx.doi.org/10.1007/978-3-319-52381-1_14]

[82]   Kim Y, Joo H, Her N, Yoon Y, Lee CH, Yoon J. Self-rotating photocatalytic system for aqueous Cr(VI) reduction on TiO2 nanotube/Ti mesh substrate. Chem Eng J 2013; 229: 66-71.
[http://dx.doi.org/10.1016/j.cej.2013.05.116]

[83]   Taghipour M, Jalali M. Effect of clay minerals and nanoparticles on chromium fractionation in soil contaminated with leather factory waste. J Hazard Mater 2015; 297: 127-33.
[http://dx.doi.org/10.1016/j.jhazmat.2015.04.067] [PMID: 25956643]

[84]   Della Puppa L, Komárek M, Bordas F, Bollinger JC, Joussein E. Adsorption of copper, cadmium, lead and zinc onto a synthetic manganese oxide. J Colloid Interface Sci 2013; 399: 99-106.
[http://dx.doi.org/10.1016/j.jcis.2013.02.029] [PMID: 23566588]

[85]   Michálková Z, Komárek M, Šillerová H, *et al.* Evaluating the potential of three Fe- and Mn-(nano)oxides for the stabilization of Cd, Cu and Pb in contaminated soils. J Environ Manage 2014; 146: 226-34.
[http://dx.doi.org/10.1016/j.jenvman.2014.08.004] [PMID: 25178528]

[86]   Baruah S, Dutta J. Nanotechnology applications in pollution sensing and degradation in agriculture: a review. Environ Chem Lett 2009; 7(3): 191-204.
[http://dx.doi.org/10.1007/s10311-009-0228-8]

[87]   Liu S, Yuan L, Yue X, Zheng Z, Tang Z. Recent advances in nanosensors for organophosphate pesticide detection. Adv Powder Technol 2008; 19(5): 419-41.
[http://dx.doi.org/10.1016/S0921-8831(08)60910-3]

[88]   Kumar P, Kim KH, Deep A. Recent advancements in sensing techniques based on functional materials for organophosphate pesticides. Biosens Bioelectron 2015; 70: 469-81.
[http://dx.doi.org/10.1016/j.bios.2015.03.066] [PMID: 25864041]

[89]   Cesarino I, Moraes FC, Lanza MRV, Machado SAS. Electrochemical detection of carbamate pesticides in fruit and vegetables with a biosensor based on acetylcholinesterase immobilised on a composite of polyaniline–carbon nanotubes. Food Chem 2012; 135(3): 873-9.
[http://dx.doi.org/10.1016/j.foodchem.2012.04.147] [PMID: 22953799]

[90]   Wu S, Huang F, Lan X, Wang X, Wang J, Meng C. Electrochemically reduced graphene oxide and Nafion nanocomposite for ultralow potential detection of organophosphate pesticide. Sens Actuators B Chem 2013; 177: 724-9.
[http://dx.doi.org/10.1016/j.snb.2012.11.069]

[91]   Dong J, Fan X, Qiao F, Ai S, Xin H. A novel protocol for ultra-trace detection of pesticides: Combined electrochemical reduction of Ellman's reagent with acetylcholinesterase inhibition. Anal Chim Acta 2013; 761: 78-83.
[http://dx.doi.org/10.1016/j.aca.2012.11.042] [PMID: 23312317]

[92]   Neha Agarwal, Vijendra Singh Solanki, Brijesh Pare, Neetu Singh, Sreekantha B. Jonnalagadda, Current trends in nanocatalysis for green chemistry and its applications- a mini-review, Current, Opinion in Green and Sustainable Chemistry, Volume 41, 2023, 100788, ISSN 2452-2236.
[http://dx.doi.org/10.1016/j.cogsc.2023.100788]

[93]   Fernández FG, Hoeft RG. Managing soil pH and crop nutrients. Ill Agron Handb 2009; 24: 91-112.

CHAPTER 16

Nanotechnology in Remediation of Persistent Organic Pollutants

Priyanka Singh[1], Mithlesh Kumar[2,*], Reenu Gill[2] and Amlesh Yadav[3]

[1] *Department of Botany, University of Lucknow, Lucknow, India*

[2] *Department of Botany, Rajkeeya Mahavidhyalaya Todarpur, Hardoi 241125, Uttar Pradesh, India*

[3] *Department of Botany, Govt. PG college, Hardoi, Uttar Pradesh, India*

Abstract: The growing concern over environmental pollution caused by toxic organic materials has led to intensive research on innovative and sustainable remediation methods. Among the emerging technologies, the use of nanomaterials (NMs) has gained significant consideration because of their exceptional properties as well as high efficiency for the degradation of various pollutants. Contaminated organic materials, including various industrial chemicals, pesticides, pharmaceuticals, and persistent organic pollutants (POPs), pose severe threats to the environment and the health of human beings and other animals. The conventional methods used in the treatment of these pollutants often exhibit limited effectiveness, high costs, and may generate harmful by-products. The use of NMs in degradation processes often requires less energy compared to conventional remediation methods, leading to the overall process being more energy-efficient and environmentally sustainable. In a nanotechnology-based remediation strategy, engineered NMs are used to clean polluted locations because of their efficient, cost-effective, sustainable as well as eco-friendly nature. Nanoparticles (NPs) are very sensitive, have catalytic behavior, high surface area to volume ratio, and excellent electronic properties. NPs have the ability to diffuse in small spaces, which promotes their use as agents for the redressal of polluted soil and water. This chapter highlights the pivotal role of NPs in the degradation of toxic organic materials by leveraging their unique properties, making NMs a promising solution for addressing environmental pollution and promoting sustainable remediation practices.

Keywords: Environmental pollution, Nanoparticles, Nanomaterials, Persistent organic pollutants, Sustainable remediation.

---

* **Corresponding author Mithlesh Kumar:** Department of Botany, Rajkeeya Mahavidhyalaya Todarpur, Hardoi 241125, Uttar Pradesh, India; E-mail: sirmithleshkumar@gmail.com

Neha Agarwal, Vijendra Singh Solanki, Neetu Singh & Maulin P. Shah (Eds.)

INTRODUCTION

Soil is the most important and vital component of Earth, which provides a crucial medium for the growth of plants by providing them with essential nutrient elements, water, and oxygen [1, 2]. The optimal physical and chemical properties of soil are very crucial for seed germination, seedling emergence, establishment, growth, and development of different crop plants, and trees as well as global food security and economic sustainability. This valuable natural resource is being adulterated by the addition of many toxic, harmful industrial and domestic waste products. These harmful substances negatively affect the growth and development of plants in such soils. The addition of harmful substances in the soil is carried out by both natural as well as anthropogenic sources. However, anthropogenic sources like industrialization and urbanization are major factors responsible for intense soil pollution at a very rapid rate, which is much higher than natural soil pollution [3]. The important soil pollutants are toxic organic pollutants (such as pharmaceuticals, flame-retardants, pesticides, polycyclic aromatic hydrocarbon (PAHs) biocides, polychlorinated biphenyls (PCBs), surfactants, polychlorinated dibenzofurans and polychlorinated dibenzo-p-dioxins), heavy metals, and radioactive substances due to the excess use of pesticides, herbicides and industrial effluents. These organic pollutants are introduced into soil mainly due to anthropogenic activities [4]. The soil organic pollutants (OPs) include a great diversity of chemical substances that exhibit different chemical properties and lack analytical standards for a whole set of OPs [5]. There is a great environmental threat associated with soil pollution. There are estimations that approximately 30% of land is contaminated and degraded by different pollutants [2]. Maintenance and restoration of soil fertility are essential for the conservation of biodiversity and for the maintenance of equilibrium between earth and environmental processes such as biogeochemical cycling, soil temperature, and soil reactions [6, 7]. The main processes responsible for land degradation are prompted by many factors like urbanization, soil erosion, dumping of wastes, overgrazing, and deforestation [8]. Besides this, chemical contamination and pollution due to the introduction of pesticides, toxic metals/metalloids, PAHs, and POPs are major anthropogenic factors leading to land degradation. Disproportionate application of pesticides, chemical fertilizers, and other contaminants leads to their excessive accumulation in such lands [9, 10].

Certainly, continuous industrialization leads to activities such as transportation, manufacturing, construction, petroleum refining, and mining, which result in the depletion of natural resources. These processes generate a substantial volume of hazardous waste, leading to the pollution of soil, water, and air and disturbing the biotic and abiotic components of ecosystems. The widespread application of pesticides and herbicides by farmers has led to significant environmental pollution. These chemicals often contain nitrogen-based compounds and other substances that are non-degradable in nature and persist in the environment for a long period of time. A reported statistical data collected in 2020 indicated that usually 51.90% of land in India is contaminated with different pollutants of serious concern [11].

Traditional pollution remediation methods often relied on specialized equipment and chemicals, which drove up costs significantly. Moreover, these methods often inadvertently caused more harm to the environment and were not economically sustainable in the long term. This is why there is an urgent need for the development of alternative remediation techniques that are inexpensive, simple, less time-consuming, require minimum labor input, eco-friendly, and sustainable. Consequently, a substantial amount of research is currently underway to develop reliable, versatile, and efficient techniques for the degradation and transformation of such toxic environmental pollutants. The application of NMs for the remediation of toxic environmental pollutants has gained greater attention due to their specific and versatile features like sensitivity, selectivity, cost-effectiveness, environment friendliness, excellent electronic properties, and superior catalytic properties [12, 13].

NMs designed for the protection and remediation of the environment have been effectively developed over the last few decades. In other words, NMs possess exceptional characteristics like electrochemical, and magnetic properties, large surface area-to-volume ratio, and size-dependent physical and chemical traits, which provide these NMs significant advantages in pollution control. The application of NMs in catalytic oxidation, chemical reduction, adsorption, photocatalysis, electrochemical, and filtration processes holds great promise for effectively addressing environmental challenges. Nanotechnology offers innovative solutions that can contribute to cleaner soil, water, and air as well as improved environmental sustainability and human health.

FATE, ACCUMULATION AND TOXIC IMPACTS OF POPS

Soil OPs have been divided into many classes according to their unique toxicological modes of action and physicochemical properties [5, 14]. The most commonly found OPs in soil include chlorinated compounds (PCBs, PCDFs, and PCDDs), monomeric aromatic hydrocarbons (toluene, benzene, xylene, and ethylbenzene), oil hydrocarbons (alkenes, alkanes, and cycloalkanes), PAHs (chrysene, benzo [a] pyrene and fluoranthene), numerous fungicides (lindane, penconazole, metalaxyl, and procymidone), pesticides, herbicides (atrazine, alachlor, acetochlor, and bifenox), insecticides (endosulfan, captan, heptachlor, benomyl, endrin), and their degradation products [14, 15]. Many of these OPs are toxic, persistent, bioaccumulative, and disposed to long-range transport. The Stockholm Convention mentioned the reduction and elimination of the release of such toxic POPs in the environment and ecosystem [16]. The invention of versatile and sophisticated analytical techniques like gas chromatography coupled with tandem mass spectroscopy, together with the speedy economic development has allowed the identification and quantification of undetected substances of very low concentrations. The toxic OPs cover a wide range of chemical classes including pharmaceuticals, PCPs, hormones and sterols, flame retardants, dioxin-like compounds, nitrosamines, phthalates (used in plasticizers and generally found in home furnishings), building materials, plastic toys, clothing, medical products, food packaging, pesticides, metabolites and transformation products of artificial chemicals [17 - 19].

Due to the rapidly increasing human population, industrialization, and urbanization, numerous organic pollutants are passed into the different matrices of our environment. Being carcinogenic, mutagenic, and teratogenic in nature, such organic pollutants when discharged into the different water bodies adversely affect aquatic life. These pollutants block sewage treatment plants leading to an increase in biochemical oxygen demand (BOD), which interrupts the organization of different water bodies. These harmful organic pollutants interfere with the action and mechanism of the endocrine system. Globally, different countries and organizations have taken initiatives for the removal of these pollutants from contaminated sites through remediation and efficient management practices on a priority basis. Out of the 12 restricted and banned POPs, which are recorded as *dirty dozen*, eight of these are pesticides and the other POPs include industrial chemicals, solvents, and pharmaceuticals [20]. Many organic pollutants and pesticides have dangerous properties of bioaccumulation and biomagnification and contaminate the food chain. Being lipophilic in nature; POPs accumulate in fat-storing adipose tissue in different organisms including human beings. At present, most of the harmful and toxic pesticides/chemicals have been banned across the globe but the hazardous effects of these chemicals are still present.

The POPs are able to contaminate new sites across the world by the process of deposition and evaporation. Thus, remediation of soil polluted with such types of pollutants is necessary for the elimination of hazardous effects of such pollutants in the food chain and to fulfill the need for food grain and bioenergy production for the ever-increasing human population [21, 22].

These contaminated soil and water resources are not suitable for food production and have the threat of contamination of the food chain at different trophic levels [23]. When the livestock feed on such degraded and polluted land and water-grown crops, food, and fodder; it poses fatal and serious outcomes to the health of human beings and animals. To prevent the entry of such pollutants *via* the food chain to the human body, such lands have been abandoned and thus they are responsible for huge losses in agricultural production. The area of contaminated lands affected by POPs is increasing year by year, resulting in the reduction of arable lands in different parts of the world. So there is an urgent need for the reduction of further expansion of degraded land as well as restoration of fertility of previously degraded land. This chapter describes the current advancement in the field of remediation of soil contaminated with POPs with the use of a versatile ecofriendly, sustainable, cost-effective technology of engineered NPs (ENPs). Thus, nanotechnology opens new dimensions for the remediation of POPs because of this eco-friendly, cost-effective, and sustainable tool for the remediation of such degraded lands.

NANOTECHNOLOGY IN REMEDIATION OF CONTAMINATED LAND

Nanotechnology has emerged as a novel and innovative technique with a great deal of application in many fields like health, medicine, and agriculture biotechnology. Nanotechnology incorporates the design, measurement, and manipulation of matter at the atomic, macromolecular, and micromolecular scale due to which the refined material has completely different, specific, and unique features from that of the original bulk material [24]. These particles are called NPs and/ or engineered NPs (ENPs). ENPs can be referred to as particles with at least one external dimension ranging between 1-100 nm [25, 26]. Nanotechnology has vast applications in many sectors like medical diagnostics, agriculture, pharmaceutical, food production, food packaging, genetic-material transfer in plants, nano-based epitomization of pesticides, drug delivery in human beings, and the treatment of many diseases [27]. At present, nanotechnology and ENPs have many applications in the remediation of different contaminants of soil and water for their restoration to optimum levels [28]. Conventional technologies are time-consuming, less efficient, costly, and not environmentally friendly. So, there is an urgent requirement for those technologies that are affordable, cost-effective, faster, and have no additional load of clear-out process in the form of environmental persistence and harmful intermediates/residue during the process of remediation [29].

The use of NPs for the redress of lands contaminated with various pollutants like pesticides, POPs, and heavy metals has opened new frontiers and provided an efficient and safe tool for remediation. NM-based technology includes the following different strategies;

1. Nanotechnology-based materials are helpful in the degradation of POPs and pesticides.

2. Nanotechnology-based materials convert toxic pollutants to their less harmful forms.
3. Nanotechnology-based materials are used as sensors for the detection of pesticide residues in soil.
4. Nanotechnology-assisted bioremediation and phytoremediation are efficient in the remediation of contaminated lands.

The most commonly used NPs for remediation of soil include nanoscale zerovalent iron (nZVI), titanium dioxide (TiO_2), multiwalled carbon nanotubes (MWCNTs), zinc oxide (ZnO), fullerenes, bimetallic NPs and stabilized NPs [29]. Soil remediation by using nanotechnology relies on the application of different NPs with a reactive nature on the contaminated sites resulting in the transformation of these toxic and polluted soils into non-toxic compounds [30]. Due to their size and other specific properties like greater surface area; and greater sorption sites, NPs behave as excellent absorbents [31]. Some other specific properties of NPs include lower thermal modification, more tunable pore size, shorter interparticle diffusion space, and wide surface chemistry [32]. The above-mentioned properties enable NPs to act as outstanding catalysts and reducing agents for the remediation of different toxic contaminants and pollutants. This is why the NPs have gathered significant attention and they are broadly used for detoxification and remediation of contaminated lands.

DIFFERENT MECHANISMS OF DEGRADATION OF POPS

POPs are naturally degraded by reduction and catalytic degradation reactions where adsorption plays a significant role in the reclamation of heavy metals and metalloid-contaminated sites. However, ENPs have shown greater importance in the remediation of contaminated sites by instantaneously eliminating various toxic pollutants through sequential treatments [33, 34]. These ENPs contain a diverse group of NPs including carbon NMs, Fe_3O_4, ZnO, TiO_2, nanocomposites, and nZVI. Some important mechanisms involved in the detoxification of POPs are described in the following sub-sections.

Photocatalytic Degradation

Photocatalysis is the process of degradation of POPs in the presence of sunlight with the help of NPs. The process of photocatalysis proves to be efficient and cost-effective for the degradation of POPs where nano-photocatalysts transform toxic POPs into simple and relatively less harmful products like water, and carbon dioxide. In this direction, TiO_2 and ZnO NPs have attracted substantial consideration as semiconductor nano photocatalysts under the influence of UV radiation [35 - 40].

The photocatalytic technique relies on the reaction of toxic organic pollutants with powerful oxidizing and reducing agents (h^+ and e^-) generated by either UV/visible light interacting with the surface of photocatalysts [41]. For instance, carbon quantum dots were utilized to support AgI/ZnO/phosphorus-doped graphitic carbon nitride in the photodegradation of 2, 4-dinitrophenol [42]. Nano-TiO_2 has received much attention and primary interest in the degradation of POPs (such as pesticides, herbicides, dyes, and phenolic compounds) because of its resilient nature, minimal toxicity, cost-effectiveness, remarkable chemical and photochemical stability, and super-hydrophilicity [43]. Nano-TiO_2 has been used for degrading three types of chlorinated pesticides α-, β-, γ-, and δ-hexachlorobenzene, cypermethrin, and dicofol [44]. TiO_2 initiates the photolytic degradation of these organochlorine pesticides on its surface by facilitating electron transfer as peroxide or hydroxyl radicals.

Rhenium doping with Nano-TiO_2 induces photocatalytic degradation of organophosphorus and carbamate pesticides present in tomato plant leaves and soil [45]. Nano-TiO_2 also decomposed a POP called carbofuran, achieving a degradation rate of 55%, which exceeded the natural degradation rate by 30%.

Altered TiO_2 NMs have also been found to decrease the decay rates of different pesticides. Metolachlor (MTLC) is a widely used herbicide, which was degraded by ozonation catalysis using MWCNTs [46]. Here carbon catalyst significantly increased the mineralization of MTLC by reducing its toxicity. The normal degradation of MTLC could produce numerous harmful by-products like aromatic compounds and organic acids but in the presence of carbon nanofibers, it resulted in its complete mineralization. In an experimental investigation focusing on the degradation of DDT by nZVI, it was observed that the degradation rate was enhanced by 50% [47].

The semiconductor NPs synthesized from nanoscale composites are characterized by a core-shell architecture comprising different metals like TiO_2, ZnO, Au/TiO_2, and Au/ZnO. When used as catalysts for the photodegradation of malathion, it was found that nanocomposites increased the light-induced catalytic degradation of malathion [48]. Another investigation reported the effect of nano-TiO_2 on the photocatalytic degradation of phenanthrene and concluded that the half-life of phenanthrene was decreased from 46 to 31 hours [49].

In a ground-based approach of remediating soil contaminated with PCBs using nZVI and Pd/Fe bimetallic NPs, it was observed that the Pd/Fe bimetallic NPs resulted in faster and more thorough hydro-dechlorination of 2,2,4,4,5,5-hexachlorobiphenyl compared to nZVI [50]. This enhanced performance may be attributed to the presence of palladium (Pd) loading. Remediation of pentachlorophenol (PCP) contaminated soil was investigated using CMC-stabilized nano-Pd/Fe. It was found that PCPs present in soil were dechlorinated to phenol by the electrokinetic transport mechanism of nano-Pd/Fe [50]. Another study on the catalytic dichlorination of lindane (an organochlorine pesticide) utilized a combined nanobiotechnological strategy, integrating FeS NPs, and subsequent microbial degradation [51]. The stabilized NPs mentioned above efficiently degraded 5mg/L of lindane with a 94% efficiency within 8 hours. Afterward, the remaining lindane and its partially degraded intermediates (along with the stabilizing polymer) were completely degraded during microbiological treatment within 1 hour. Through this process, they successfully removed all 5mg/L of lindane within a total span of 9 hours. A bimetallic iron biocomposite (BioCAT slurry) was also synthesized by coating biologically captured NP agents onto the surface of ZVI containing palladium NPs (PdNPs). This composite was designed for the degradation of pentachlorophenol, a commonly used organochlorine pesticide. Over the course of 21 days of treatment, the pentachlorophenol was degraded by 90% [52].

Immobilization

Presently, the majority of remedial procedures rely on the immobilization process to remove contaminants from land and water. This method is efficient, cost-effective, and relatively environmentally friendly, especially for cleaning up metal-contaminated soils [53]. Selecting suitable materials and/or coating compounds is crucial in immobilization remediation. This selection depends on the specific contaminants present and the conditions of the soil being treated. Thus, a wide range of nano additives, including carbon-based NMs (such as CNTs, fullerene, graphene, *etc.*), metal oxide NMs (like ZnO, TiO_2), nanocomposites, and other engineered NMs have been extensively used in immobilizing complex pollutants within the soil. Carbon NMs absorb organic pollutants *via* van der Waals interactions, and π-π stacking while metal oxide NMs and other nanocomposites capture trace metals and organic compounds through surface complexation mechanisms.

Carbon NMs

There is a vast application of carbon NMs for the remediation of organic contaminants mainly present in water and soils. This technology relies mainly on the adsorption and surface hydrophobicity capacities of these NPs [54]. The adsorption and desorption process and mechanism have been extensively used and studied in the remediation of contaminated water [55]. These organic compounds show relatively more adsorption affinity to CNTs as compared to the soil particles so CNTs have a significant role in the remediation of such organic pollutants [56].

Furthermore, these CNTs have been reported to adsorb and capture ionizable POPs because of the hydrogen bonds facilitated by charge interaction, low-barrier, and cation-π assisted π-π interactions [57] The sorption of PAHs compounds like naphthalene, and phenanthrene, fluorene, and pyrene by CNTs have been extensively evaluated in diverse environmental situations [58]. CNTs in soils could obstruct the mobilization and mineralization of benzo [a] pyrene and phenanthrene thus they reduce the bioavailability and extractability of these POPs to flora and soil microflora. Carbon nanotubes demonstrate exceptional efficacy in adsorbing and immobilizing pyrene [56, 59, 60].

For phenolic compounds, the adsorption capacity and affinity of CNTs rely on their physicochemical attributes, including morphology, size, layering, and the abundance of hydroxyl groups ($^-$OH). The incorporation of OH$^-$ in the aromatic ring in these phenolic compounds resulted in the augmentation of π-π interactions between their aromatic rings and CNTs facilitating the simultaneous attraction of both π-receptor and π-donor molecules to their surface, thereby enhancing the adsorption process [60].

Metal Oxide NMs

NPs based on different metal oxides like Fe_3O_4 and TiO_2 *etc.* are extensively used and studied. ENMs have great potential for the adsorption of numerous pollutants present in the soil. NPs based on metal oxides like Fe_3O_4 are capable of internalizing and fixing change. NPs have a strong capability to adsorb and immobilize trace metals like cadmium and arsenic [61, 62]. The maximum cadmium sorption efficiency of 37.03 mg/g is shown by Fe_3O_4 under the condition of 10–20°C and pH 6.0. The findings indicate that Fe_3O_4 exhibits potential for remediating metal-contaminated soils and mitigating metal-induced stress in terrestrial plants through various effective methods [61, 63]. Coating TiO_2 with dissolved organic matter, such as tannic acid, lignin, and capsorubin, markedly enhances its sorption capacity for PAHs. This leads to a substantial reduction in phenanthrene mobility within the soil [64].

Nanocomposites

Nanocomposites have been testified to have greater effectiveness as compared to single materials in the degradation of pollutants and have shown better efficiency after numerous adsorption-desorption cycles [65, 66]. A composite of nano-$Fe/Ca/CaO/[PO_4]$ material was synthesized for dual separation and immobilization of cesium (Cs) in polluted soils and found 100% immobilization of stable Cs (133Cs) after simple crushing under parched conditions [67]. Studies have shown that hydroxyapatite (nHAP) NPs effectively mitigate the leachability, mobility, and bioavailability of Zn, Cu, and Pb in heavy metal-contaminated soils [68, 69].

Fenton-like Reaction

For the remediation of PAHs contaminated soils, Fenton oxidation is extensively used. The Fenton reagent is a mixture of hydrogen peroxide (H_2O_2+ and iron Fe $_{II}$), which is used for the removal of organic wastes from contaminated soil and water. In this method, the hydroxyl radicals ($^.OH$) generated from hydrogen peroxidase help in the oxidation of organic compounds in the presence of ferrous ions, which function as a catalyst in the reaction [70]. A wide range of materials act as catalysts in the Fenton oxidation reaction including iron metal and some other related engineered NMs like Fe_3O_4 and nanocomposites, which are the most prevalent materials and gained significant consideration [71]. Fe_3O_4 acts as a promising agent for in-situ remediation of contaminants from soil and groundwater due to its specific features like abundance in the environment, low cost, ease of synthesis and reuse, and environmentally friendly characteristics of iron oxides in the soil [71]. For example, Fe_3O_4-catalyzed Fenton-like reactions act as potential agents for the remediation of PAHs contaminated soils with no need for pH adjustment as well as without the formation of harmful byproducts [72].

Reduction Reactions

For the removal of heavy metals and organic contaminants from soil, water, and air; reduction reactions are followed as these reactions have a high potential for the removal of these pollutants. nZVI has attracted significant attention [73 - 75] in such reactions. It has a negative reduction potential and a free electron. It is used for the removal of polychlorinated biphenyls, chlorinated organic solvents, and organochlorine pesticides *via* redox transformation strategies [76]. Due to specific properties of nZVI like large surface area, these NPs have better efficiency to remediate contaminated lands by direct contact with pollutants. Due to some other properties of nZVI like high reduction capacity and better absorption efficiency, these particles can convert toxic contaminants into relatively lesser toxic substances. For example, nZVI can reduce Cr (VI) to Cr (III) and mostly results in the formation of ferrous chromite [77]. Furthermore, the combination of biochar with ZVI NPs has been observed to improve the reduction capacity of nZVI as well as improve its effectiveness towards the removal of Cr (VI) by reinforcing the distribution of Fe- particles and reducing the movement within the soil [78, 79].

There are reports that the combined use of carboxymethyl cellulose stabilizer with nZVI had shown a high capacity for conversion of Cr (VI) pollutants into iron-manganese and carbonate-bound compounds [80]. For the removal of organic contaminants like trichloroethylene (TCE), tetra-bromo-bisphenol A (TBBPA), dichlorodiphenyltrichloroethane (DDT), and 2,4-dichlorophenoxyacetic acid (2,4-D), combined application of carboxymethyl cellulose stabilizer with nZVI has shown better response [81, 82]. The effectiveness of nZVI in the degradation of DDT can be greatly improved and influenced by specific properties like reactivity, particle size, and suspension stability, as well as the pH condition of the soil [83].

Nanofiltration

A recent breakthrough in pressure-driven membrane filtration has been utilized effectively for separating molecules from the liquid phase. Nanofiltration (NF) possesses a unique property for heavy metals treatment and removal, attributed to its constant charge resulting from the dissociation of surface charges. The presence of sulfonate and carboxylate groups in these charge groups allows for their application in treating pharmaceutical waste, recovering metals, and removing organic and inorganic pollutants from surface water [84, 85]. Another key advantage of nanofiltration membranes is their smaller pore size, which enhances their efficiency. With high recovery rates and low costs, nanofiltration membranes are used for the investigation of their efficacy in the elimination of heavy metals. Studies have shown that nanofiltration membranes can achieve up to 92% removal of copper from high volumes of contaminated water.

Nanobioremediation

Nanobioremediation is being widely used for the removal and treatment of contaminants in soil, water, and air. This method combines nanotechnology and bioremediation to speed up the treatment process, achieving efficient results in minimal time. Therefore, this method offers significant advantages in eliminating pollutants, organic compounds, and toxic metals [86]. Anaerobic bacteria, including organohalide respiring bacteria, sulfate-reducing bacteria (SRB), and iron-reducing bacteria (IRB), were integrated with nZVI, yielding promising results in the removal of both inorganic and organic pollutants.

CONCLUSION AND FUTURE PERSPECTIVE

In summary, the present state of nanotechnology and ENMs in soil remediation along with recent advancements and risk assessments has been discussed. The use of ENMs and nanotechnology in soil remediation involves reduction reactions, immobilization mechanisms, fusion of different technologies, and their incorporation into bioremediation strategies such as phytoremediation and micro-remediation.

Carbon NMs (especially CNTs), nZVI, and metal oxide NMs (such as Fe_3O_4, TiO_2, and ZnO) are the most efficient NMs in removing heavy metal and organic contaminants from the soils. A contaminant's fate in the soil environment is determined by soil microbe, adsorption onto soil particles, aggregation, and dissolution into underground water reservoirs. These dynamics are regulated by both soil conditions and the physicochemical properties of ENMs. Utilizing nanotechnology can significantly increase crop yield by promoting growth.

It is a valid inquiry to explore the potential impact of ENMs on terrestrial plants, soil organisms (including microorganisms and invertebrates), and human health. Evaluating the effects of nanotechnology is essential to determine whether the advantages of its application outweigh the associated environmental risks. Therefore, further investigation is needed to advance the application of nanotechnology in soil remediation and promote environmental conservation efforts on a global scale. Perspectives on future research directions may encompass a wide range of topics that include-

- Customized synthesis and fabrication of innovative ENMs with specific capabilities.
- Developing diverse forms of nano-composites and integrating them with alternative technologies, such as bioremediation.
- Constant supervision and controlling the use of ENMs in real-time, while evaluating how effectively they remove pollutants.
- Studying the prolonged exposure and lasting impacts of ENMs utilized in soil remediation.

REFERENCES

[1]   Young IM, Crawford JW. Interactions and self-organization in the soil-microbe complex. Science 2004; 304(5677): 1634-7.
        [http://dx.doi.org/10.1126/science.1097394] [PMID: 15192219]

[2]   Abhilash PC, Dubey RK, Tripathi V, Srivastava P, Verma JP, Singh HB. Remediation and management of POPs-contaminated soils in a warming climate: challenges and perspectives. Environ Sci Pollut Res Int 2013; 20(8): 5879-85.
        [http://dx.doi.org/10.1007/s11356-013-1808-5] [PMID: 23677754]

[3]   Smith P, House JI, Bustamante M, *et al.* Global change pressures on soils from land use and management. Glob Change Biol 2016; 22(3): 1008-28.
        [http://dx.doi.org/10.1111/gcb.13068] [PMID: 26301476]

[4]   Schaeffer A, Amelung W, Hollert H, *et al.* The impact of chemical pollution on the resilience of soils under multiple stresses: A conceptual framework for future research. Sci Total Environ 2016; 568: 1076-85.
        [http://dx.doi.org/10.1016/j.scitotenv.2016.06.161] [PMID: 27372890]

[5]   Andreu V, Picó Y. Determination of pesticides and their degradation products in soil: critical review and comparison of methods. Trends Analyt Chem 2004; 23(10-11): 772-89.
[http://dx.doi.org/10.1016/j.trac.2004.07.008]

[6]   Lal R. Restoring soil quality to mitigate soil degradation. Sustainability (Basel) 2015; 7(5): 5875-95.
[http://dx.doi.org/10.3390/su7055875]

[7]   Lal R. Soil degradation as a reason for inadequate human nutrition. Food Secur 2009; 1(1): 45-57.
[http://dx.doi.org/10.1007/s12571-009-0009-z]

[8]   León JD, Osorio NW. Role of litter turnover in soil quality in tropical degraded lands of Colombia. ScientificWorldJournal 2014; 2014: 1-11.
[http://dx.doi.org/10.1155/2014/693981] [PMID: 24696656]

[9]   Ghormade V, Deshpande MV, Paknikar KM. Perspectives for nano-biotechnology enabled protection and nutrition of plants. Biotechnol Adv 2011; 29(6): 792-803.
[http://dx.doi.org/10.1016/j.biotechadv.2011.06.007] [PMID: 21729746]

[10]   Maharia RS, Dutta RK, Acharya R, Reddy AVR. Heavy metal bioaccumulation in selected medicinal plants collected from Khetri copper mines and comparison with those collected from fertile soil in Haridwar, India. J Environ Sci Health B 2010; 45(2): 174-81.
[http://dx.doi.org/10.1080/03601230903472249] [PMID: 20390948]

[11]   Bora J, Imam S, Vaibhav V, Malik S. Use of Genetic Engineering Approach in Bioremediation of Wastewater. In: Shah MP, Ed. Modern Approaches in Waste Bioremediation. Switzerland: Springer Nature A.G 2023; pp. 485-513.
[http://dx.doi.org/10.1007/978-3-031-24086-7_23]

[12]   Ghasemzadeh G, Momenpour M, Omidi F, Hosseini MR, Ahani M, Barzegari A. Applications of nanomaterials in water treatment and environmental remediation. Front Environ Sci Eng 2014; 8(4): 471-82.
[http://dx.doi.org/10.1007/s11783-014-0654-0]

[13]   Khan I, Saeed K, Khan I. Nanoparticles: Properties, applications and toxicities. Arab J Chem 2019; 12(7): 908-31.
[http://dx.doi.org/10.1016/j.arabjc.2017.05.011]

[14]   Stokes JD, Paton GI, Semple KT. Behaviour and assessment of bioavailability of organic contaminants in soil: relevance for risk assessment and remediation. Soil Use Manage 2005; 21(s2): 475-86.
[http://dx.doi.org/10.1079/SUM2005347]

[15]   Tsibart AS, Gennadiev AN. Polycyclic aromatic hydrocarbons in soils: Sources, behavior, and indication significance (a review). Eurasian Soil Sci 2013; 46(7): 728-41.
[http://dx.doi.org/10.1134/S1064229313070090]

[16]   Wang J, Chen B, Xing B. Wrinkles and folds of activated graphene nanosheets as fast and efficient adsorptive sites for hydrophobic organic contaminants. Environ Sci Technol 2016; 50(7): 3798-808.
[http://dx.doi.org/10.1021/acs.est.5b04865] [PMID: 26938576]

[17]   Vodyanitskii YN, Yakovlev AS. Contamination of soils and groundwater with new organic micropollutants: A review. Eurasian Soil Sci 2016; 49(5): 560-9.
[http://dx.doi.org/10.1134/S1064229316050148]

[18]   Wu C, Spongberg AL, Witter JD, Fang M, Ames A, Czajkowski KP. Detection of pharmaceuticals and personal care products in agricultural soils receiving biosolids application. Clean (Weinh) 2010; 38(3): 230-7.
[http://dx.doi.org/10.1002/clen.200900263]

[19]   Careghini A, Mastorgio AF, Saponaro S, Sezenna E. Bisphenol A, nonylphenols, benzophenones, and benzotriazoles in soils, groundwater, surface water, sediments, and food: a review. Environ Sci Pollut Res Int 2015; 22(8): 5711-41.
[http://dx.doi.org/10.1007/s11356-014-3974-5] [PMID: 25548011]

[20]   Allinson G, Allinson M, Bui A, *et al.* Pesticide and trace metals in surface waters and sediments of rivers entering the Corner Inlet Marine National Park, Victoria, Australia. Environ Sci Pollut Res Int 2016; 23(6): 5881-91.
[http://dx.doi.org/10.1007/s11356-015-5795-6] [PMID: 26593725]

[21]   Tripathi V, Fraceto LF, Abhilash PC. Sustainable clean-up technologies for soils contaminated with multiple pollutants: Plant-microbe-pollutant and climate nexus. Ecol Eng 2015; 82: 330-5.
[http://dx.doi.org/10.1016/j.ecoleng.2015.05.027]

[22]   Zhang X, Wang H, He L, *et al.* Using biochar for remediation of soils contaminated with heavy metals and organic pollutants. Environ Sci Pollut Res Int 2013; 20(12): 8472-83.
[http://dx.doi.org/10.1007/s11356-013-1659-0] [PMID: 23589248]

[23]   Lu Y, Song S, Wang R, *et al.* Impacts of soil and water pollution on food safety and health risks in China. Environ Int 2015; 77: 5-15.
[http://dx.doi.org/10.1016/j.envint.2014.12.010] [PMID: 25603422]

[24]   Khan MN, Mobin M, Abbas ZK, AlMutairi KA, Siddiqui ZH. Role of nanomaterials in plants under challenging environments. Plant Physiol Biochem 2017; 110: 194-209.
[http://dx.doi.org/10.1016/j.plaphy.2016.05.038] [PMID: 27269705]

[25]   Shi J, Abid AD, Kennedy IM, Hristova KR, Silk WK. To duckweeds (Landoltia punctata), nanoparticulate copper oxide is more inhibitory than the soluble copper in the bulk solution. Environ Pollut 2011; 159(5): 1277-82.
[http://dx.doi.org/10.1016/j.envpol.2011.01.028] [PMID: 21333422]

[26]   Thuesombat P, Hannongbua S, Akasit S, Chadchawan S. Effect of silver nanoparticles on rice (Oryza sativa L. cv. KDML 105) seed germination and seedling growth. Ecotoxicol Environ Saf 2014; 104: 302-9.
[http://dx.doi.org/10.1016/j.ecoenv.2014.03.022] [PMID: 24726943]

[27]   Gogos A, Knauer K, Bucheli TD. Nanomaterials in plant protection and fertilization: current state, foreseen applications, and research priorities. J Agric Food Chem 2012; 60(39): 9781-92.
[http://dx.doi.org/10.1021/jf302154y] [PMID: 22963545]

[28]   Shi J, Votruba AR, Farokhzad OC, Langer R. Nanotechnology in drug delivery and tissue engineering: from discovery to applications. Nano Lett 2010; 10(9): 3223-30.
[http://dx.doi.org/10.1021/nl102184c] [PMID: 20726522]

[29]   Cai C, Zhao M, Yu Z, Rong H, Zhang C. Utilization of nanomaterials for in-situ remediation of heavy metal(loid) contaminated sediments: A review. Sci Total Environ 2019; 662: 205-17.
[http://dx.doi.org/10.1016/j.scitotenv.2019.01.180] [PMID: 30690355]

[30]   Yan W, Lien HL, Koel BE, Zhang W. Iron nanoparticles for environmental clean-up: recent developments and future outlook. Environ Sci Process Impacts 2013; 15(1): 63-77.
[http://dx.doi.org/10.1039/C2EM30691C] [PMID: 24592428]

[31]   Gong X, Huang D, Liu Y, *et al.* Remediation of contaminated soils by biotechnology with nanomaterials: bio-behavior, applications, and perspectives. Crit Rev Biotechnol 2018; 38(3): 455-68.
[http://dx.doi.org/10.1080/07388551.2017.1368446] [PMID: 28903604]

[32]   Tang WW, Zeng GM, Gong JL, *et al.* Impact of humic/fulvic acid on the removal of heavy metals from aqueous solutions using nanomaterials: A review. Sci Total Environ 2014; 468-469: 1014-27.
[http://dx.doi.org/10.1016/j.scitotenv.2013.09.044] [PMID: 24095965]

[33]   Zhang T, Lowry GV, Capiro NL, *et al. in situ* remediation of subsurface contamination: opportunities and challenges for nanotechnology and advanced materials. Environ Sci Nano 2019; 6(5): 1283-302. a
[http://dx.doi.org/10.1039/C9EN00143C]

[34]   Trujillo-Reyes J, Peralta-Videa JR, Gardea-Torresdey JL. Supported and unsupported nanomaterials for water and soil remediation: Are they a useful solution for worldwide pollution? J Hazard Mater 2014; 280: 487-503.

[http://dx.doi.org/10.1016/j.jhazmat.2014.08.029] [PMID: 25203809]

[35]   Matos J, Miralles-Cuevas S, Ruíz-Delgado A, Oller I, Malato S. Development of TiO2-C photocatalysts for solar treatment of polluted water. Carbon 2017; 122: 361-73.
[http://dx.doi.org/10.1016/j.carbon.2017.06.091]

[36]   Tang X, Wang Z, Wang Y. Visible active N-doped TiO2/reduced graphene oxide for the degradation of tetracycline hydrochloride. Chem Phys Lett 2018; 691: 408-14.
[http://dx.doi.org/10.1016/j.cplett.2017.11.037]

[37]   Fujishima A, Zhang X, Tryk D. TiO2 photocatalysis and related surface phenomena. Surf Sci Rep 2008; 63(12): 515-82.
[http://dx.doi.org/10.1016/j.surfrep.2008.10.001]

[38]   Ali T, Tripathi P, Azam A, *et al.* Photocatalytic performance of Fe-doped TiO$_2$ nanoparticles under visible-light irradiation. Mater Res Express 2017; 4(1): 015022.
[http://dx.doi.org/10.1088/2053-1591/aa576d]

[39]   Sudhaik A, Raizada P, Shandilya P, Jeong DY, Lim JH, Singh P. Review on fabrication of graphitic carbon nitride based efficient nanocomposites for photodegradation of aqueous phase organic pollutants. J Ind Eng Chem 2018; 67: 28-51. a
[http://dx.doi.org/10.1016/j.jiec.2018.07.007]

[40]   Sharma S, Dutta V, Singh P, *et al.* Carbon quantum dot supported semiconductor photocatalysts for efficient degradation of organic pollutants in water: A review. J Clean Prod 2019; 228: 755-69.
[http://dx.doi.org/10.1016/j.jclepro.2019.04.292]

[41]   Hasija V, Raizada P, Sudhaik A, *et al.* Recent advances in noble metal free doped graphitic carbon nitride based nanohybrids for photocatalysis of organic contaminants in water: A review. Appl Mater Today 2019; 15: 494-524. a
[http://dx.doi.org/10.1016/j.apmt.2019.04.003]

[42]   Hasija V, Sudhaik A, Raizada P, Hosseini-Bandegharaei A, Singh P. Carbon quantum dots supported AgI /ZnO/phosphorus doped graphitic carbon nitride as Z-scheme photocatalyst for efficient photodegradation of 2, 4-dinitrophenol. J Environ Chem Eng 2019; 7(4): 103272. b
[http://dx.doi.org/10.1016/j.jece.2019.103272]

[43]   Fujishima A, Honda K. Electrochemical photolysis of water at a semiconductor electrode. Nature 1972; 238(5358): 37-8.
[http://dx.doi.org/10.1038/238037a0] [PMID: 12635268]

[44]   Yu B, Zeng J, Gong L, Zhang M, Zhang L, Chen X. Investigation of the photocatalytic degradation of organochlorine pesticides on a nano-TiO2 coated film. Talanta 2007; 72(5): 1667-74.
[http://dx.doi.org/10.1016/j.talanta.2007.03.013] [PMID: 19071814]

[45]   Zeng R, Wang J, Cui J, Hu L, Mu K. Photocatalytic degradation of pesticide residues with RE3+ -doped nano-TiO2. J Rare Earths 2010; 28: 353-6.
[http://dx.doi.org/10.1016/S1002-0721(10)60329-8]

[46]   Restivo J, Órfão JJM, Armenise S, Garcia-Bordejé E, Pereira MFR. Catalytic ozonation of metolachlor under continuous operation using nanocarbon materials grown on a ceramic monolith. J Hazard Mater 2012; 239-240: 249-56.
[http://dx.doi.org/10.1016/j.jhazmat.2012.08.073] [PMID: 23009793]

[47]   El-Temsah YS, Sevcu A, Bobcikova K, Cernik M, Joner EJ. DDT degradation efficiency and ecotoxicological effects of two types of nano-sized zero-valent iron (nZVI) in water and soil. Chemosphere 2016; 144: 2221-8.
[http://dx.doi.org/10.1016/j.chemosphere.2015.10.122] [PMID: 26598990]

[48]   Fouad DM, Mohamed MB. Studies on the photo-catalytic activity of semiconductor nanostructures and their gold core–shell on the photodegradation of malathion. Nanotechnology 2011; 22(45): 455705.

[http://dx.doi.org/10.1088/0957-4484/22/45/455705] [PMID: 22020195]

[49]   Gu J, Dong D, Kong L, Zheng Y, Li X. Photocatalytic degradation of phenanthrene on soil surfaces in the presence of nanometer anatase TiO2 under UV-light. J Environ Sci (China) 2012; 24(12): 2122-6.
[http://dx.doi.org/10.1016/S1001-0742(11)61063-2] [PMID: 23534208]

[50]   Chen X, Yao X, Yu C, *et al.* Hydrodechlorination of polychlorinated biphenyls in contaminated soil from an e-waste recycling area, using nanoscale zerovalent iron and Pd/Fe bimetallic nanoparticles. Environ Sci Pollut Res Int 2014; 21(7): 5201-10.
[http://dx.doi.org/10.1007/s11356-013-2089-8] [PMID: 24390111]

[51]   Paknikar KM, Nagpal V, Pethkar AV, Rajwade JM. Degradation of lindane from aqueous solutions using iron sulfide nanoparticles stabilized by biopolymers. Sci Technol Adv Mater 2005; 6(3-4): 370-4.
[http://dx.doi.org/10.1016/j.stam.2005.02.016]

[52]   Dien NT, De Windt W, Buekens A, Chang MB. Application of bimetallic iron (BioCAT slurry) for pentachlorophenol removal from sandy soil. J Hazard Mater 2013; 252-253: 83-90.
[http://dx.doi.org/10.1016/j.jhazmat.2013.02.029] [PMID: 23500793]

[53]   Barzegar G, Jorfi S, Soltani RDC, *et al.* Enhanced Sono-Fenton-like oxidation of PAH-contaminated soil using Nano-sized magnetite as catalyst: Optimization with response surface methodology. Soil Sediment Contam 2017; 26(5): 538-57.
[http://dx.doi.org/10.1080/15320383.2017.1363157]

[54]   Gupta VK, Saleh TA. Sorption of pollutants by porous carbon, carbon nanotubes and fullerene- An overview. Environ Sci Pollut Res Int 2013; 20(5): 2828-43.
[http://dx.doi.org/10.1007/s11356-013-1524-1] [PMID: 23430732]

[55]   Peng H, Zhang D, Pan B, Peng J. Contribution of hydrophobic effect to the sorption of phenanthrene, 9-phenanthrol and 9, 10-phenanthrenequinone on carbon nanotubes. Chemosphere 2017; 168: 739-47.
[http://dx.doi.org/10.1016/j.chemosphere.2016.10.143] [PMID: 27836280]

[56]   Zhang W, Lu Y, Sun H, *et al.* Effects of multi−walled carbon nanotubes on pyrene adsorption and desorption in soils: The role of soil constituents. Chemosphere 2019; 221: 203-11. b
[http://dx.doi.org/10.1016/j.chemosphere.2019.01.030] [PMID: 30640002]

[57]   Kah M, Sigmund G, Xiao F, Hofmann T. Sorption of ionizable and ionic organic compounds to biochar, activated carbon and other carbonaceous materials. Water Res 2017; 124: 673-92.
[http://dx.doi.org/10.1016/j.watres.2017.07.070] [PMID: 28825985]

[58]   Li S, Turaga U, Shrestha B, *et al.* Mobility of polyaromatic hydrocarbons (PAHs) in soil in the presence of carbon nanotubes. Ecotoxicol Environ Saf 2013; 96: 168-74.
[http://dx.doi.org/10.1016/j.ecoenv.2013.07.005] [PMID: 23896179]

[59]   Towell MG, Browne LA, Paton GI, Semple KT. Impact of carbon nanomaterials on the behaviour of 14C-phenanthrene and 14C-benzo-[a] pyrene in soil. Environ Pollut 2011; 159(3): 706-15.
[http://dx.doi.org/10.1016/j.envpol.2010.11.040] [PMID: 21195517]

[60]   Wu W, Jiang W, Zhang W, Lin D, Yang K. Influence of functional groups on desorption of organic compounds from carbon nanotubes into water: insight into desorption hysteresis. Environ Sci Technol 2013; 47(15)
[http://dx.doi.org/10.1021/es401567g] [PMID: 23848495]

[61]   Sebastian A, Nangia A, Prasad MNV. Cadmium and sodium adsorption properties of magnetite nanoparticles synthesized from Hevea brasiliensis Muell. Arg. bark: Relevance in amelioration of metal stress in rice. J Hazard Mater 2019; 371: 261-72.
[http://dx.doi.org/10.1016/j.jhazmat.2019.03.021] [PMID: 30856436]

[62]   Sun J, Chillrud SN, Mailloux BJ, *et al.* Enhanced and stabilized arsenic retention in microcosms through the microbial oxidation of ferrous iron by nitrate. Chemosphere 2016; 144: 1106-15.
[http://dx.doi.org/10.1016/j.chemosphere.2015.09.045] [PMID: 26454120]

[63] Liang Q, An B, Zhao D, Ahuja S. Removal and Immobilization of Arsenic in Water and Soil Using Polysaccharide-Modified Magnetite NPs Monitoring Water Quality: Pollution Assessment. Analysis, and Remediation 2013; pp. 285-97.

[64] Wang X, Ma E, Shen X, *et al.* Effect of model dissolved organic matter coating on sorption of phenanthrene by TiO 2 nanoparticles. Environ Pollut 2014; 194: 31-7.
[http://dx.doi.org/10.1016/j.envpol.2014.06.039] [PMID: 25089890]

[65] Mahmoud ME, Osman MM, Yakout AA, Abdelfattah AM. Water and soil decontamination of toxic heavy metals using aminosilica-functionalized-ionic liquid nanocomposite. Journal of Molecular Liquids 2018; 266(834)
[http://dx.doi.org/10.1016/j.molliq.2018.06.055]

[66] An B, Zhao D. Immobilization of As(III) in soil and groundwater using a new class of polysaccharide stabilized Fe–Mn oxide nanoparticles. J Hazard Mater 2012; 211-212: 332-41.
[http://dx.doi.org/10.1016/j.jhazmat.2011.10.062] [PMID: 22119304]

[67] Mallampati SR, Mitoma Y, Okuda T, Simion C, Lee BK. Solvent-free synthesis and application of nano-Fe/Ca/CaO/[PO4] composite for dual separation and immobilization of stable and radioactive cesium in contaminated soils. J Hazard Mater 2015; 297: 74-82.
[http://dx.doi.org/10.1016/j.jhazmat.2015.04.071] [PMID: 25942697]

[68] Yang Z, Fang Z, Zheng L, *et al.* Remediation of lead contaminated soil by biochar-supported nano-hydroxyapatite. Ecotoxicol Environ Saf 2016; 132: 224-30.
[http://dx.doi.org/10.1016/j.ecoenv.2016.06.008] [PMID: 27337496]

[69] Sun RJ, Chen JH, Fan TT, Zhou DM, Wang YJ. Effect of nanoparticle hydroxyapatite on the immobilization of Cu and Zn in polluted soil. Environ Sci Pollut Res Int 2018; 25(1): 73-80.
[http://dx.doi.org/10.1007/s11356-016-8063-5] [PMID: 27844320]

[70] Hou L, Wang L, Royer S, Zhang H. Ultrasound-assisted heterogeneous Fenton-like degradation of tetracycline over a magnetite catalyst. J Hazard Mater 2016; 302: 458-67.
[http://dx.doi.org/10.1016/j.jhazmat.2015.09.033] [PMID: 26521091]

[71] Garrido-Ramírez eg, Theng BKG, Mora ML. Clays and oxide minerals as catalysts and nanocatalysts in Fenton-like reactions — A review. Appl Clay Sci 2010; 47(3-4): 182-92.
[http://dx.doi.org/10.1016/j.clay.2009.11.044]

[72] Usman M, Faure P, Ruby C, Hanna K. Remediation of PAH-contaminated soils by magnetite catalyzed Fenton-like oxidation. Appl Catal B 2012; 117-118: 10-7.
[http://dx.doi.org/10.1016/j.apcatb.2012.01.007]

[73] Zhao X, Liu W, Cai Z, Han B, Qian T, Zhao D. An overview of preparation and applications of stabilized nZVINPs for soil and groundwater remediation. Water Res 2016; 100: 245-66.
[http://dx.doi.org/10.1016/j.watres.2016.05.019] [PMID: 27206054]

[74] Stefaniuk M, Oleszczuk P, Ok YS. Review on nano zerovalent iron (nZVI): From synthesis to environmental applications. Chem Eng J 2016; 287: 618-32.
[http://dx.doi.org/10.1016/j.cej.2015.11.046]

[75] Chrysochoou M, Johnston CP, Dahal G. A comparative evaluation of hexavalent chromium treatment in contaminated soil by calcium polysulfide and green-tea nanoscale zero-valent iron. J Hazard Mater 2012; 201-202: 33-42.
[http://dx.doi.org/10.1016/j.jhazmat.2011.11.003] [PMID: 22169240]

[76] Cheng X, Lai C, Li J, *et al.* Toward enhancing desalination and heavy metal removal of TFC nanofiltration membranes: a cost-effective interface temperature-regulated interfacial polymerization. ACS Appl Mater Interfaces 2021; 13(48): 57998-8010.
[http://dx.doi.org/10.1021/acsami.1c17783] [PMID: 34817167]

[77] Qian L, Shang X, Zhang B, *et al.* Enhanced removal of Cr(VI) by silicon rich biochar-supported nanoscale zero-valent iron. Chemosphere 2019; 215: 739-45.

[http://dx.doi.org/10.1016/j.chemosphere.2018.10.030] [PMID: 30347367]

[78] Qian L, Liu S, Zhang W, *et al.* Enhanced reduction and adsorption of hexavalent chromium by palladium and silicon rich biochar supported nanoscale zero-valent iron. J Colloid Interface Sci 2019; 533: 428-36. a
[http://dx.doi.org/10.1016/j.jcis.2018.08.075] [PMID: 30172153]

[79] Galdames A, Mendoza A, Orueta M, *et al.* Development of new remediation technologies for contaminated soils based on the application of nZVINPs and bioremediation with compost. Resource-Efficient Technologies 2017; 3(2): 166-76.
[http://dx.doi.org/10.1016/j.reffit.2017.03.008]

[80] Zhang R, Zhang N, Fang Z. *in situ* remediation of hexavalent chromium contaminated soil by CMC-stabilized nanoscale zero-valent iron composited with biochar. Water Sci Technol 2018; 77(6): 1622-31.
[http://dx.doi.org/10.2166/wst.2018.039] [PMID: 29595164]

[81] El-Temsah YS, Joner EJ. Effects of nano-sized zero-valent iron (nZVI) on DDT degradation in soil and its toxicity to collembola and ostracods. Chemosphere 2013; 92(1): 131-7.
[http://dx.doi.org/10.1016/j.chemosphere.2013.02.039] [PMID: 23522781]

[82] Zhang M, He F, Zhao D, Hao X. Degradation of soil-sorbed trichloroethylene by stabilized zero valent iron nanoparticles: Effects of sorption, surfactants, and natural organic matter. Water Res 2011; 45(7): 2401-14. b
[http://dx.doi.org/10.1016/j.watres.2011.01.028] [PMID: 21376362]

[83] El-Temsah YS, Joner EJ. Effects of nano-sized nZVI(nZVI) on DDT degradation in soil and residual toxicity in soil: a column experiment. Plant Soil 2012; 368: 189-200.
[http://dx.doi.org/10.1007/s11104-012-1509-8]

[84] Bowen WR, Welfoot JS. Modelling the performance of membrane nanofiltration—critical assessment and model development. Chem Eng Sci 2002; 57(7): 1121-37.
[http://dx.doi.org/10.1016/S0009-2509(01)00413-4]

[85] Abdel-Fatah MA. Nanofiltration systems and applications in wastewater treatment: Review article. Ain Shams Eng J 2018; 9(4): 3077-92.
[http://dx.doi.org/10.1016/j.asej.2018.08.001]

[86] Umair Azhar, Huma Ahmad, Hafsa Shafqat, Muhammad Babar, Hafiz Muhammad Shahzad Munir, Muhammad Sagir, Muhammad Arif, Afaq Hassan, Nova Rachmadona, Saravanan Rajendran, Muhammad Mubashir, Kuan Shiong Khoo, Remediation techniques for elimination of heavy metal pollutants from soil: A review, Environmental Research, Volume 214, Part 4, 2022, 113918, ISSN 0013-9351.
[http://dx.doi.org/10.1016/j.envres.2022.113918]

SUBJECT INDEX

A

Adsorbing toxic gases 315
Adsorption 22, 49, 70, 73, 75, 76, 77, 78, 79, 81, 82, 83, 121, 123, 124, 153, 154, 174, 175, 211, 212, 232, 241, 311, 315, 328, 351
  activated carbon 49, 174
  gas molecule 311
  ion 81, 123
  pollutant 241
  efficacy 73, 75, 76, 81, 82
  isotherm 77, 78
  methods 49, 79, 82
  pathways 82
Advanced oxidation processes (AOPs) 22, 168, 169, 172, 266, 269, 328, 329
Air contaminated 43
Air filtration 212, 306, 313, 316
  applications 313
  processes 306
  systems 212, 313, 316
Air pollution 189, 198, 212, 213, 223, 230, 306, 307, 309, 316
  remediation applications 306
Air purification 175, 198, 239, 240, 249, 277, 306, 309
  applications 249
  systems 249, 309
Airborne 189, 250
  microbes 250
  pollutants 189
Algae processing processes 14
Alzheimer's disease 42
Anaerobic 12, 150, 354
  bacteria 150, 354
  environments 12
Anesthesiologists 293
Antimicrobial properties 144, 145, 186, 228
Arsenic metals 330
Atomic 20, 21
  absorption spectroscopy (AAS) 20, 21

emission spectrometry (AES) 21
fluorescence spectroscopy (AFS) 21

B

Bacillus subtilis transformation 153
Bacteria 9, 11, 12, 13, 18, 19, 49, 52, 139, 147, 182, 202, 203, 228, 325, 354
  iron-reducing 354
Bacterial degradation 331
Biobased green synthesis techniques 9
Biochemical oxygen demand (BOD) 81, 347
Biodegradation 176, 179
  processes 176
  reactions 179
Biomagnification 347
Biomass 13, 15, 16, 60, 109, 110, 113, 114, 115, 116, 185, 229, 233, 234
  burning 233
  dry 110
  plant-based 185
  pyrolyzing 109, 115
Biopolymers, natural 142
Biosensing systems 335
Biosensors, novel laccase-based 143
Biosynthesis, microbial 138
Brome mosaic virus (BMV) 140

C

Chemical 124, 273, 310
  degradation techniques 273
  pollutants 124
  vapour deposition 310
Chronic obstructive pulmonary disease (COPD) 293
Coatings 144, 146, 212, 249, 351
  antimicrobial 144
  protein 212
Combustion 110, 137, 170, 171, 172, 269, 299
  coal 171
  emissions 137

spectroscopy 21, 40
-visible spectroscopy method 205

V

Vapor deposition synthesis 137
Volatile organic compounds (VOCs) 22, 111,
      170, 173, 230, 232, 233, 240, 306, 308,
      311, 329

W

Waste 3, 33, 110, 118, 122, 142, 146, 169,
      175, 176, 178, 198, 223, 270, 302, 322,
      332, 345, 353
   industrial 223, 270, 322
   inorganic 176
   municipal 110
   organic 118, 353
   plant fiber 142
   treatment method 169
Water 32, 246
      decontamination techniques 32
      treatment technologies 246
Wireless communication technologies 300

X

X-ray 13, 20, 149, 150, 205, 207
      diffraction (XRD) 13, 20, 149, 150, 205,
         207
      photoelectron spectroscopy 20

www.ingramcontent.com/pod-product-compliance
Lightning Source LLC
Chambersburg PA
CBHW050801220326
41598CB00006B/87